T0331644

FRACTALS IN PROBABILITY AND ANALYSIS

A mathematically rigorous introduction to fractals which emphasizes examples and fundamental ideas. Building up from basic techniques of geometric measure theory and probability, central topics such as Hausdorff dimension, self-similar sets and Brownian motion are introduced, as are more specialized topics, including Kakeya sets, capacity, percolation on trees and the Traveling Salesman Theorem. The broad range of techniques presented enables key ideas to be highlighted, without the distraction of excessive technicalities. The authors incorporate some novel proofs which are simpler than those available elsewhere. Where possible, chapters are designed to be read independently so the book can be used to teach a variety of courses, with the clear structure offering students an accessible route into the topic.

Christopher J. Bishop is a professor in the Department of Mathematics at Stony Brook University. He has made contributions to the theory of function algebras, Kleinian groups, harmonic measure, conformal and quasiconformal mapping, holomorphic dynamics and computational geometry.

Yuval Peres is a Principal Researcher at Microsoft Research in Redmond, WA. He is particularly known for his research in topics such as fractals and Hausdorff measure, random walks, Brownian motion, percolation and Markov chain mixing times. In 2016 he was elected to the National Academy of Science.

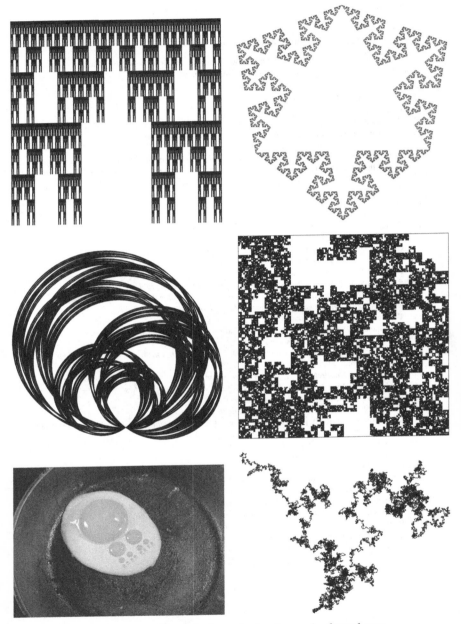

A selection of fractals from the book – and a fractal egg,
reproduced with the permission of Kevin Van Aelst.

Fractals in Probability
and Analysis

CHRISTOPHER J. BISHOP
Stony Brook University, Stony Brook, NY

YUVAL PERES
Microsoft Research, Redmond, WA

CAMBRIDGE
UNIVERSITY PRESS

CAMBRIDGE
UNIVERSITY PRESS

Shaftesbury Road, Cambridge CB2 8EA, United Kingdom

One Liberty Plaza, 20th Floor, New York, NY 10006, USA

477 Williamstown Road, Port Melbourne, VIC 3207, Australia

314–321, 3rd Floor, Plot 3, Splendor Forum, Jasola District Centre, New Delhi – 110025, India

103 Penang Road, #05–06/07, Visioncrest Commercial, Singapore 238467

Cambridge University Press is part of Cambridge University Press & Assessment,
a department of the University of Cambridge.

We share the University's mission to contribute to society through the pursuit of
education, learning and research at the highest international levels of excellence.

www.cambridge.org
Information on this title: www.cambridge.org/9781107134119

10.1017/9781316460238

First published 2017

A catalogue record for this publication is available from the British Library

ISBN   978-1-107-13411-9   Hardback

Cambridge University Press & Assessment has no responsibility for the persistence
or accuracy of URLs for external or third-party internet websites referred to in this
publication and does not guarantee that any content on such websites is, or will
remain, accurate or appropriate.

Contents

Preface

The aim of this book is to acquaint readers with some fractal sets that arise naturally in probability and analysis, and the methods used to study them. The book is based on courses taught by the authors at Yale, Stony Brook University, the Hebrew University and UC Berkeley. We owe a great debt to our advisors, Peter Jones and Hillel Furstenberg; thus the book conveys some of their perspectives on the subject, as well as our own.

We have made an effort to keep the book self-contained. The only prerequisite is familiarity with measure theory and probability at the level acquired in a first graduate course. The book contains many exercises of varying difficulty. We have indicated with a "•" those for which a solution, or a hint, is given in Appendix C. A few sections are technically challenging and not needed for subsequent sections, so could be skipped in the presentation of a given chapter. We mark these with a "*" in the section title.

Acknowledgments: We are very grateful to Tonći Antunović, Subhroshekhar Ghosh and Liat Kessler for helpful comments and crucial editorial work. We also thank Ilgar Eroglu, Hrant Hakobyan, Michael Hochman, Nina Holden, Pertti Mattila, Elchanan Mossel, Boris Solomyak, Perla Sousi, Ryokichi Tanaka, Tatiana Toro, Bálint Virág, Samuel S. Watson, Yimin Xiao and Alex Zhai for useful comments. Richárd Balka carefully read the entire manuscript and provided hundreds of detailed corrections and suggestions. Many thanks to David Tranah and Sam Harrison at Cambridge University Press for numerous helpful suggestions.

Finally, we dedicate this book to our families: Cheryl, David and Emily Bishop, and Deborah, Alon and Noam Peres; without their support and understanding, it would have taken even longer to write.

1

Minkowski and Hausdorff dimensions

In this chapter we will define the Minkowski and Hausdorff dimensions of a set and will compute each in a few basic examples. We will then prove Billingsley's Lemma and the Law of Large Numbers. These allow us to deal with more sophisticated examples: sets defined in terms of digit frequencies, random slices of the Sierpiński gasket, and intersections of random translates of the middle thirds Cantor set with itself. Both Minkowski and Hausdorff dimensions measure how efficiently a set K can be covered by balls. Minkowski dimension requires that the covering be by balls all of the same radius. This makes it easy to compute, but it lacks certain desirable properties. In the definition of Hausdorff dimension we will allow coverings by balls of different radii. This gives a better behaved notion of dimension, but (as we shall see) it is usually more difficult to compute.

1.1 Minkowski dimension

A subset K of a metric space is called **totally bounded** if for any $\varepsilon > 0$, it can be covered by a finite number of balls of diameter ε. For Euclidean space, this is the same as being a bounded set. For a totally bounded set K, let $N(K, \varepsilon)$ denote the minimal number of sets of diameter at most ε needed to cover K. We define the **upper Minkowski dimension** as

$$\overline{\dim}_{\mathscr{M}}(K) = \limsup_{\varepsilon \to 0} \frac{\log N(K, \varepsilon)}{\log 1/\varepsilon},$$

and the **lower Minkowski dimension**

$$\underline{\dim}_{\mathscr{M}}(K) = \liminf_{\varepsilon \to 0} \frac{\log N(K, \varepsilon)}{\log 1/\varepsilon}.$$

1

If the two values agree, the common value is simply called the **Minkowski dimension** of K and denoted by $\dim_{\mathscr{M}}(K)$. When the Minkowski dimension of a set K exists, the number of sets of diameter ε needed to cover K grows like $\varepsilon^{-\dim_{\mathscr{M}}(K)+o(1)}$ as $\varepsilon \to 0$.

We get the same values of $\overline{\dim}_{\mathscr{M}}(K)$ and $\underline{\dim}_{\mathscr{M}}(K)$ if we replace $N(K,\varepsilon)$ by $N_B(K,\varepsilon)$, which is the number of closed balls of radius ε needed to cover K. This is because $N_B(K,\varepsilon) \leq N(K,\varepsilon) \leq N(K,\varepsilon/2)$ (any set is contained in a ball of at most twice the diameter and any ball of radius $\varepsilon/2$ has diameter at most ε; strict inequality could hold in a metric space). For subsets of Euclidean space we can also count the number of axis-parallel squares of side length ε needed to cover K, or the number of such squares taken from a grid. Both possibilities give the same values for upper and lower Minkowski dimension, and for this reason Minkowski dimension is sometimes called the **box counting dimension**. It is also easy to see that a bounded set A and its closure $\overline{A}$ satisfy $\overline{\dim}_{\mathscr{M}}(A) = \overline{\dim}_{\mathscr{M}}(\overline{A})$ and $\underline{\dim}_{\mathscr{M}}(A) = \underline{\dim}_{\mathscr{M}}(\overline{A})$.

If X is a set and $x,y \in X$ implies $|x-y| \geq \varepsilon$, we say X is ε-**separated**. Let $N_{\text{sep}}(K,\varepsilon)$ be the number of elements in a maximal ε-separated subset X of K. Clearly, any set of diameter $\varepsilon/2$ can contain at most one point of an ε-separated set X, so $N_{\text{sep}}(K,\varepsilon) \leq N(K,\varepsilon/2)$. On the other hand, every point of K is within ε of a maximal ε-separated subset X (otherwise add that point to X). Thus $N(K,\varepsilon) \leq N_{\text{sep}}(K,\varepsilon)$. Therefore replacing $N(K,\varepsilon)$ by $N_{\text{sep}}(K,\varepsilon)$ in the definition of upper and lower Minkowski dimension gives the same values (and it is often easier to give a lower bound in terms of separated sets).

Example 1.1.1 Suppose that K is a finite set. Then $N(K,\varepsilon)$ is bounded and $\dim_{\mathscr{M}}(K)$ exists and equals 0.

Example 1.1.2 Suppose $K = [0,1]$. Then at least $1/\varepsilon$ intervals of length ε are needed to cover K and clearly $\varepsilon^{-1}+1$ suffice. Thus $\dim_{\mathscr{M}}(K)$ exists and equals 1. Similarly, any bounded set in $\mathbb{R}^d$ with interior has Minkowski dimension d.

Example 1.1.3 Let $\mathbf{C}$ be the usual middle thirds Cantor set obtained as follows. Let $\mathbf{C}^0 = [0,1]$ and define $\mathbf{C}^1 = [0,\frac{1}{3}] \cup [\frac{2}{3},1] \subset \mathbf{C}^0$ by removing the central interval of length $\frac{1}{3}$. In general, $\mathbf{C}^n$ is a union of 2^n intervals of length 3^{-n} and $\mathbf{C}^{n+1}$ is obtained by removing the central third of each. This gives a decreasing nested sequence of compact sets whose intersection is the desired set $\mathbf{C}$.

The construction gives a covering of $\mathbf{C}$ that uses 2^n intervals of length 3^{-n}. Thus for $3^{-n} \leq \varepsilon < 3^{-n+1}$ we have

$$N(\mathbf{C},\varepsilon) \leq 2^n,$$

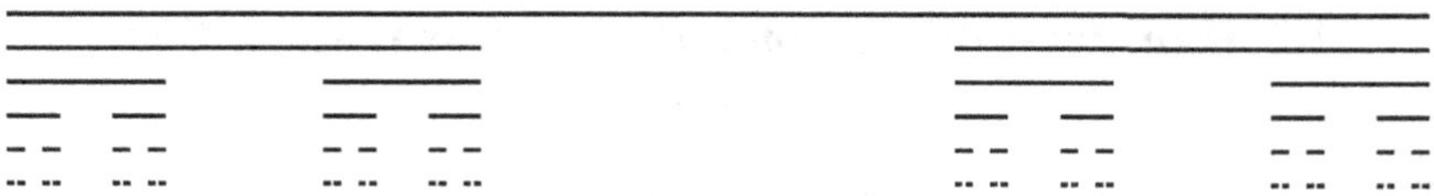

Figure 1.1.1 The Cantor middle thirds construction.

and hence

$$\overline{\dim}_{\mathcal{M}}(\mathbf{C}) \leq \frac{\log 2}{\log 3}.$$

Conversely, the centers of the nth generation intervals form a 3^{-n}-separated set of size 2^n, so $N_{\text{sep}}(\mathbf{C}, 3^{-n}) \geq 2^n$. Thus

$$\underline{\dim}_{\mathcal{M}}(\mathbf{C}) \geq \frac{\log 2}{\log 3} = \log_3 2.$$

Therefore the Minkowski dimension exists and equals this common value. If at each stage we remove the middle α $(0 < \alpha < 1)$ we get a Cantor set $\mathbf{C}_\alpha$ with Minkowski dimension $\log 2 / (\log 2 + \log \frac{1}{1-\alpha})$.

Example 1.1.4 Consider $K = \{0\} \cup \{1, \frac{1}{2}, \frac{1}{3}, \frac{1}{4}, \ldots\}$. Observe that

$$\frac{1}{n-1} - \frac{1}{n} = \frac{1}{n(n-1)} > \frac{1}{n^2}.$$

So, for $\varepsilon > 0$, if we choose n so that $(n+1)^{-2} < \varepsilon \leq n^{-2}$, then $n \leq \varepsilon^{-1/2}$ and n distinct intervals of length ε are needed to cover the points $1, \frac{1}{2}, \ldots, \frac{1}{n}$. The interval $[0, \frac{1}{n+1}]$ can be covered by $n+1$ additional intervals of length ε. Thus

$$\varepsilon^{-1/2} \leq N(K, \varepsilon) \leq 2\varepsilon^{-1/2} + 1.$$

Hence $\dim_{\mathcal{M}}(K) = 1/2$.

This example illustrates a drawback of Minkowski dimension: finite sets have dimension zero, but countable sets can have positive dimension. In particular, it is not true that $\dim_{\mathcal{M}}(\bigcup_n E_n) = \sup_n \dim_{\mathcal{M}}(E_n)$, a useful property for a dimension to have. In the next section, we will introduce Hausdorff dimension, which does have this property (Exercise 1.6). In the next chapter, we will introduce packing dimension, which is a version of upper Minkowski dimension forced to have this property.

1.2 Hausdorff dimension and the Mass Distribution Principle

Given any set K in a metric space, we define the α-**dimensional Hausdorff content** as

$$\mathscr{H}^\alpha_\infty(K) = \inf\Big\{ \sum_i |U_i|^\alpha : K \subset \bigcup_i U_i \Big\},$$

where $\{U_i\}$ is a countable cover of K by any sets and $|E|$ denotes the diameter of a set E.

Definition 1.2.1 The **Hausdorff dimension** of K is defined to be

$$\dim(K) = \inf\{\alpha : \mathscr{H}^\alpha_\infty(K) = 0\}.$$

More generally we define

$$\mathscr{H}^\alpha_\varepsilon(K) = \inf\Big\{ \sum_i |U_i|^\alpha : K \subset \bigcup_i U_i, |U_i| < \varepsilon \Big\},$$

where each U_i is now required to have diameter less than ε. The α-**dimensional Hausdorff measure** of K is defined as

$$\mathscr{H}^\alpha(K) = \lim_{\varepsilon \to 0} \mathscr{H}^\alpha_\varepsilon(K).$$

This is an outer measure; an **outer measure** on a non-empty set X is a function μ^* from the family of subsets of X to $[0, \infty]$ that satisfies

- $\mu^*(\emptyset) = 0$,
- $\mu^*(A) \le \mu^*(B)$ if $A \subset B$,
- $\mu^*(\bigcup_{j=1}^\infty A_j) \le \sum_{j=1}^\infty \mu^*(A_j)$.

For background on real analysis see Folland (1999). The α-dimensional Hausdorff measure is even a Borel measure in $\mathbb{R}^d$; see Theorem 1.2.4 below. When $\alpha = d \in \mathbb{N}$, then $\mathscr{H}^\alpha$ is a constant multiple of $\mathscr{L}_d$, d-dimensional Lebesgue measure.

If we admit only open sets in the covers of K, then the value of $\mathscr{H}^\alpha_\varepsilon(K)$ does not change. This is also true if we only use closed sets or only use convex sets. Using only balls might increase $\mathscr{H}^\alpha_\varepsilon$ by at most a factor of 2^α, since any set K is contained in a ball of at most twice the diameter. Still, the values for which $\mathscr{H}^\alpha(K) = 0$ are the same whether we allow covers by arbitrary sets or only covers by balls.

Definition 1.2.2 Let μ^* be an outer measure on X. A set K in X is μ^*-**measurable**, if for Every set $A \subset X$ we have

$$\mathscr{H}^\alpha(A) = \mu^*(A \cap K) + \mu^*(A \cap K^c).$$

Definition 1.2.3 Let (Ω, d) be a metric space. An outer measure μ on Ω is called a **metric outer measure** if $\text{dist}(A, B) > 0 \implies \mu(A \cup B) = \mu(A) + \mu(B)$, where A and B are two subsets of Ω.

Since Hausdorff measure $\mathscr{H}^{\alpha}$ is clearly a metric outer measure, the following theorem shows that all Borel sets are $\mathscr{H}^{\alpha}$-measurable. This implies that $\mathscr{H}^{\alpha}$ is a Borel measure (see Folland (1999)).

Theorem 1.2.4 *Let μ be a metric outer measure. Then all Borel sets are μ-measurable.*

Proof It suffices to show any closed set K is μ-measurable, since the measurable sets form a σ-algebra. So, let K be a closed set. We must show for any set $A \subset \Omega$ with $\mu(A) < \infty$,

$$\mu(A) \geq \mu(A \cap K) + \mu(A \cap K^c). \tag{1.2.1}$$

Let $B_0 = \emptyset$ and for $n \geq 1$ define $B_n = \{x \in A : \text{dist}(x, K) > \frac{1}{n}\}$, so that

$$\bigcup_{n=1}^{\infty} B_n = A \cap K^c$$

(since K is closed). Since μ is a metric outer measure and $B_n \subset A \setminus K$,

$$\mu(A) \geq \mu[(A \cap K) \cup B_n] = \mu(A \cap K) + \mu(B_n). \tag{1.2.2}$$

For all $m \in \mathbb{N}$, the sets $D_n = B_n \setminus B_{n-1}$ satisfy

$$\sum_{j=1}^{m} \mu(D_{2j}) = \mu\left(\bigcup_{j=1}^{m} D_{2j}\right) \leq \mu(A), \forall m,$$

since if $x \in B_n$, and $y \in D_{n+2}$, then

$$\text{dist}(x, y) \geq \text{dist}(x, K) - \text{dist}(y, K) \geq \frac{1}{n} - \frac{1}{n+1}.$$

Similarly $\sum_{j=1}^{m} \mu(D_{2j-1}) \leq \mu(A)$. So $\sum_{j=1}^{\infty} \mu(D_j) < \infty$. The inequality

$$\mu(B_n) \leq \mu(A \cap K^c) \leq \mu(B_n) + \sum_{j=n+1}^{\infty} \mu(D_j)$$

implies that $\mu(B_n) \to \mu(A \cap K^c)$ as $n \to \infty$. Thus letting $n \to \infty$ in (1.2.2) gives (1.2.1). □

The construction of Hausdorff measure can be made a little more general by considering a positive, increasing function φ on $[0,\infty)$ with $\varphi(0) = 0$. This is called a **gauge function** and we may associate to it the Hausdorff content

$$\mathcal{H}_\infty^\varphi(K) = \inf\left\{ \sum_i \varphi(|U_i|) : K \subset \bigcup_i U_i \right\};$$

then $\mathcal{H}_\varepsilon^\varphi(K)$, and $\mathcal{H}^\varphi(K) = \lim_{\varepsilon\to 0} \mathcal{H}_\varepsilon^\varphi(K)$ are defined as before. The case $\varphi(t) = t^\alpha$ is just the case considered above. We will not use other gauge functions in the first few chapters, but they are important in many applications, e.g., see Exercise 1.59 and the Notes for Chapter 6.

Lemma 1.2.5 *If $\mathcal{H}^\alpha(K) < \infty$ then $\mathcal{H}^\beta(K) = 0$ for any $\beta > \alpha$.*

Proof It follows from the definition of $\mathcal{H}_\varepsilon^\alpha$ that

$$\mathcal{H}_\varepsilon^\beta(K) \le \varepsilon^{\beta-\alpha} \mathcal{H}_\varepsilon^\alpha(K),$$

which gives the desired result as $\varepsilon \to 0$. $\square$

Thus if we think of $\mathcal{H}^\alpha(K)$ as a function of α, the graph of $\mathcal{H}^\alpha(K)$ versus α shows that there is a critical value of α where $\mathcal{H}^\alpha(K)$ jumps from ∞ to 0. This critical value is equal to the Hausdorff dimension of the set. More generally we have:

Proposition 1.2.6 *For every metric space E we have*

$$\mathcal{H}^\alpha(E) = 0 \quad \Leftrightarrow \quad \mathcal{H}_\infty^\alpha(E) = 0$$

and therefore

$$\dim E = \inf\{\alpha : \mathcal{H}^\alpha(E) = 0\} = \inf\{\alpha : \mathcal{H}^\alpha(E) < \infty\}$$
$$= \sup\{\alpha : \mathcal{H}^\alpha(E) > 0\} = \sup\{\alpha : \mathcal{H}^\alpha(E) = \infty\}.$$

Proof Since $\mathcal{H}^\alpha(E) \ge \mathcal{H}_\infty^\alpha(E)$, it suffices to prove "$\Leftarrow$". If $\mathcal{H}_\infty^\alpha(E) = 0$, then for every $\delta > 0$ there is a covering of E by sets $\{E_k\}$ with $\sum_{k=1}^\infty |E_k|^\alpha < \delta$. These sets have diameter less than $\delta^{1/\alpha}$, hence $\mathcal{H}_{\delta^{1/\alpha}}^\alpha(E) < \delta$. Letting $\delta \downarrow 0$ yields $\mathcal{H}^\alpha(E) = 0$, proving the claimed equivalence. The equivalence readily implies that $\dim E = \inf\{\alpha : \mathcal{H}^\alpha(E) = 0\} = \sup\{\alpha : \mathcal{H}^\alpha(E) > 0\}$. The other conclusions follow from Lemma 1.2.5. $\square$

The following relationship to Minkowski dimension is clear

$$\dim(K) \le \underline{\dim}_\mathcal{M}(K) \le \overline{\dim}_\mathcal{M}(K). \tag{1.2.3}$$

Indeed, if $B_i = B(x_i, \varepsilon/2)$ are $N(K, \varepsilon)$ balls of radius $\varepsilon/2$ and centers x_i in K that cover K, then consider the sum

$$S_\varepsilon = \sum_{i=1}^{N(K,\varepsilon)} |B_i|^\alpha = N(K, \varepsilon)\varepsilon^\alpha = \varepsilon^{\alpha - R_\varepsilon},$$

where $R_\varepsilon = \frac{\log N(K,\varepsilon)}{\log(1/\varepsilon)}$. If $\alpha > \liminf_{\varepsilon \to 0} R_\varepsilon = \underline{\dim}_{\mathscr{M}}(K)$ then $\inf_{\varepsilon > 0} S_\varepsilon = 0$. Strict inequalities in (1.2.3) are possible.

Example 1.2.7 Example 1.1.4 showed that $K = \{0\} \bigcup_n \{\frac{1}{n}\}$ has Minkowski dimension $\frac{1}{2}$. However, any countable set has Hausdorff dimension 0, for if we enumerate the points $\{x_1, x_2, \dots\}$ and cover the nth point by a ball of diameter $\delta_n = \varepsilon 2^{-n}$ we can make $\sum_n \delta_n^\alpha$ as small as we wish for any $\alpha > 0$. Thus K is a compact set for which the Minkowski dimension exists, but is different from the Hausdorff dimension.

Lemma 1.2.8 (Mass Distribution Principle) *If E supports a strictly positive Borel measure μ that satisfies*

$$\mu(B(x,r)) \le Cr^\alpha,$$

for some constant $0 < C < \infty$ and for every ball $B(x,r)$, then $\mathscr{H}^\alpha(E) \ge \mathscr{H}^\alpha_\infty(E) \ge \mu(E)/C$. In particular, $\dim(E) \ge \alpha$.

Proof Let $\{U_i\}$ be a cover of E. For $\{r_i\}$, where $r_i > |U_i|$, we look at the following cover: choose x_i in each U_i, and take open balls $B(x_i, r_i)$. By assumption,

$$\mu(U_i) \le \mu(B(x_i, r_i)) \le Cr_i^\alpha.$$

We deduce that $\mu(U_i) \le C|U_i|^\alpha$, whence

$$\sum_i |U_i|^\alpha \ge \sum_i \frac{\mu(U_i)}{C} \ge \frac{\mu(E)}{C}.$$

Thus $\mathscr{H}^\alpha(E) \ge \mathscr{H}^\alpha_\infty(E) \ge \mu(E)/C$. $\qquad \square$

We note that upper bounds for Hausdorff dimension usually come from finding explicit coverings of the set, but lower bounds are proven by constructing an appropriate measure supported on the set. Later in this chapter we will generalize the Mass Distribution Principle by proving Billingsley's Lemma (Theorem 1.4.1) and will generalize it even further in later chapters. As a special case of the Mass Distribution Principle, note that if $A \subseteq \mathbb{R}^d$ has positive Lebesgue d-measure then $\dim(A) = d$.

Example 1.2.9 Consider the Cantor set E obtained by replacing the unit square in the plane by four congruent sub-squares of side length $\alpha < 1/2$ and continuing similarly. See Figure 1.2.1. We can cover the set by 4^n squares of diameter $\sqrt{2} \cdot \alpha^n$. Thus

$$\overline{\dim}_{\mathscr{M}}(E) \leq \lim_{n \to \infty} \frac{\log 4^n}{-\log(\sqrt{2}\alpha^n)} = \frac{\log 4}{-\log \alpha}.$$

On the other hand, it is also easy to check that at least 4^n sets of diameter α^n are needed, so

$$\underline{\dim}_{\mathscr{M}}(E) \geq \frac{\log 4}{-\log \alpha}.$$

Thus the Minkowski dimension of this set equals $\beta = -\log 4/\log \alpha$.

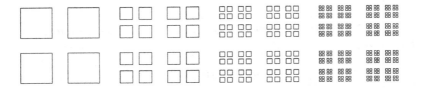

Figure 1.2.1 Four generations of a Cantor set.

We automatically get $\dim(E) \leq \beta$ and we will prove the equality using Lemma 1.2.8. Let μ be the probability measure defined on E that gives each nth generation square the same mass (namely 4^{-n}). We claim that

$$\mu(B(x,r)) \leq Cr^{\beta},$$

for all disks and some $0 < C < \infty$. To prove this, suppose $B = B(x,r)$ is some disk hitting E and choose n so that $\alpha^{n+1} \leq r < \alpha^n$. Then B can hit at most 4 of the nth generation squares and so, since $\alpha^{\beta} = 1/4$,

$$\mu(B \cap E) \leq 4 \cdot 4^{-n} = 4\alpha^{n\beta} \leq 16r^{\beta}.$$

Example 1.2.10 Another simple set for which the two dimensions agree and are easy to compute is the von Koch snowflake. To construct this we start with an equilateral triangle. At each stage we add to each edge an equilateral triangle pointing outward of side length $1/3$ the size of the current edges and centered on the edge. See Figure 1.2.2 for the first four iterations of this process. The boundary of this region is a curve with dimension $\log 4/\log 3$ (see Theorem 2.2.2). We can also think of this as a replacement construction, in which at each stage, a line segment is replaced by an appropriately scaled copy of a polygonal curve.

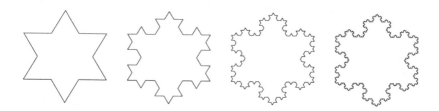

Figure 1.2.2 Four generations of the von Koch snowflake.

Even for some relatively simple sets the Hausdorff dimension is still unknown. Consider the Weierstrass function (Figure 1.2.3)

$$f_{\alpha,b}(x) = \sum_{n=1}^{\infty} b^{-n\alpha} \cos(b^n x),$$

where $b > 1$ is real and $0 < \alpha < 1$. It is conjectured that the Hausdorff dimension of its graph is $2 - \alpha$, and this has been proven when b is an integer; see the discussion in Example 5.1.7. On the other hand, some sets that are more difficult to define, such as the graph of Brownian motion (Figure 1.2.4), will turn out to have easier dimensions to compute ($3/2$ by Theorem 6.4.3).

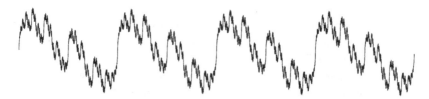

Figure 1.2.3 The Weierstrass function with $b = 2$, $\alpha = 1/2$. This graph has Minkowski dimension $3/2$ and is conjectured to have the same Hausdorff dimension.

1.3 Sets defined by digit restrictions

In this section we will consider some more complicated sets for which the Minkowski dimension is easy to compute, but the Hausdorff dimension is not

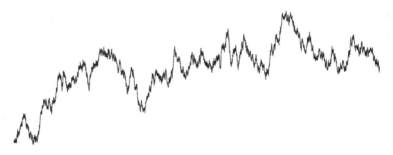

Figure 1.2.4 1-dimensional Brownian motion. This graph has dimension 3/2 almost surely.

so obvious, and will be left to later sections. These subsets of $[0,1]$ will be defined by restricting which digits can occur at a certain position of a number's b-ary expansion. In a later section we will consider sets defined by the asymptotic distribution of the digits. We start by adapting Hausdorff measures to b-adic grids.

Let $b \geq 2$ be an integer and consider b-adic expansions of real numbers, i.e., to each sequence $\{x_n\} \in \{0, 1, \ldots, b-1\}^{\mathbb{N}}$ we associate the real number

$$x = \sum_{n=1}^{\infty} x_n b^{-n} \in [0,1].$$

b-adic expansions give rise to Cantor sets by restricting the digits we are allowed to use. For example, if we set $b = 3$ and require $x_n \in \{0, 2\}$ for all n we get the middle thirds Cantor set **C**.

For each integer n let $I_n(x)$ denote the unique half-open interval of the form $[\frac{k-1}{b^n}, \frac{k}{b^n})$ containing x. Such intervals are called b-**adic intervals of generation** n (**dyadic** if $b = 2$).

It has been observed (by Frostman (1935) and Besicovitch (1952)) that we can restrict the infimum in the definition of Hausdorff measure to coverings of the set that involve only b-adic intervals and only change the value by a bounded factor. The advantage of dealing with these intervals is that they are nested, i.e., two such intervals either are disjoint or one is contained in the other. In particular, any covering by b-adic intervals always contains a subcover by disjoint intervals (just take the maximal intervals). Furthermore, the b-adic intervals can be given the structure of a tree, an observation that we will use extensively in later chapters.

We define the **grid Hausdorff content** by

$$\tilde{\mathcal{H}}_{\infty}^{\alpha}(A) = \inf\{\sum_i |J_i|^{\alpha}, A \subset \bigcup_i J_i\}, \qquad (1.3.1)$$

and the **grid Hausdorff measures** by

$$\tilde{\mathcal{H}}_{\varepsilon}^{\alpha}(A) = \inf\{\sum_i |J_i|^{\alpha}, A \subset \bigcup_i J_i, |J_i| < \varepsilon\}, \qquad (1.3.2)$$

$$\tilde{\mathcal{H}}^{\alpha}(A) = \lim_{\varepsilon \to 0} \tilde{\mathcal{H}}_{\varepsilon}^{\alpha}(A), \qquad (1.3.3)$$

where the infimums are over all coverings of $A \subset \mathbb{R}$ by collections $\{J_i\}$ of b-adic intervals. The grid measure depends on b, but we omit it from the notation; usually the value of b is clear from context.

Clearly

$$\tilde{\mathcal{H}}^{\alpha}(A) \geq \mathcal{H}^{\alpha}(A),$$

since we are taking the infimum over a smaller set of coverings. However, the two sides are almost of the same size, i.e.,

$$\tilde{\mathcal{H}}^{\alpha}(A) \leq (b+1)\mathcal{H}^{\alpha}(A),$$

since any interval I can be covered by at most $(b+1)$ shorter b-adic intervals; just take the smallest n so that $b^{-n} \leq |I|$ and take all b-adic intervals of length b^{-n} that hit I. Thus $\tilde{\mathcal{H}}^{\alpha}(A) = 0$ if and only if $\mathcal{H}^{\alpha}(A) = 0$. Similarly, we can define $\tilde{N}(K, \varepsilon)$ to be the minimal number of closed b-adic intervals of length ε needed to cover K. We get

$$N(K, \varepsilon) \leq \tilde{N}(K, \varepsilon) \leq (b+1)N(K, \varepsilon).$$

Hence, the Hausdorff and the Minkowski dimensions are not changed.

One can define grid Hausdorff content $\tilde{\mathcal{H}}_{\infty}^{\varphi}(A)$ and grid Hausdorff measures $\tilde{\mathcal{H}}_{\varepsilon}^{\varphi}(A)$ and $\tilde{\mathcal{H}}^{\varphi}(A)$ with respect to any gauge function φ by replacing $|J_i|^{\alpha}$ with $\varphi(|J_i|)$ in (1.3.1) and (1.3.2). Furthermore, the definitions can be extended to subsets of $\mathbb{R}^n$ by replacing the b-adic intervals J_i in (1.3.1) and (1.3.2) with b-adic cubes (products of b-adic intervals of equal length).

Definition 1.3.1 For $S \subset \mathbb{N}$ the **upper density** of S is

$$\overline{d}(S) = \limsup_{N \to \infty} \frac{\#(S \cap \{1, \ldots, N\})}{N}.$$

Here and later $\#(E)$ denotes the number of elements in E, i.e., $\#$ is counting measure. The **lower density** is

$$\underline{d}(S) = \liminf_{N \to \infty} \frac{\#(S \cap \{1, \ldots, N\})}{N}.$$

If $\overline{d}(S) = \underline{d}(S)$, then the limit exists and is called $d(S)$, the **density** of S.

Example 1.3.2   Suppose $S \subset \mathbb{N}$, and define

$$A_S = \left\{ x = \sum_{k \in S} x_k 2^{-k} : x_k \in \{0,1\} \right\}.$$

The set A_S is covered by exactly

$$2^{\sum_{k=1}^n \mathbf{1}_S(k)} = 2^{\#(S \cap \{1,\ldots,n\})}$$

closed dyadic intervals of generation n, where $\mathbf{1}_S$ is the characteristic function of S, i.e., $\mathbf{1}_S(n) = 1$ for $n \in S$, and $\mathbf{1}_S(n) = 0$ for $n \notin S$. So

$$\log_2 \tilde{N}(A_S, 2^{-n}) = \sum_{k=1}^n \mathbf{1}_S(k).$$

Thus

$$\frac{\log \tilde{N}(A_S, 2^{-n})}{\log 2^n} = \frac{1}{n} \sum_{k=1}^n \mathbf{1}_S(k),$$

which implies

$$\overline{\dim}_{\mathscr{M}}(A_S) = \overline{d}(S),$$

$$\underline{\dim}_{\mathscr{M}}(A_S) = \underline{d}(S).$$

It is easy to construct sets S where the liminf and limsup differ and hence we get compact sets where the Minkowski dimension does not exist (see Exercise 1.17). The Hausdorff dimension of A_S is equal to the lower Minkowski dimension. We shall prove this in the next section as an application of Billingsley's Lemma.

We can also construct A_S as follows. Start with the interval $[0,1]$ and subdivide it into two equal length subintervals $[0,1/2]$ and $[1/2,1]$. If $1 \in S$ then keep both intervals and if $1 \notin S$ then keep only the leftmost, $[0,1/2]$. Cut each of the remaining intervals in half, keeping both subintervals if $2 \in S$ and only keeping the left interval otherwise. In general, at the nth step we have a set $A_n^S \subset A_{n-1}^S$ that is a finite union of intervals of length 2^{-n} (some may be adjoining). We cut each of the intervals in half, keeping both subintervals if $n \in S$ and throwing away the right-hand one if $n \notin S$. The limiting set is $A_S = \bigcap_{n=1}^\infty A_n^S$. In Figure 1.3.1 we have drawn the first generations of the construction corresponding to $S = \{1,3,4,6,8,10,\ldots\}$.

Example 1.3.3   The **shifts of finite type** are defined by restricting which

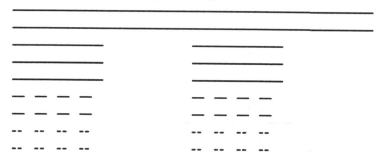

Figure 1.3.1 First 8 generations of A_S for $S = \{1,3,4,6,8,10,\dots\}$.

digits can follow other digits. Let $A = (A_{ij}), 0 \le i, j < b$, be a $b \times b$ matrix of 0s and 1s, and define

$$X_A = \left\{ \sum_{n=1}^{\infty} x_n b^{-n} : A_{x_n x_{n+1}} = 1 \text{ for all } n \ge 1 \right\}.$$

We will also assume that if the jth column of A is all zeros, so is the jth row; this implies every finite sequence $\{x_n\}_1^n$ satisfying the condition above can be extended to at least one infinite sequence satisfying the condition. Thus such finite strings correspond to b-adic closed dyadic intervals that intersect X_A.

Notice that a b-ary rational number r will belong to X_A if either of its two possible b-ary expansions does (one is eventually all 0s, the other eventually all $(b-1)$s). The condition on $\{x_n\}$ described in the example is clearly shift invariant and is the intersection of countably many "closed" conditions, so the set X_A is compact and invariant under the map T_b, where $T_b(x) = (bx) \mod 1$. For example, if

$$A = \begin{pmatrix} 1 & 1 \\ 1 & 0 \end{pmatrix},$$

then $x \in X_A$ if and only if the binary representation of x has no two consecutive 1s. The first ten generations in the construction of the Cantor set X_A are shown in Figure 1.3.3. Any such matrix can be represented by a directed graph where the vertices represent the numbers $\{0,\dots,b-1\}$ and a directed edge connects i to j if $A_{ij} = 1$. For example, the 2×2 matrix A above is represented by Figure 1.3.2.

An element of $x = \sum_n x_n b^{-n} \in X_A$ corresponds in a natural way to an infinite path $\{x_1, x_2, \dots\}$ on this graph and a b-adic interval hitting X_A corresponds to a finite path. Conversely, any finite path of length n corresponds to a b-adic interval of length b^{-n} hitting X_A. Thus $N_n(X_A) \equiv \tilde{N}(X_A, b^{-n})$ is the number of

Figure 1.3.2 The directed graph representing A.

Figure 1.3.3 Approximations of X_A.

distinct paths of length n in the graph (this uses our assumption about the rows and columns of A). By the definition of matrix multiplication, the number of paths of length n from i to j is $(A^n)_{ij}$ (i.e., the (ith, jth) entry in the matrix A^n). (See Exercise 1.23.) Thus the total number of length n paths is $N_n(X_A) = \|A^n\|$, where the norm of a matrix is defined as $\|B\| = \sum_{i,j} |B_{ij}|$. (However, since any two norms on a finite-dimensional space are comparable, the precise norm won't matter.) Thus

$$\dim_{\mathscr{M}}(X_A) = \lim_{n\to\infty} \frac{\log N_n(X_A)}{n\log b} = \lim_{n\to\infty} \frac{\log(N_n(X_A))^{1/n}}{\log b} = \frac{\log \rho(A)}{\log b},$$

where

$$\rho(A) = \lim_{n\to\infty} \|A^n\|^{1/n} \qquad (1.3.4)$$

is, by definition, the **spectral radius** of the matrix A, and is equal to the absolute value of the largest eigenvalue. That this limit exists is a standard fact about matrices and is left to the reader (Exercise 1.24).

For the matrix

$$A = \begin{pmatrix} 1 & 1 \\ 1 & 0 \end{pmatrix},$$

the spectral radius is $(1 + \sqrt{5})/2$ (Exercise 1.25), so the Minkowski dimension of X_A is

$$\frac{\log \rho(A)}{\log 2} = \frac{\log(1 + \sqrt{5}) - \log 2}{\log 2}.$$

We shall see in Example 2.3.3, that for shifts of finite type the Hausdorff and Minkowski dimensions agree.

Example 1.3.4 Now we consider some sets in the plane. Suppose A is a $b \times b$ matrix of 0s and 1s with rows labeled by 0 to $b - 1$ **from bottom to top** (the unusual labeling ensures the matrix corresponds to the picture of the set) and columns by 0 to $b - 1$ from left to right. Let

$$Y_A = \{(x,y) : A_{y_n x_n} = 1 \text{ for all } n\},$$

where $\{x_n\}, \{y_n\}$ are the b-ary expansions of x and y. For example, if

$$A = \begin{pmatrix} 1 & 0 \\ 1 & 1 \end{pmatrix},$$

then $x_n = 0$ implies that y_n can be either 0 or 1, but if $x_n = 1$ then y_n must be 0. This matrix A gives a Sierpiński gasket. See Figure 1.3.4.

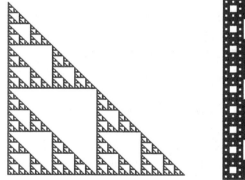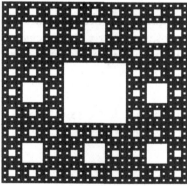

Figure 1.3.4 The Sierpiński gasket (8th generation) and the Sierpiński carpet (4th generation).

If

$$A = \begin{pmatrix} 1 & 0 & 1 \\ 0 & 0 & 0 \\ 1 & 0 & 1 \end{pmatrix},$$

then we get $\mathbf{C}_{1/3} \times \mathbf{C}_{1/3}$, the product of the middle thirds Cantor set with itself.

For

$$A = \begin{pmatrix} 1 & 1 & 1 \\ 1 & 0 & 1 \\ 1 & 1 & 1 \end{pmatrix},$$

we get the Sierpiński carpet. See the right side of Figure 1.3.4. If the matrix A has r 1s in it, then we can cover Y_A by r^n squares of side b^{-n} and it is easy to see this many are needed. Thus $\dim_{\mathcal{M}}(Y_A) = \log r / \log b = \log_b r$. Defining a measure on Y_A which gives equal mass to each of the nth generational squares and using the Mass Distribution Principle, we get that the Hausdorff dimension of the set Y_A is also $\log_b r$.

The Sierpiński gasket (as well as the other examples so far) are "self-similar" in the sense that they are invariant under a certain collection of maps that are constructed from isometries and contractive dilations. Self-similar sets are discussed in Section 2.1.

If we replace the square matrices of the previous example by rectangular ones (i.e., use different base expansions for x and y) we get a much more difficult class of sets that are invariant under affine maps. These sets will be studied in Chapter 4, but we give an example here.

Example 1.3.5   Self-affine sets. We modify the previous example by taking a non-square matrix. Suppose $m < n$ and suppose A is an $m \times n$ matrix of 0s and 1s with rows and columns labeled as in the previous example. Let

$$Y_A = \{(x,y) : A_{y_k x_k} = 1 \text{ for all } k\},$$

where $\{x_k\}$ is the n-ary expansion of x and $\{y_k\}$ is the m-ary expansion of y. For example, if

$$A = \begin{pmatrix} 0 & 1 & 0 \\ 1 & 0 & 1 \end{pmatrix},$$

we obtain the McMullen set in Figure 1.3.5. To construct the set, start with a rectangle Q and divide it into 6 sub-rectangles by making one horizontal and 2 vertical subdivisions. Choose one of the rectangles on the top row and two rectangles from the bottom row according to the pattern A. Now subdivide the chosen rectangles in the same way and select the corresponding sub-rectangles. Figure 1.3.5 illustrates the resulting set after a few iterations.

The Minkowski dimension of this set exists and is equal to $1 + \log_3 \frac{3}{2} \approx 1.36907$ (Theorem 4.1.1). We will prove later in this chapter that the Hausdorff dimension of the McMullen set is strictly larger than 1 (see Example

Figure 1.3.5 Four generations of the McMullen set.

1.6.5). We shall eventually prove C. McMullen's (1984) result that the Hausdorff dimension of this set is $\log_2(2^{\log_3 2} + 1) \approx 1.34968$, which is strictly less than the Minkowski dimension (Theorem 4.2.1).

1.4 Billingsley's Lemma and the dimension of measures

In this section we prove a refinement of the Mass Distribution Principle. The measure in this version need not satisfy uniform estimates on all balls as in the Mass Distribution Principle, but only estimates in a neighborhood of each point, where the size of that neighborhood may vary from point to point. An even more general result was proved by Rogers and Taylor (1959), see Proposition 4.3.3. Let $b \geq 2$ be an integer and for $x \in [0,1]$ let $I_n(x)$ be the nth generation, half-open b-adic interval of the form $[\frac{j-1}{b^n}, \frac{j}{b^n})$ containing x. The following is due to Billingsley (1961).

Lemma 1.4.1 (Billingsley's Lemma)   *Let $A \subset [0,1]$ be Borel and let μ be a finite Borel measure on $[0,1]$. Suppose $\mu(A) > 0$. If*

$$\alpha_1 \leq \liminf_{n \to \infty} \frac{\log \mu(I_n(x))}{\log |I_n(x)|} \leq \beta_1, \qquad (1.4.1)$$

for all $x \in A$, then $\alpha_1 \leq \dim(A) \leq \beta_1$.

Proof   Let $\alpha < \alpha_1 < \beta_1 < \beta$. The inequalities (1.4.1) yield that

$$\text{for all } x \in A, \quad \limsup_{n \to \infty} \frac{\mu(I_n(x))}{|I_n(x)|^\beta} \geq 1, \qquad (1.4.2)$$

and

$$\text{for all } x \in A, \quad \limsup_{n \to \infty} \frac{\mu(I_n(x))}{|I_n(x)|^\alpha} \leq 1. \qquad (1.4.3)$$

We will show that (1.4.2) implies $\tilde{\mathscr{H}}^\beta(A) \leq \mu([0,1])$ and that (1.4.3) implies $\tilde{\mathscr{H}}^\alpha(A) \geq \mu(A)$. The lemma follows from these two claims.

We start with the first assertion. Assume (1.4.2) holds. For $0 < c < 1$, fix $\varepsilon > 0$. For every $x \in A$ we can find integers n as large as we wish satisfying

$$\frac{\mu(I_n(x))}{|I_n(x)|^\beta} > c.$$

Take $n(x)$ to be the minimal integer satisfying this condition and such that $b^{-n(x)} < \varepsilon$.

Now $\{I_{n(x)}(x) : x \in A\}$ is a cover of A, and suppose $\{J_k\}$ is a subcover by disjoint intervals. This covering of A has the property that $|J_k| < \varepsilon$ for all k (by our choice of $n(x)$) and

$$\sum_k |J_k|^\beta \le c^{-1} \sum_k \mu(J_k) \le c^{-1} \mu([0,1]). \tag{1.4.4}$$

This implies

$$\tilde{\mathcal{H}}_\varepsilon^\beta(A) \le c^{-1} \mu([0,1]).$$

Taking $c \to 1$ and $\varepsilon \to 0$ gives the first assertion. Therefore $\tilde{\mathcal{H}}^\beta(A) \le \mu([0,1])$.

Next we prove the second assertion. Assume (1.4.3) holds. For a fixed $C > 1$ and a positive integer m, let

$$A_m = \{x \in A : \mu(I_n(x)) < C|I_n(x)|^\alpha \text{ for all } n > m\}.$$

Since $A = \bigcup_m A_m$ and $A_{m+1} \supset A_m$, we have $\mu(A) = \lim_{m \to \infty} \mu(A_m)$.

Fix $\varepsilon < b^{-m}$ and consider any cover of A by b-adic intervals $\{J_k\}$ such that $|J_k| < \varepsilon$. Then

$$\sum_k |J_k|^\alpha \ge \sum_{k: J_k \cap A_m \ne \emptyset} |J_k|^\alpha \ge C^{-1} \sum_{k: J_k \cap A_m \ne \emptyset} \mu(J_k) \ge C^{-1} \mu(A_m).$$

This shows $\tilde{\mathcal{H}}_\varepsilon^\alpha(A) \ge C^{-1} \mu(A_m)$. Taking $\varepsilon \to 0, m \to \infty$ and $C \to 1$ gives the desired result. $\square$

The above proof of Billingsley's Lemma can be generalized in several ways:

(a) It is clear that the proof generalizes directly to subsets $A \subset [0,1]^d$. In the proof one needs to replace b-adic intervals with cubes that are products of intervals of the form $\prod_{i=1}^d \left[\frac{j_i-1}{b^n}, \frac{j_i}{b^n}\right)$.

(b) A covering $\mathcal{S} = \{S_i\}_i$ of A is called a **Vitali** covering if for every point $x \in A$ and all $\delta > 0$ there is a set S_i such that $x \in S_i$ and such that $0 < |S_i| < \delta$. We say $\mathcal{S}$ has the **bounded subcover property** if there exists a constant C such that whenever a set E is covered by a subcover $\mathcal{S}_E \subset \mathcal{S}$ then there is a further subcover $\tilde{\mathcal{S}}_E \subset \mathcal{S}_E$ of E such that $\sum_{D \in \tilde{\mathcal{S}}_E} \mathbf{1}_D \le C$. For example, by Besicovitch's covering theorem (see Mattila, 1995, Theorem 2.6) the family of open balls in $\mathbb{R}^d$ enjoys the bounded subcover property. Replacing the family of dyadic intervals by a Vitali covering with bounded subcover property in (1.3.2) and (1.3.3) and in the statement of Billingsley's Lemma one can proceed as in the above proof to conclude that $\tilde{\mathscr{H}}^\alpha(A) \geq \mu(A)$ and $\tilde{\mathscr{H}}^\beta(A) \leq C\mu([0,1])$. In particular (1.4.4) is replaced by

$$\sum_{D \in \mathscr{F}_A} |D|^\beta \leq c^{-1} \sum_{D \in \mathscr{F}_A} \mu(D) = c^{-1} \int \sum_{D \in \mathscr{F}_A} 1_D \, d\mu \leq c^{-1} C\mu([0,1]).$$

However, in general, $\tilde{\mathscr{H}}^\gamma(A)$ will not be comparable to $\mathscr{H}^\gamma(A)$. To replace $\tilde{\mathscr{H}}^\alpha(A)$ and $\tilde{\mathscr{H}}^\beta(A)$ by $\mathscr{H}^\alpha(A)$ and $\mathscr{H}^\beta(A)$ and prove the corresponding version of Billingsley's Lemma, one needs additional assumptions. For example, it suffices to assume that there exists a constant C such that any ball B can be covered by no more than C elements of the cover of diameter less than $C|B|$. An example of a covering satisfying all the above assumptions are approximate squares as in Definition 4.2.2; these will be used to compute dimensions of self-affine sets in Chapter 4.

(c) The assumptions can be further weakened. For example, instead of the last assumption we can assume that there is a function $\Psi \colon \mathbb{R}^+ \to \mathbb{R}^+$ satisfying $\lim_{r \to 0} \frac{\log \Psi(r)}{\log 1/r} = 0$ and such that any ball of radius r can be covered by $\Psi(r)$ elements of the cover of diameter at most $\Psi(r)r$.

Example 1.4.2 For $S \subset \mathbb{N}$, recall

$$A_S = \left\{ x = \sum_{k \in S} x_k 2^{-k} : x_k \in \{0,1\} \right\}. \tag{1.4.5}$$

We computed the upper and lower Minkowski dimensions of this set in Example 1.3.2 and claimed that

$$\dim(A_S) = \liminf_{N \to \infty} \frac{\#(S \cap \{1, \dots, N\})}{N}.$$

To prove this, let μ be the probability measure on A_S that gives equal measure to the nth generation covering intervals. This measure makes the digits $\{x_k\}_{k \in S}$ in (1.4.5) independent identically distributed (i.i.d.) uniform random bits. For any $x \in A_S$,

$$\frac{\log \mu(I_n(x))}{\log |I_n(x)|} = \frac{\log 2^{-\#(S \cap \{1, \dots, n\})}}{\log 2^{-n}} = \frac{\#(S \cap \{1, \dots, n\})}{n}.$$

Thus the liminf of the left-hand side is the liminf of the right-hand side. By Billingsley's Lemma, this proves the claim.

Definition 1.4.3 If μ is a Borel measure on $\mathbb{R}^n$ we define

$$\dim(\mu) = \inf\{\dim(A) : \mu(A^c) = 0, A \subset \mathbb{R}^n \text{ is Borel }\}.$$

Observe that the infimum is really a minimum, because if $\dim(A_n) \to \dim(\mu)$ and $\mu(A_n^c) = 0$ for all n then $A = \bigcap_n A_n$ satisfies $\dim(A) = \dim(\mu)$. An equivalent definition is to write

$$\dim(\mu) = \inf\{\alpha : \mu \perp \mathscr{H}^\alpha\}$$

where $\mu \perp \nu$ means the two measures are **mutually singular**, i.e., there is a set $A \subset \mathbb{R}^n$ such that $\mu(A^c) = \nu(A) = 0$.

Lemma 1.4.4 *Let b be a positive integer. Given $x \in [0, 1]$, let $I_n(x)$ denote the b-adic interval of the form $[\frac{j-1}{b^n}, \frac{j}{b^n})$ containing x. Let*

$$\alpha_\mu = \operatorname{ess\,sup}\left\{ \liminf_{n \to \infty} \frac{\log \mu(I_n(x))}{\log |I_n(x)|} \right\},$$

where $\operatorname{ess\,sup}(f) = \min\{\alpha : \mu(\{x : f(x) > \alpha\}) = 0\}$. Then $\dim(\mu) = \alpha_\mu$.

Proof   First take $\alpha > \alpha_\mu$, and set

$$A = \left\{ x : \liminf_{n \to \infty} \frac{\log \mu(I_n(x))}{\log |I_n(x)|} \leq \alpha \right\}.$$

By definition of essential supremum $\mu(A^c) = 0$. Hence, $\dim(\mu) \leq \dim(A)$. By Billingsley's Lemma, $\dim(A) \leq \alpha$. Taking $\alpha \to \alpha_\mu$ gives $\dim(\mu) \leq \alpha_\mu$.

To prove the other direction, let $\alpha < \alpha_\mu$ and consider

$$B = \left\{ x : \liminf_{n \to \infty} \frac{\log \mu(I_n(x))}{\log |I_n(x)|} \geq \alpha \right\}.$$

By the definition of α_μ, we have $\mu(B) > 0$. If $\mu(E^c) = 0$, then we also have $0 < \mu(B) = \mu(E \cap B)$ and

$$\liminf_{n \to \infty} \frac{\log \mu(I_n(x))}{\log |I_n(x)|} \geq \alpha$$

on $E \cap B$. Billingsley's Lemma shows $\dim(E) \geq \dim(E \cap B) \geq \alpha$. Therefore $\dim(\mu) \geq \alpha$ for all $\alpha < \alpha_\mu$. We deduce $\dim(\mu) = \alpha_\mu$. $\square$

Example 1.4.5   Consider the measure μ on the middle thirds Cantor set **C** that gives equal mass to each nth generation interval in the construction. This is called the **Cantor singular measure**. If we consider a 3-adic interval I of length 3^{-n} then

$$\mu(I) = 2^{-n} = |I|^{\log_3 2},$$

if I^o (the interior of I) hits **C** and is 0 otherwise. Thus

$$\lim_{n \to \infty} \frac{\log \mu(I_n(x))}{\log |I_n(x)|} = \log_3 2,$$

for all $x \in \mathbf{C}$, and hence for μ almost every x. Therefore Lemma 1.4.4 implies $\dim(\mu) = \log_3 2$.

1.5 Sets defined by digit frequency

We previously considered sets with restrictions on what the nth digit of the b-ary expansion could be. In this section we do not restrict particular digits, but will require that each digit occurs with a certain frequency. The resulting sets are dense in $[0,1]$, so we need only consider their Hausdorff dimension.

Example 1.5.1 $A_p = \{x = \sum_{n=1}^{\infty} x_n 2^{-n} : x_n \in \{0,1\}, \lim_{j\to\infty} \frac{1}{j} \sum_{k=1}^{j} x_k = p\}$.

Thus A_p is the set of real numbers in $[0,1]$ in which a 1 occurs in the binary expansion with asymptotic frequency p. For "typical" real numbers we expect a 1 to occur about half the time, and indeed, $A_{1/2}$ is a set of full Lebesgue measure in $[0,1]$ (see below). In general we will show

$$\dim(A_p) = h_2(p) = -p\log_2 p - (1-p)\log_2(1-p).$$

The quantity h_2 is called the **entropy** of p and is strictly less than 1 except for $p = 1/2$. It represents the uncertainty associated to the probability (if $p = 0$ or 1 the entropy is 0; it is maximized when $p = 1/2$). See Cover and Thomas (1991).

In addition to Billingsley's Lemma we will need

Theorem 1.5.2 (Strong Law of Large Numbers)   *Let (X, dv) be a probability space and $\{f_n\}$, $n = 1, 2 \ldots$ a sequence of orthogonal functions in $L^2(X, dv)$. Suppose $E(f_n^2) = \int |f_n|^2 dv \leq 1$, for all n. Then*

$$\frac{1}{n}S_n = \frac{1}{n}\sum_{k=1}^{n} f_k \to 0,$$

a.e. (with respect to v) as $n \to \infty$.

Proof   We begin with the simple observation that if $\{g_n\}$ is a sequence of functions on a probability space (X, dv) such that

$$\sum_n \int |g_n|^2 \, dv < \infty,$$

then $\sum_n |g_n|^2 < \infty$ v-a.e. and hence $g_n \to 0$ v-a.e.

Using this, it is easy to verify the Strong Law of Large Numbers (LLN) for

$n \to \infty$ along the sequence of squares. Specifically, since the functions $\{f_n\}$ are orthogonal,

$$\int \left(\frac{1}{n} S_n\right)^2 dv = \frac{1}{n^2} \int |S_n|^2 dv = \frac{1}{n^2} \sum_{k=1}^{n} \int |f_k|^2 dv \le \frac{1}{n}.$$

Thus if we set $g_n = \frac{1}{n^2} S_{n^2}$, we have

$$\int g_n^2 \, dv \le \frac{1}{n^2}.$$

Since the right-hand side is summable, the observation made above implies that $g_n = n^{-2} S_{n^2} \to 0$ v-a.e. To handle the limit over all positive integers, suppose that $m^2 \le n < (m+1)^2$. Then

$$\int \left|\frac{1}{m^2} S_n - \frac{1}{m^2} S_{m^2}\right|^2 dv = \frac{1}{m^4} \int \left| \sum_{k=m^2+1}^{n} f_k \right|^2 dv$$

$$= \frac{1}{m^4} \int \sum_{k=m^2+1}^{n} |f_k|^2 dv$$

$$\le \frac{2}{m^3},$$

since the sum has at most $2m$ terms, each of size at most 1. Set $m(n) = \lfloor \sqrt{n} \rfloor$ and

$$h_n = \frac{S_n}{m(n)^2} - \frac{S_{m(n)^2}}{m(n)^2}.$$

Now each integer m equals $m(n)$ for at most $2m + 1$ different choices of n. Therefore,

$$\sum_{n=1}^{\infty} \int |h_n|^2 d\mu \le \sum_{n=1}^{\infty} \frac{2}{m(n)^3} \le \sum_m (2m+1) \frac{2}{m^3} < \infty,$$

so by the initial observation, $h_n \to 0$ a.e. with respect to v. This yields that

$$\frac{1}{m(n)^2} S_n \to 0 \text{ a.e.},$$

that, in turn, implies that $\frac{1}{n} S_n \to 0$ a.e., as claimed.   □

Theorem 1.5.2 is called a *Strong* Law of Large Numbers because it gives a.e. convergence, as opposed to the *Weak* Law of Large Numbers, that refers to convergence in measure. We frequently make use of the following

Corollary 1.5.3   *If $\{X_k\}$ are independent identically distributed (i.i.d.) random variables and $\mathbb{E}[X_k^2] < \infty$ then $\lim_{n\to\infty} \frac{1}{n} \sum_{k=1}^{n} X_k = \mathbb{E}[X_1]$.*

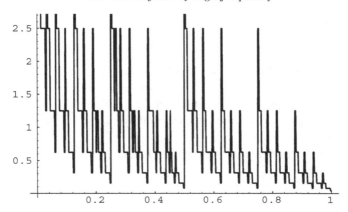

Figure 1.5.1 A histogram of $\mu_{1/3}$ applied to dyadic intervals of length 2^{-7}.

Proof Note that $\{X_k - \mathbb{E}[X_k]\}$ are orthogonal and apply Theorem 1.5.2 a.e.
□

Next, define a measure μ_p on $[0,1]$ as follows. By Caratheodory's Extension Theorem it suffices to define μ_p on dyadic intervals. Let

$$\mu_p\left(\left[\frac{j}{2^n}, \frac{j+1}{2^n}\right)\right) = p^{k(j)}(1-p)^{n-k(j)},$$

where $k(j)$ is the number of 1s in the binary expression of j. We can also write μ_p as

$$\mu_p(I_n(x)) = p^{\sum_{k=1}^{n} x_k}(1-p)^{n-\sum_{k=1}^{n} x_k},$$

where $x = \sum_{k=1}^{\infty} x_k 2^{-k}$.

An alternative way of describing μ_p is that it gives the whole interval $[0,1]$ mass 1, and gives the two subintervals $[0,1/2]$, $[1/2,1]$ mass $1-p$ and p respectively. In general, if a dyadic interval I has measure $\mu_p(I)$, then the left half has measure $(1-p)\mu_p(I)$ and the right half has measure $p\mu_p(I)$. Observe that if $p = 1/2$ then this is exactly Lebesgue measure on $[0,1]$. In Figure 1.5.1 we have graphed the density of the measure μ_p for $p = 1/3$ on all dyadic intervals of length 2^{-7} in $[0,1]$.

Lemma 1.5.4 $\dim(A_p) = \dim(\mu_p) = h_2(p)$

Proof We start by proving that

$$\mu_p(A_p) = 1.$$

To see this, let

$$f_n(x) = (x_n - p)$$

(where $\sum_k x_k 2^{-k}$ is the binary expansion of x). If $S_n = \sum_{k=1}^n f_k$, then unwinding definitions shows that A_p is exactly the set where $\frac{1}{n} S_n \to 0$, so we will be done if we can apply the Law of Large Numbers to $\{f_n\}$ and the measure μ_p.

Clearly, $\int f_n^2 \, d\mu_p \leq 1$. It is easy to check orthogonality, $\int f_n f_m \, d\mu_p = 0$ for $n \neq m$. Thus $\mu_p(A_p) = 1$.

Now, we show that $\dim(\mu_p) = h_2(p)$. By Lemma 1.4.4,

$$\dim(\mu_p) = \operatorname{ess\,sup} \left\{ \liminf_{n \to \infty} \frac{\log \mu(I_n(x))}{\log |I_n(x)|} \right\}.$$

Note that

$$\frac{\log \mu_p(I_n(x))}{\log |I_n(x)|} = \frac{1}{\log 2} \left(\left(\frac{1}{n} \sum_{k=1}^n x_k \right) \log \frac{1}{p} + \left(\frac{1}{n} \sum_{k=1}^n (1 - x_k) \log \frac{1}{1-p} \right) \right).$$

Since μ_p-almost every point is in A_p, we see that

$$\lim_{n \to \infty} \frac{1}{n} \sum_{k=1}^n x_k = p,$$

for μ_p-a.e. x. Therefore,

$$\frac{\log \mu_p(I_n(x))}{\log |I_n(x)|} \to \frac{p \log \frac{1}{p} + (1 - p) \log \frac{1}{1-p}}{\log 2} = h_2(p),$$

for a.e. x with respect to μ_p. Thus $\dim(\mu_p) = h_2(p)$. By Definition 1.4.3 of $\dim(\mu_p)$, we deduce from $\mu_p(A_p) = 1$ that $\dim(A_p) \geq \dim(\mu_p) = h_2(p)$. However, since

$$A_p \subset \left\{ x \in [0,1] : \liminf_{n \to \infty} \frac{\log \mu_p(I_n(x))}{\log |I_n(x)|} = h_2(p) \right\},$$

and $\mu_p(A_p) = 1 > 0$, Billingsley's Lemma implies $\dim(A_p) \leq h_2(p)$, hence equality. $\qquad\square$

Example 1.5.5   Consider the related sets

$$\widetilde{A_p} = \left\{ x \in [0,1] : \limsup_{n \to \infty} \frac{1}{n} \sum_{k=1}^n x_k \leq p \right\},$$

$$\widehat{A_p} = \left\{ x \in [0,1] : \liminf_{n \to \infty} \frac{1}{n} \sum_{k=1}^n x_k \leq p \right\}.$$

Clearly, $A_p \subset \widetilde{A_p} \subset \widehat{A_p}$. Since $\dim(A_{1/2}) = 1$, we have $\dim(\widetilde{A_p}) = \dim(\widehat{A_p}) = 1$ if $p \geq 1/2$. If $p < 1/2$, then by Lemma 1.5.4,

$$h_2(p) = \dim(A_p) \leq \dim(\widetilde{A_p}) \leq \dim(\widehat{A_p}).$$

On the other hand, we saw before that

$$\frac{\log \mu_p(I_n(x))}{\log |I_n(x)|} = \frac{1}{\log 2}\left(\log \frac{1}{1-p} + \left(\log \frac{1-p}{p}\right) \cdot \frac{1}{n} \sum_{k=1}^{n} x_k\right).$$

We have $\log \frac{1-p}{p} > 0$ since $p < 1/2$, so for $x \in \widehat{A_p}$, we get

$$\liminf_{n \to \infty} \frac{\log \mu_p(I_n(x))}{\log |I_n(x)|} \leq h_2(p).$$

Therefore, for $p < 1/2$, Billingsley's Lemma implies that

$$\dim(A_p) = \dim(\widetilde{A_p}) = \dim(\widehat{A_p}) = h_2(p).$$

Example 1.5.6 The same argument used in Lemma 1.5.4 to compute the dimension of A_p works in a more general setting. Suppose $\mathbf{p} = (p_0, \ldots, p_{b-1})$ is a probability vector, i.e., $\sum_{k=0}^{b-1} p_k = 1$ and define a measure $\mu_{\mathbf{p}}$ by

$$\mu_{\mathbf{p}}(I_n(x)) = \prod_{j=1}^{n} p_{x_j},$$

where $\{x_n\}$ is the b-ary expansion of x. Then repeating the proof of the lemma shows

$$\dim(\mu_{\mathbf{p}}) = h_b(\mathbf{p}) = \sum_{k=0}^{b-1} p_k \log_b \frac{1}{p_k}.$$

Similarly, the set of x's in $[0, 1]$ such that

$$\lim_{N \to \infty} \frac{\#(\{n \in [0, N] : x_n = k\})}{N} = p_k$$

for each $k = 0, \ldots, b-1$ has dimension $h_b(\mathbf{p})$.

The following is a variant we will need later in the proof of Proposition 1.7.7 about intersections of random translates of Cantor sets.

Fix an integer $b \geq 2$ and a set $E \subset \{0, 1, \ldots, b-1\}$. Let $\{x_n\}$ be the b-ary expansion of the real number x.

Lemma 1.5.7 *The set*

$$X_p^E = \left\{x \in [0, 1) : \lim_{N \to \infty} \frac{1}{N} \sum_{n=1}^{N} \mathbf{1}_E(x_n) = p\right\}$$

has Hausdorff dimension

$$-p\log_b\Big(\frac{p}{\#(E)}\Big) - (1-p)\log_b\Big(\frac{1-p}{b-\#(E)}\Big).$$

Proof Let $I_n(x)$ denote the b-ary interval of generation n containing x. Define a Borel measure on $[0,1)$ by

$$\mu(I_n(x)) = \Big(\frac{p}{\#(E)}\Big)^{\Sigma_{k\leq n}\mathbf{1}_E(x_k)}\Big(\frac{1-p}{b-\#(E)}\Big)^{\Sigma_{k\leq n}(1-\mathbf{1}_E(x_k))}.$$

The proof of Lemma 1.5.4 shows that

$$\dim(X_p^E) = \dim(\mu) = -p\log_b\Big(\frac{p}{\#(E)}\Big) - (1-p)\log_b\Big(\frac{1-p}{b-\#(E)}\Big). \qquad \square$$

1.6 Slices

If $A \subset \mathbb{R}^2$ has dimension α, what should we expect the dimension of $A \cap L$ to be, where L is a "typical" line? For the present, let us consider only vertical lines and set

$$A_x = \{y : (x,y) \in A\}.$$

Theorem 1.6.1 (Marstrand Slicing Theorem) *Let $A \subset \mathbb{R}^2$ and suppose that* $\dim(A) \geq 1$. *Then*

$$\dim(A_x) \leq \dim(A) - 1,$$

for (Lebesgue) almost every x.

If $\dim(A) < 1$, then the slice A_x is empty for almost every x (in fact, it is empty except for a set of exceptional x of dimension at most $\dim(A)$; see Exercise 1.9). On the other hand, it is possible that $\dim(A_x) = \dim(A)$ for some values of x, e.g., if A is a vertical line segment.

The inequality can be strict for every x, e.g., there are real-valued functions on $\mathbb{R}$ whose graphs have dimension strictly larger than one, even though all the vertical slices are singletons. See Chapter 5. For example, we will prove that the graph of a 1-dimensional Brownian motion has dimension $3/2$ almost surely (see Theorem 6.4.3). We shall give several other examples of strict inequality later in this section.

Proof We start with the following claim: for $1 \leq \alpha \leq 2$

$$\mathscr{H}^\alpha(A) \geq c_2 \int_{\mathbb{R}} \mathscr{H}^{\alpha-1}(A_x)\, dx.$$

(Measurability of the integrand follows from a monotone class argument that we omit.) In fact, we can take $c_2 = 1$. (But in higher dimensions the analogous constant is $c_n < 1$.) It suffices to prove this claim because for $\alpha > \dim(A)$,

$$0 = \mathscr{H}^\alpha(A) \geq \int_{\mathbb{R}} \mathscr{H}^{\alpha-1}(A_x)\, dx$$

implies $\mathscr{H}^{\alpha-1}(A_x) = 0$ for Lebesgue almost every x. Thus $\dim(A_x) \leq \alpha - 1$ for Lebesgue almost every x.

To prove the claim fix $\varepsilon > 0$ and $\delta > 0$ and let $\{D_j\}$ be a cover of A with $|D_j| < \varepsilon$ and

$$\sum_j |D_j|^\alpha < \mathscr{H}^\alpha_\varepsilon(A) + \delta.$$

Enclose each D_j in a square Q_j with sides parallel to the axes and with side length $s_j \leq |D_j|$. Let I_j be the projection of Q_j onto the x-axis. For each x, the slices $\{(Q_j)_x\}$ form a cover of A_x and have length

$$|(Q_j)_x| = \begin{cases} s_j, & x \in I_j \\ 0, & x \notin I_j \end{cases}.$$

(In the higher-dimensional case, this is the point where the constant $c_n < 1$ appears. For example, if $n = 3$, then a slice of a cube Q_j has diameter either $\sqrt{2}s_j$ or 0.)

We now have an ε-cover of A_x and

$$\mathscr{H}^{\alpha-1}_\varepsilon(A_x) \leq \sum_j |(Q_j)_x|^{\alpha-1} = \sum_{j:x\in I_j} s_j^{\alpha-1}.$$

Therefore

$$\int_{\mathbb{R}} \mathscr{H}^{\alpha-1}_\varepsilon(A_x)\, dx \leq \int_{\mathbb{R}} \Big(\sum_{j:x\in I_j} s_j^{\alpha-1} \Big) dx$$
$$= \sum_j s_j^\alpha$$
$$\leq \mathscr{H}^\alpha_\varepsilon(A) + \delta.$$

Taking $\delta \to 0$ gives

$$\int_{\mathbb{R}} \mathscr{H}^{\alpha-1}_\varepsilon(A_x)\, dx \leq \mathscr{H}^\alpha_\varepsilon(A).$$

As $\varepsilon \to 0$, $\mathscr{H}^{\alpha-1}_\varepsilon(A_x) \nearrow \mathscr{H}^{\alpha-1}(A_x)$, so the Monotone Convergence Theorem implies

$$\int_{\mathbb{R}} \mathscr{H}^{\alpha-1}(A_x)\, dx \leq \mathscr{H}^\alpha(A). \qquad \square$$

There is a generalization of the Slicing Theorem to higher dimensions.

Theorem 1.6.2   *Let $A \subset \mathbb{R}^n$ be a Borel set with $\dim(A) > n - m$ and let E_m be an m-dimensional subspace. Then for almost every $x \in E_m^\perp$ (or equivalently, almost every $x \in \mathbb{R}^n$)*

$$\dim(A \cap (E_m + x)) \leq \dim(A) - (n - m),$$

and for $\alpha > n - m$,

$$\mathscr{H}^\alpha(A) \geq c_n \int_{E_m^\perp} \mathscr{H}^{\alpha - (n-m)}(A \cap (E_m + x))\, dx.$$

The proof is almost the same as of Theorem 1.6.1, so we omit it. As before, the equality can be strict. There are also generalizations that replace the "almost everywhere with respect to Lebesgue measure on $\mathbb{R}^n$" with a more general measure. See Theorem 3.3.1.

Example 1.6.3   Let $A = \mathbf{C} \times \mathbf{C}$ be the product of the middle thirds Cantor set with itself. We saw in Example 1.3.4 that $\dim(A) = \log_3 4$. The vertical slices of A are empty almost surely, so $\dim(A_x) = 0 < (\log_3 4) - 1$ almost surely. This gives strict inequality in Theorem 1.6.1 for almost all slices.

Example 1.6.4   A set with more interesting slices is the Sierpiński gasket G (see Example 1.3.4). We claim that (Lebesgue) almost every vertical slice G_x of G has dimension $1/2$ (note that this is strictly smaller than the estimate in the slicing theorem, $\dim(G) - 1 = \log_2 3 - 1 = .58496\ldots$).

Proof   For each $x \in [0, 1]$, let

$$S(x) = \{n : x_n = 0\},$$

where $\{x_n\}$ is the binary expansion of x. Then it is easy to see

$$G_x = A_{S(x)},$$

where A_S is the set defined in Example 1.3.2 and discussed in Example 1.4.2. The Law of Large Numbers says that in almost every x the digits 0 and 1 occur with equal frequency in its binary expansion. We deduce $\dim(G_x) = 1/2$ for almost every x.   □

More generally, $\dim(G_x) = p$ if and only if $\limsup \frac{\sum_{k=1}^n x_k}{n} = 1 - p$. It follows from the proof of Lemma 1.5.4 and Example 1.5.5 that the set of such xs has dimension $h_2(p)$. Thus

$$\dim(\{x : \dim(G_x) = p\}) = h_2(p).$$

Example 1.6.5   Consider the self-affine sets of Example 1.3.5. More precisely, consider the McMullen set X obtained by taking

$$A = \begin{pmatrix} 0 & 1 & 0 \\ 1 & 0 & 1 \end{pmatrix}.$$

The intersection of X with any vertical line is a single point. On the other hand the intersection of the set with a horizontal line of height $y = \sum_n y_n 2^{-n}$ is a set described as follows: at the nth generation replace each interval by its middle third if $y_n = 1$ but remove the middle third if $y_n = 0$. By Lemma 1.4.4 the dimension of such a set is

$$(\log_3 2) \liminf_{N \to \infty} \frac{1}{N} \sum_{n=1}^{N} (1 - y_n).$$

Since for almost every $y \in [0,1]$ the limit exists and equals $\frac{1}{2}$, we see that almost every horizontal cross-section of the McMullen set has dimension $\frac{1}{2} \log_3 2$. See Exercise 1.42 for a refinement.

While it is easy to compute the dimension of the cross-sections, it is more challenging to compute the dimension of the McMullen set itself. Marstrand's Slicing Theorem implies that its dimension is at least $1 + \frac{1}{2} \log_3 2 \approx 1.31546$ (we shall use the Law of Large Numbers in Chapter 4 to prove that its dimension is exactly $\log_2(1 + 2^{\log_3 2}) \approx 1.34968$).

1.7 Intersecting translates of Cantor sets *

In the previous section we considered intersecting a set in the plane with random lines. In this section we will intersect a set in the line with random translates of itself. It turns out that this is easiest when the set has a particular arithmetic structure, such as the middle thirds Cantor set. We include the discussion here, because this special case behaves very much like the "digit restriction sets" considered earlier. This section also serves to illustrate a principle that will be more apparent later in the book: it is often easier to compute the expected dimension of a random family of sets than to compute the dimension of any particular member of the family.

J. Hawkes (1975) proved the following:

Theorem 1.7.1   *Let* **C** *be the middle thirds Cantor set. Then*

$$\dim((\mathbf{C} + t) \cap \mathbf{C}) = \frac{1}{3} \frac{\log 2}{\log 3}$$

for Lebesgue–a.e. $t \in [-1, 1]$.

We start by sketching a simplified proof Theorem 1.7.1 for Minkowski dimension. The proof for Hausdorff dimension will be given in greater generality in Theorem 1.7.3. Hawkes' original proof was more elaborate, as he proved a result about general Hausdorff gauge functions and derived Theorem 1.7.1 as a corollary.

Proof of Theorem 1.7.1 for Minkowski dimension   The key observation is that the digits $\{-2,0,2\}$ can be used to represent any $t \in [-1,1]$ in base 3, and the representation is unique if t is not a ternary rational. Fix a t that is not a ternary rational, and let $\mathbf{C}$ denote the middle thirds Cantor set. Represent

$$t = \sum_{n=1}^{\infty} t_n 3^{-n}$$

with $t_n \in \{-2,0,2\}$. We have

$$(\mathbf{C}+t) \cap \mathbf{C} = \{y \in \mathbf{C} : \exists x \in \mathbf{C}, y - x = t\}$$

$$= \left\{ \sum_{1}^{\infty} y_n 3^{-n} : \exists x = \sum_{1}^{\infty} x_n 3^{-n}, \ y_n, x_n \in \{0,2\}, \ y_n - x_n = t_n \right\}$$

$$= \left\{ \sum_{1}^{\infty} y_n 3^{-n} : y_n \in \{0,2\}, \ y_n = 2 \text{ if } t_n = 2, \ y_n = 0 \text{ if } t_n = -2 \right\}.$$

Thus for $y \in (\mathbf{C}+t) \cap \mathbf{C}$ the nth ternary digit is determined unless $t_n = 0$, in which case it may take two values. For almost all $t \in [-1,1]$ with respect to Lebesgue measure, the set of indices $\{n : t_n = 0\}$ has density $1/3$, so we get $N((\mathbf{C}+t) \cap \mathbf{C}, 3^{-n}) = 2^{n/3+o(n)}$. This gives

$$\dim_{\mathscr{M}}((\mathbf{C}+t) \cap \mathbf{C}) = \frac{\log 2^{\frac{1}{3}}}{\log 3} = \frac{1}{3} \log_3 2,$$

for such t.   □

Next we discuss the cases to which (essentially) Hawkes' method extends. Let $b > 1$ be an integer, and D a finite set of integers. We denote

$$\Lambda(D,b) = \left\{ \sum_{n=1}^{\infty} d_n b^{-n} : d_n \in D \right\}.$$

As before, we denote the number of elements in D by $\#(D)$. When $b = 3$ and $D = \{0,2\}$, the set $\Lambda(D,b)$ is the middle thirds Cantor set.

Definition 1.7.2   Say that two finite sets of integers D_1, D_2 satisfy the b-**difference-set condition** if the difference-set $D_2 - D_1 \subset \mathbb{Z}$ is contained in an arithmetic progression of length b.

If D_1, D_2 satisfy the b-difference-set condition, then the representation in base b with digits from $D = D_2 - D_1$ "behaves" like the standard representation in base b: we claim that the representation is unique except at countably many points. This follows from the b-difference-set condition that guarantees that for some $a_0, d_0 \in \mathbb{Z}$ the inclusion

$$D \subset \widetilde{D} = \{a_0 + jd_0 : 0 \le j < b\}$$

holds. Let $s, t \in \mathbb{R}$ with $t = a_0/(b-1) + sd_0$. Then $s \in \Lambda(\{0, 1, \ldots, b-1\}, b)$ if and only if $t \in \Lambda(\widetilde{D}, b)$, and this equivalence establishes the claim above.

Theorem 1.7.3 *Fix an integer $b > 1$. Suppose D_1, D_2 are finite sets of integers satisfying the b-difference-set condition. For each integer $i \in D = D_2 - D_1$, let $M_i = \#((D_1 + i) \cap D_2)$. Denote $\Lambda_1 = \Lambda(D_1, b)$ and $\Lambda_2 = \Lambda(D_2, b)$. Then*

1. For all $t \in \mathbb{R}$, $\dim((\Lambda_1 + t) \cap \Lambda_2) = \dim_{\mathscr{M}}((\Lambda_1 + t) \cap \Lambda_2)$.
2. If $\#(D) = b$ then

$$\dim((\Lambda_1 + t) \cap \Lambda_2) = \frac{1}{b} \sum_{i \in D} \log_b M_i,$$

for Lebesgue almost all t in the set $\Lambda(D, b) = \Lambda_2 - \Lambda_1$.

In the examples below, we show how Theorem 1.7.3 implies Hawkes' Theorem as well as Proposition 1.7.7. The proof of Theorem 1.7.3 depends on the following lemma.

Lemma 1.7.4 *Fix an integer $b > 1$. Assume D_1 and D_2 are finite sets of integers that satisfy the b-difference-set condition. Let $D = D_2 - D_1$, and for each $i \in D$, let $M_i = \#((D_1 + i) \cap D_2)$. If $t \in \Lambda(D, b)$ has a unique representation $t = \sum_{n=1}^{\infty} t_n b^{-n}$ with $t_n \in D$, define*

$$\varphi_1(t) = \liminf_{N \to \infty} \frac{1}{N} \sum_{n=1}^{N} \log_b M_{t_n}. \qquad (1.7.1)$$

If t has two such representations, or $t \notin \Lambda(D, b)$, define $\varphi_1(t) = 0$. Then

$$\dim((\Lambda_1 + t) \cap \Lambda_2) = \underline{\dim}_{\mathscr{M}}((\Lambda_1 + t) \cap \Lambda_2) = \varphi_1(t), \qquad (1.7.2)$$

for all $t \in \mathbb{R}$, where $\Lambda_1 = \Lambda(D_1, b)$ and $\Lambda_2 = \Lambda(D_2, b)$.

The assumption that $t_n \in D$ implies that $M_{t_n} \ge 1$. Therefore the right-hand side of (1.7.1) is well defined.

Proof of Lemma 1.7.4 Assume first that t has a unique representation

$$t = \sum_{n=1}^{\infty} t_n b^{-n}$$

with $t_n \in D$. Then

$$(\Lambda_1 + t) \cap \Lambda_2 = \left\{ \sum_{n=1}^{\infty} d_n b^{-n} : d_n \in (D_1 + t_n) \cap D_2 \text{ for all } n \right\}$$

and it follows immediately from the definition of Minkowski dimension and φ_1 that

$$\dim_{\mathscr{M}}((\Lambda_1 + t) \cap \Lambda_2) = \varphi_1(t).$$

To compute the Hausdorff dimension, define a probability measure μ_t supported on $(\Lambda_1 + t) \cap \Lambda_2$ as follows. For any sequence $d_1^*, d_2^*, \dots, d_N^*$ such that $d_j^* \in (D_1 + t_j) \cap D_2$ let

$$\mu_t \left\{ \sum_{j=1}^{N} d_j^* b^{-j} + \sum_{j=N+1}^{\infty} d_j b^{-j} : d_j \in (D_1 + t_j) \cap D_2 \text{ for all } j > N \right\} = \prod_{j=1}^{N} M_{t_j}^{-1}.$$

Fix $y \in (\Lambda_1 + t) \cap \Lambda_2$, and let $C_N(y)$ denote the subset of $(\Lambda_1 + t) \cap \Lambda_2$ consisting of points for which the first N digits in the base b representation (with digit set D_2) agree with the first N digits of y. Then

$$\liminf_{N \to \infty} \frac{-\log \mu_t(C_N(y))}{N \log b} = \varphi_1(t),$$

so by Billingsley's Lemma, $\dim((\Lambda_1 + t) \cap \Lambda_2) = \varphi_1(t)$. Similar considerations appear in Cajar (1981).

If $t \in \Lambda(D, b)$ has two distinct representations

$$t = \sum_{n=1}^{\infty} t_n b^{-n} = \sum_{n=1}^{\infty} t_n' b^{-n}$$

then (as we noted earlier in the chapter) we eventually have $t_n = \max D$ and $t_n' = \min D$ or vice versa. For such a t, we can show the set $(\Lambda_1 + t) \cap \Lambda_2$ is finite. Finally, for $t \notin \Lambda(D, b)$ the set $(\Lambda_1 + t) \cap \Lambda_2$ is empty. $\square$

Proof of Theorem 1.7.3 Statement (i) is contained in Lemma 1.7.4. Next we prove (ii). When representing numbers in $\Lambda(D, b)$, the digits are independent, identically distributed random variables with respect to the normalized Lebesgue measure on the closed set $\Lambda(D, b)$ (we have used this several times already). Thus, by the Law of Large Numbers,

$$\lim_{N \to \infty} \frac{1}{N} \sum_{n=1}^{N} \log_b M_{t_n} = \frac{\int_{\Lambda(D,b)} \log_b(M_t) \, dt}{\int_{\Lambda(D,b)} 1 \, dt} = \frac{1}{b} \sum_{i \in D} \log_b M_i$$

for a.e. $t = \sum_{n=1}^{\infty} t_n b^{-n} \in \Lambda(D, b)$. Lemma 1.7.4 now gives the desired conclusion. $\square$

Example 1.7.5   If $b = 3$ and $D_1 = D_2 = \{0, 2\}$, then $D = D_1 - D_2 = \{-2, 0, 2\}$, $\Lambda_1, \Lambda_2 = \mathbf{C}$, the middle thirds Cantor set, and $\{M_j\}_{j \in D} = \{1, 2, 1\}$. Thus Theorem 1.7.1 is contained in Theorem 1.7.3.

Example 1.7.6   Fix an integer $b > 1$. Extending the previous example, assume D_1, D_2 are arithmetic progressions with the same difference of length n_1, n_2 respectively, where $n_1 + n_2 = b + 1$. Without loss of generality, $n_2 \le n_1$,

$$D_1 = \{1, \ldots, n_1\} \text{ and } D_2 = \{1, \ldots, n_2\}.$$

Then $D = D_2 - D_1 = \{1 - n_1, 2 - n_1, \ldots, n_2 - 1\}$ and

$$M_j = \begin{cases} j + n_1 & -n_1 < j \le n_2 - n_1, \\ n_2 & n_2 - n_1 < j < 0, \\ n_2 - j & 0 \le j < n_2. \end{cases}$$

Thus Theorem 1.7.3 (ii) shows that for a.e. $t \in \Lambda(D, b)$

$$\dim((\Lambda_1 + t) \cap \Lambda_2) = \frac{1}{b} \log_b((n_2!)^2 n_2^{n_1 - n_2 - 1}).$$

Theorem 1.7.3 gives the dimension of intersections for typical translates. It is possible to go further and compute certain iterated dimensions, just as we did for slices of the Sierpiński gasket. Such iterated dimensions are closely related to the popular "multi-fractal analysis".

Proposition 1.7.7

$$\dim\{t : \dim((\mathbf{C} + t) \cap \mathbf{C}) = \alpha \log_3 2\} = h_3\left(\frac{1 - \alpha}{2}, \alpha, \frac{1 - \alpha}{2}\right), \quad (1.7.3)$$

where $\mathbf{C}$ is the middle thirds Cantor set, $0 < \alpha < 1$ and

$$h_3(p_1, p_2, p_3) = -\sum_{i=1}^{3} p_i \log_3 p_i$$

is the ternary entropy function.

Proof   From the considerations in the proofs of Lemma 1.7.4 and Theorem 1.7.3 we see that we want to find the dimension of the set of ts such that

$$\liminf_{N \to \infty} \frac{1}{N} \sum_{n=1}^{N} \log_b M_{t_n} = \alpha \log_3 2,$$

where $\{t_n\}$ is the ternary expansion of t. Equivalently (since $M_{t_n} = 2$ if $t_n = 0$ and 1 otherwise), we want the set of t so that

$$\liminf_{N \to \infty} \frac{1}{N} \sum_{n=1}^{N} \mathbf{1}_{\{t_n = 0\}} = \alpha.$$

Thus Lemma 1.5.7 implies the dimension in (1.7.3) is

$$-\alpha \log_3 \alpha - (1 - \alpha) \log_3 \frac{(1 - \alpha)}{2} = h_3 \left(\frac{1 - \alpha}{2}, \alpha, \frac{1 - \alpha}{2} \right),$$

as desired. $\qquad\qquad\qquad\qquad\qquad\qquad\qquad\qquad\qquad\qquad\qquad$ $\square$

1.8 Notes

Hausdorff dimension was invented by Felix Hausdorff in his 1918 paper *Dimension und äußeres Maß*. Hausdorff starts his paper with the comment (from the English translation in Edgar (2004)):

Mr. Carathéodory has defined an exceptionally simple and general measure theory, that contains Lebesgue's theory as a special case, and which, in particular, defines the p-dimensional measure of a point set in q-dimensional space. In this paper we shall add a small contribution to this work. . . . we introduce an explanation of p-dimensional measure which can be immediately be extended to non-integer values of p and suggests the existence of sets of fractional dimension, and even of sets whose dimensions fill out the scale of positive integers to a more refined, e.g., logarithmic scale.

In addition to the definition of Hausdorff dimension, the paper contains its computation for various Cantor sets and the construction of Jordan curves with all dimensions between 1 and 2. The definition of Hausdorff dimension we give using content is equivalent to Hausdorff's original 1918 definition, but he used Hausdorff measure, as in Proposition 1.2.6.

If $A = (a_{ij})$ has real, non-negative entries then the Perron–Frobenius theorem says that $\rho(A)$ (defined in(1.3.4)) is an eigenvalue of A of maximal modulus. Moreover, if A is primitive, then all the other eigenvalues are strictly smaller in modulus. **Primitive** means that some power A^n of A has all positive entries. Thus the dimension of X_A can (in theory) be computed from A. Such a formula for $\dim_{\mathscr{H}}(X_A)$ was first given by Parry (1964) and Furstenberg (1967) proved the equality for Hausdorff dimension in 1967. We will give the proof in Section 2.3.

The dimension of sets defined by digit expansions was first determined by Besicovitch (1935) for binary expansions and later by Eggleston (1949) for general bases. The latter paper appeared just a year after Shannon's seminal paper on information theory, Shannon (1948). In the Math Review by J.L. Doob of Shannon's paper, he wrote: "The discussion is suggestive throughout, rather than mathematical, and it is not always clear that the author's mathematical intentions are honorable."

In the Strong Law of Large Numbers (Theorem 1.5.2), better estimates for the decay of S_n are possible if we assume that the functions $\{f_n\}$ are independent with respect to the measure v. This means that for any n and any collection of measurable sets $\{A_1, \ldots, A_n\}$ we have

$$v(\{x \in X : f_j(x) \in A_j, j = 1, \ldots, n\}) = \prod_{j=1}^{n} v(\{x \in X : f_j(x) \in A_j\}).$$

Roughly, this says that knowing the values at x for any subset of the $\{f_j\}$ does not give us any information about the values of the remaining functions there.

By 1915 Hausdorff had proved that if $\{f_n\}$ are independent and satisfy $\int f_n dv = 0$ and $\int f_n^2 \, dv = 1$, then

$$\lim_{N \to \infty} \frac{1}{N^{\frac{1}{2}+\varepsilon}} \sum_{n=0}^{N} f_n(x) = 0 \text{ for a.e. } x$$

and for every $\varepsilon > 0$. After that Hardy–Littlewood, and independently Khinchin, proved

$$\lim_{N \to \infty} \frac{1}{\sqrt{N \log N}} \sum_{n=0}^{N} f_n(x) = 0 \text{ for a.e. } x.$$

The "final" result, found by Khinchin for a special case in 1928 and proved in general by Hartman–Wintner in 1941, says

$$\limsup_{N \to \infty} \frac{1}{\sqrt{2N \log \log N}} \sum_{n=0}^{N} f_n(x) = 1 \text{ for a.e. } x.$$

It is natural to expect that Theorem 1.7.1 should extend to other Cantor sets defined by digit restrictions, but this extension, given in Kenyon and Peres (1991), turns out to depend on the theory of random matrix products, which is beyond the scope of this volume. The dimensions that arise in general also seem different, as they do not appear to be ratios of logarithms of rational numbers. For example, if $\Lambda = \{\sum_{n=1}^{\infty} d_n 4^{-n} : d_n \in \{0,1,2\}\}$ then for almost every $t \in [-1,1]$,

$$\dim((\Lambda+t) \cap \Lambda) = \frac{1}{6} \log_4 \frac{2}{3} + \sum_{k=0}^{\infty} 4^{-k-1} \log_4 \frac{(3 \cdot 2^k)!}{(2^{k+1})!} \approx 0.575228.$$

Proposition 1.7.7 is due to Kenyon and Peres (1991).

The Sierpiński carpet (Figure 1.3.4) is a special case of a more general class of planar sets called carpets: any planar set that is compact, connected, nowhere dense and locally connected, and so that any complementary domains are bounded by Jordan curves, is homeomorphic to the Sierpiński carpet. Examples of carpets arise naturally as Julia sets in rational dynamics and as limit sets of Kleinian groups. Although any two planar carpets are homeomorphic, they exhibit strong rigidity with respect to other classes of maps. For example, if we define a carpet S_p for odd p by iteratively omitting the center square from a $p \times p$ grid, Bonk and Merenkov (2013) showed S_p cannot be mapped to S_q by any quasisymmetric mapping and all the quasisymmetric self-maps of S_3 are isometries. See also Bonk et al. (2009), Bonk (2011).

1.9 Exercises

Exercise 1.1   For $0 < \alpha, \beta < 1$, let $K_{\alpha,\beta}$ be the Cantor set obtained as an intersection of the following nested compact sets. $K^0_{\alpha,\beta} = [0,1]$. The set $K^1_{\alpha,\beta}$ is obtained by leaving the first interval of length α and the last interval of length β, and removing the interval in between. To get $K^n_{\alpha,\beta}$, for each interval I in $K^{n-1}_{\alpha,\beta}$, leave the first interval of length $\alpha|I|$ and the last interval of length $\beta|I|$, and remove the subinterval in between. Compute the Minkowski dimension of $K_{\alpha,\beta}$.

Exercise 1.2   For $\alpha > 0$, let $E_\alpha = \{0\} \cup \{n^{-\alpha}\}_{n=1}^\infty$. Find $\dim_{\mathcal{M}}(E)$.

Exercise 1.3   A function f that satisfies $|f(x) - f(y)| \le C|x - y|^\gamma$ for some $C < \infty$, $\gamma > 0$ and for all $x, y \in E$ is called a Hölder function of order γ on E. Show that if E_α is defined as in Exercise 1.2, and f is Hölder of order γ on E, then $\overline{\dim}_{\mathcal{M}}(f(E_\alpha)) \le \frac{1}{1+\gamma\alpha}$.

Exercise 1.4   Construct a set K so that
$$\dim(K) < \underline{\dim}_{\mathcal{M}}(K) < \overline{\dim}_{\mathcal{M}}(K).$$

Exercise 1.5   Construct a countable compact set K so that
$$\underline{\dim}_{\mathcal{M}}(K) < \overline{\dim}_{\mathcal{M}}(K).$$

Exercise 1.6   Prove that if $A = \cup_{n=1}^\infty A_n$, then $\dim(A) = \sup_n \dim(A_n)$.

Exercise 1.7   If $E_1 \supset E_2 \supset \cdots$, is it true that
$$\dim\left(\bigcap_n E_n\right) = \lim_{n\to\infty} \dim(E_n)?$$

Exercise 1.8   Suppose φ is a continuous, increasing function on $[0,\infty)$ with $\varphi(0) = 0$ and assume $\liminf_{r\to 0} r^{-d}\varphi(r) > 0$. Construct a set $E \subset \mathbb{R}^d$ so that $0 < \mathcal{H}^\varphi(E) < \infty$.

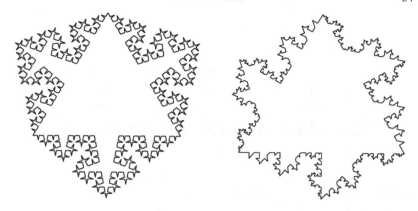

Figure 1.9.1 Curves for Exercise 1.12.

Exercise 1.9 Suppose $E \subset \mathbb{R}^n$ and $P : \mathbb{R}^n \to V$ is an orthogonal projection onto a subspace V. Prove that $\dim(P(E)) \leq \dim(E)$. More generally, prove this if $P : \mathbb{R}^n \to \mathbb{R}^n$ is any map satisfying the Lipschitz condition

$$\|P(x) - P(y)\| \leq A\|x - y\|,$$

for some $A < \infty$ and all $x, y \in \mathbb{R}^n$.

Exercise 1.10 Suppose $f : X \to Y$ is a γ-Hölder mapping between metric spaces. Show that $\dim(f(X)) \leq \dim(X)/\gamma$.

Exercise 1.11 Consider $[0,1]$ with $d(x,y) = \sqrt{|x-y|}$. Show the interval has dimension 2 with this metric.

• **Exercise 1.12** Estimate the dimension of the curves shown in Figure 1.9.1 assuming they are constructed like the von Koch snowflake (see Example 1.2.10) by replacing intervals by the same polygonal arc each time. (You may need a ruler to do this one.)

Exercise 1.13 Create a set in the complex plane by starting with the line segment $[0,1]$ and at each stage replacing each interval $I = [x,y]$ by the union of intervals $[x,z] \cup [z,w] \cup [w,z] \cup [z,y]$ where $z = \frac{1}{2}(x+y)$ and $w = z + i\beta(y-x)$, and we take $0 \leq \beta \leq 1/2$. See Figure 1.9.2. What is the dimension of the resulting set? This set is sometimes called the "antenna set".

Exercise 1.14 In the previous exercise, show the set is a solid triangle when $\beta = 1/2$. Use this to construct a continuous function f on the interval that maps the interval to a set that has non-empty interior. This is called a **Peano curve**.

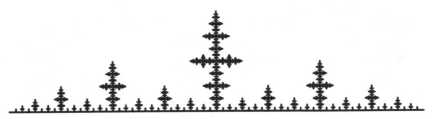

Figure 1.9.2 Set in Exercise 1.13.

Exercise 1.15   What is the dimension of almost every vertical slice of the antenna set in Exercise 1.13? Start with $\beta = 1/4$.

Exercise 1.16   Given two compact sets A, B, we define the **Hausdorff distance** between them as

$$d_H(A,B) = \max\{\max_{a \in A} \text{dist}(a,B), \max_{b \in B} \text{dist}(b,A)\}.$$

If $A_n \to A$ in the Hausdorff metric, does $\dim(A_n) \to \dim(A)$?

Exercise 1.17   Let $S = \bigcup_{n=1}^{\infty} [(2n)!, (2n+1)!) \subset \mathbb{N}$. Show that $\underline{d}(S) = 0$ and $\overline{d}(S) = 1$.

Exercise 1.18   When does A_S (defined in Example 1.4.2) have non-zero Hausdorff measure in its dimension? Finite measure?

Exercise 1.19   Suppose $S \subset \mathbb{N}$, and we are given $E, F \subset \{0,1,2\}$. Define $B_S = \{x = \sum_{k=1}^{\infty} x_k 2^{-k}\}$ where

$$x_k \in \begin{cases} E, & k \in S \\ F, & k \notin S \end{cases}.$$

Find $\dim(B_S)$ in terms of E, F and S.

Exercise 1.20   Consider the sets described in Example 1.3.2. What conditions on S_1, S_2 ensure that $A_{S_1} + A_{S_2} = [0, 2]$?

Exercise 1.21   Characterize which directed graphs correspond to sets of dimension 0 in Example 1.3.3 (of shifts of finite type).

Exercise 1.22   Construct a directed graph so that the number of paths of length n is $\sim n^2$.

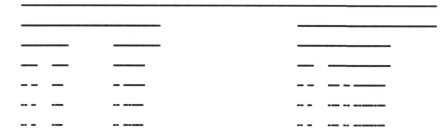

Figure 1.9.3 Set in Exercise 1.27.

Exercise 1.23 Suppose a finite graph is represented by the $0 - 1$ matrix A (with a 1 in position (i, j) if i and j are joined by an edge). Show that the number of paths of length n from i to j is $(A^n)_{ij}$ (i.e., the $(i$th, jth) entry in the matrix A^n).

Exercise 1.24 If $a_n > 0$ and $a_{n+m} \leq a_n + a_m$ then $\lim_{n \to \infty} \frac{a_n}{n} = \inf_n \frac{a_n}{n}$ exists. Use this to prove that the spectral radius of a matrix A is well defined in Section 1.3, Example 1.3.3. Also show that the spectral radius does not depend on the particular norm that is used.

• **Exercise 1.25** Show that the spectral radius of $A = \begin{pmatrix} 1 & 1 \\ 1 & 0 \end{pmatrix}$ is $(1 + \sqrt{5})/2$.

Exercise 1.26 Compute $\dim_{\mathcal{M}}(X_A)$ (see Example 1.3.3) when

$$A = \begin{pmatrix} 0 & 1 & 0 \\ 0 & 1 & 0 \\ 1 & 1 & 0 \end{pmatrix}.$$

Exercise 1.27 Figure 1.9.3 shows the first few approximations to a Cantor set of finite type X_A, corresponding to a 3×3 matrix A. What is the matrix A? What is the dimension of the set?

Exercise 1.28 Compute $\dim(X_A)$ for $A = \begin{pmatrix} 1 & 1 \\ 1 & 0 \end{pmatrix}$.

Exercise 1.29 Compute $\dim(X_A)$ for

$$A = \begin{pmatrix} 1 & 1 & 0 & 0 \\ 0 & 0 & 1 & 1 \\ 1 & 1 & 0 & 0 \\ 0 & 0 & 1 & 1 \end{pmatrix}.$$

Exercise 1.30 Compute the dimension of the set of numbers x with no two consecutive occurrences of the digit 2 in the base 3 expansion of x (for example, $x = .33212312111\ldots$ is allowed, but $y = .311212231131$ is not).

Exercise 1.31 Show that the following two constructions define the Sierpiński gasket G, described in Example 1.3.4:

(1) $G = \{\sum_{n=1}^{\infty} a_n 2^{-n} : a_n \in \{0,1,i\}\} \subset \mathbb{C}$.

(2) Let G_0 be the solid triangle with vertices $\{0,1,i\}$. Find the midpoints of each edge of G_0 and remove from G_0 the triangle with these as vertices. What remains is the union of three triangles and we denote it G_1. In general, G_n is the union of 3^n right triangles with legs of length 2^{-n}. For each triangle we find the midpoints of each edge and remove the corresponding triangle. This gives a nested sequence of sets. The limiting set is G.

Exercise 1.32 In Example 1.3.4, which matrices give connected sets?

Exercise 1.33 Show that the two definitions of $\dim(\mu)$ in Section 1.4 are equivalent.

Exercise 1.34 A set is called **meager** or **of first category** if it is a countable union of nowhere dense sets. The complement of a meager set is called **residual**. Prove that A_p is meager.

Exercise 1.35 If $p < 1/2$ what can be said of $\widehat{A}_p + \widehat{A}_p$?

Exercise 1.36 Show that for all $p \in (0,1)$, the sumset $\widehat{A}_p + \widehat{A}_p$ contains $[0,1]$.

Exercise 1.37 Construct a residual set of dimension 0.

Exercise 1.38 Let $E,F \subset \{0,1,\ldots,b-1\}$ and let $\{x_n\}$ be the b-ary expansion of $x \in [0,1]$. Let $S \subset \mathbb{N}$. What is the dimension of

$$A_{E,F,S} = \left\{ \sum_{n=1}^{\infty} x_n b^{-n} : x_n \in \left\{ \begin{array}{ll} E, & n \in S \\ F, & n \notin S \end{array} \right. \right\}?$$

Exercise 1.39 Given $\{s_n\} \in \{0,1,\ldots,b-1\}^{\mathbb{N}}$ and sets $E_i \subset \{0,1,\ldots,b-1\}$, $0 \le i \le b-1$, find the dimension of the set

$$\left\{ \sum_{n=1}^{\infty} x_n b^{-n} : x_n \in E_{s_n} \right\}.$$

Exercise 1.40 Let G be the Sierpiński gasket. Let L_c be the line in the plane $\{(x,y) : x+y = c\}$. For almost every $c \in [0,1]$ find $\dim(L_c \cap G)$.

Exercise 1.41 Let C be the Sierpiński carpet. Find $\dim(\{x : \dim(C_x) = \alpha\})$.

Exercise 1.42 Let S denote the McMullen set of Example 1.3.5 and let S_y denote the intersection of this set with the horizontal line at height y. Prove

$$\dim(\{y : \dim(S_y) = \alpha\}) = h_2\left(\frac{\alpha}{\log_3 2}\right),$$

where h_2 is the entropy function of Section 1.5 .

Exercise 1.43 Let $\mathbf{C}_{1/2}$ be the middle half Cantor set. Let $X = \mathbf{C}_{1/2} \times \mathbf{C}_{1/2}$ and let L_c be as in Exercise 1.40. Find $\dim(\{c : \dim(L_c \cap X) = \alpha\})$.

Exercise 1.44 Let t_n denote the ternary expansion of $t \in [0, 1]$. For $0 \le p \le 2$, what is the dimension of the set

$$\left\{t \in [0, 1] : \lim_{N\to\infty} \frac{1}{N} \sum_{n=1}^{N} t_n = p\right\}?$$

Exercise 1.45 Given a real number x, consider its b-adic expansion and define a sequence in $\mathbb{R}^b$ by

$$\mathbf{p}^N(x) = \{p_0^N(x), \dots, p_{b-1}^N(x)\}$$
$$= \frac{1}{N}\{\#(\{n \le N : x_n = 0\}), \dots, \#(\{n \le N : x_n = b-1\})\}.$$

Let $V(x)$ be the accumulation set of the sequence $\{\mathbf{p}^N(x)\}$. Show that $V(x)$ is a closed and connected set.

Exercise 1.46 Let $V(x)$ be as in the previous exercise. Given a subset V_0 of the space of probability vectors, show

$$\dim(\{x : V(x) \subset V_0\}) = \sup_{p\in V_0} h_b(p).$$

This is from Volkmann (1958). See also Colebrook (1970) and Cajar (1981).

Exercise 1.47 Let $\{x_n\}$ be the b-ary expansion of x and suppose f is a function from $\{0, 1 \dots, b-1\}$ to the reals. Show

$$\dim\left(\left\{x : \lim_{N\to\infty} \frac{1}{N} \sum_{n=1}^{N} f(x_n) = \alpha\right\}\right) = \max_{\mathbf{p}:\Sigma_j p_j f(j)=\alpha} h_b(\mathbf{p}).$$

See Eggleston (1949) and Colebrook (1970).

Exercise 1.48 Given a closed connected set $E \subset \mathbb{R}^b$ of probability vectors show that

$$\dim(\{x : V(x) = E\}) = \min_{\mathbf{p}\in E} h_b(\mathbf{p}).$$

See Colebrook (1970) and Cajar (1981).

Exercise 1.49 Let **C** be the middle thirds Cantor set. What is

$$\dim(\mathbf{C} \cap (\mathbf{C} + t) \cap (\mathbf{C} + s)),$$

for almost every $(s,t) \in [0,1]^2$?

Exercise 1.50 Construct a Cantor set that is disjoint from all of its non-zero rational translates. (By a Cantor set we mean a set that is compact, totally disconnected and has no isolated points.)

Exercise 1.51 Consider the self-affine type set Y_A (defined in Example 1.3.5) corresponding to the matrix

$$A = \begin{pmatrix} 1 & 0 & 1 & 0 & 1 \\ 0 & 0 & 0 & 0 & 0 \\ 1 & 0 & 1 & 0 & 1 \end{pmatrix}.$$

What is $\dim(Y_A \cap (Y_A + t))$ for almost every $t \in [0,1]^2$?

Exercise 1.52 Find $\dim(\{x \in [0,1] : \lim_{n \to \infty} \sin(2^n x) = 0\})$.

Exercise 1.53 Suppose $S \subset \mathbb{N}$ is a set of density α. Let $E \subset [0,1]$ be the collection of numbers x whose binary expansions satisfy

$$\lim_{n \to \infty} \frac{1}{\#([0,n] \cap S)} \sum_{k \in [0,n] \cap S} x_k = \beta,$$

$$\lim_{n \to \infty} \frac{1}{\#([0,n] \setminus S)} \sum_{k \notin [0,n] \setminus S} x_k = \gamma.$$

What is $\dim(E)$ in terms of α, β and γ?

Exercise 1.54 A closed set $K \subset \mathbb{R}^d$ is called **uniformly perfect** if there is a constant $M < \infty$ so that for any $0 < r \le |K|$ and $x \in K$, there is $y \in K$ with $r \le |x - y| \le Mr$. Show that any uniformly perfect set has positive Hausdorff dimension.

Exercise 1.55 A closed set $K \subset \mathbb{R}^d$ is called **porous** if there is a constant $M < \infty$ so that for any $0 < r \le |K|$ and $x \in K$, there is a y with $r \le |x - y| \le Mr$ and $B(y, \frac{r}{M}) \cap K = \emptyset$. Show that any porous set in $\mathbb{R}^d$ has upper Minkowski dimension $< d$.

Exercise 1.56 A Jordan curve $\gamma \subset \mathbb{R}^2$ satisfies **Ahlfors' 3-point property** if there is an $M < \infty$ so that $|x - y| \le M|x - z|$ for any y on the smaller diameter arc between $x, z \in \gamma$. Show that any such curve is porous, so has dimension < 2. Such arcs are also called "bounded turning" or "quasiarcs" (since this condition characterizes quasiconformal images of lines and circles).

Exercise 1.57 A homeomorphism f of $\mathbb{R}$ to itself is called M-**quasisymmetric** if $M^{-1} \leq (f(x+t) - f(x))/(f(x) - f(x-t)) \leq M$, for all $x \in \mathbb{R}, t > 0$. If f is quasisymmetric for some M, show that it is bi-Hölder, i.e., it satisfies

$$|x - y|^\alpha / C \leq |f(x) - f(y)| \leq C|x - y|^{1/\alpha}$$

for some $C < \infty$ and $\alpha \geq 1$. Hence f maps any set of positive Hausdorff dimension to a set of positive Hausdorff dimension.

• **Exercise 1.58** Construct a quasisymmetric homeomorphism that maps a set of zero Lebesgue measure to positive Lebesgue measure.

• **Exercise 1.59** Let $x_k(x)$ denote the kth binary digit of $x \in [0,1]$. Then the step functions $s_n = \sum_{k=1}^n (2x_k - 1), n \geq q$ model a random walk on the integers. For Lebesgue almost every x, $\{s_n(x)\}$ takes every integer value infinitely often, but there is a subset x for which $s_n(x) \to \infty$ and hence takes each value finitely often. Show this set has Hausdorff dimension 1. (In fact, it has positive $\mathscr{H}^\varphi$-measure for the function $\varphi(t) = t \log \frac{1}{t}$.)

• **Exercise 1.60** (Kolmogorov's maximal inequality) Let X_i be independent with mean zero and finite variance. Write $S_n = \sum_{k=1}^n X_i$. Prove that

$$\mathbb{P}[\max_{1 \leq k \leq n} |S_k| \geq h] \leq \frac{\mathrm{Var}\, S_n}{h^2}.$$

Exercise 1.61 Use the inequality above to give another proof of the strong law for i.i.d. variables with finite variance.

Exercise 1.62 Let X_i be i.i.d. with mean zero and finite variance. Hsu and Robbins (1947) proved that $\sum_n \mathbb{P}(S_n > na)$ converges for any $a > 0$. A converse was proven by Erdős (1949) and further refinements can be found in Chow and Teicher (1997). Fill in the details of the following sketched proof of the Hsu–Robbins Theorem.

(1) By scaling we may assume that $\mathrm{Var}(X_i) = 1$. Set $X_n^* = \max_{1 \leq i \leq n} X_i$ and $S_n^* = \max_{1 \leq i \leq n} S_i$. For $h > 0$, the stopping time $\tau_h = \min\{k : S_k \geq h\}$ satisfies

$$\mathbb{P}(S_n > 3h \text{ and } X_n^* \leq h) \leq \mathbb{P}(\tau_h \leq n)^2.$$

This uses the inequality $S_{\tau_h} \in [h, 2h)$, which holds on the event in the left-hand side.

(2) Kolmogorov's maximal inequality implies that

$$\mathbb{P}(\tau_h \leq n) = \mathbb{P}(S_n^* \geq h) \leq \mathrm{Var}(S_n)/h^2.$$

Apply this with $h = an/3$ and the previous step to show

$$\mathbb{P}(S_n > na \text{ and } X_n^* \leq an/3) \leq n^2/h^4 = 81/(a^4 n^2).$$

(3) Deduce that

$$\mathbb{P}(S_n > na) \leq \mathbb{P}(X_n^* > an/3) + 81/(a^4 n^2)$$
$$\leq n\mathbb{P}(X_1 > an/3) + O(1/n^2).$$

The right-hand side is summable if and only if X_1 has finite variance.

2

Self-similarity and packing dimension

We saw in the preceding chapter that the Minkowski dimension of a general set need not exist, and that when it does exist, it need not agree with the Hausdorff dimension. In this chapter we will consider various conditions on a compact set K which ensure that the Minkowski dimension of K exists and equals the Hausdorff dimension. The main idea is that the sets should "look the same at all scales". We start with the simplest class where this holds, the self-similar sets, and then consider weaker versions of self-similarity, as in Furstenberg's Lemma. We also introduce packing dimension; this is a variation of the upper Minkowski dimension, defined to have a number of better properties.

2.1 Self-similar sets

Many familiar compact sets can be written as a finite union of their images by contraction maps. Consider a family of **contracting self-maps** (or **contractions**) $\{f_i\}_{i=1}^{\ell}$ of a metric space (X,d), i.e., for all $x,y \in X$,

$$d(f_i(x), f_i(y)) \leq r_i d(x,y) \text{ with } r_i < 1$$

for $1 \leq i \leq \ell$. We always assume $\ell > 1$. A non-empty compact set $K \subset X$ is called an **attractor** for the family $\{f_i\}_{i=1}^{\ell}$ if

$$K = \bigcup_{i=1}^{\ell} f_i(K). \tag{2.1.1}$$

The most celebrated attractors are
- the middle thirds Cantor set **C**, which is an attractor for the maps of $\mathbb{R}$

$$f_1(x) = \frac{x}{3} \text{ and } f_2(x) = \frac{x+2}{3}; \tag{2.1.2}$$

45

- the Sierpiński gasket (Figure 1.3.4), which is an attractor for $f_1(x,y) = \frac{1}{2}(x,y)$, $f_2(x,y) = \frac{1}{2}(x+1,y)$ and $f_3(x,y) = \frac{1}{2}(x,y+1)$;
- the Sierpiński carpet (Figure 1.3.4) and the top third of the von Koch snowflake (Figure 1.2.2), for which we leave to the reader to write down the appropriate maps. The following theorem is well known; the elegant proof is from Hutchinson (1981).

Theorem 2.1.1 *Let $\{f_i\}_{i=1}^{\ell}$ be a family of contracting self-maps of a complete metric space (X,d). Then:*

(i) *There exists a unique non-empty compact set $K \subset X$ (the* **attractor***) that satisfies*

$$K = \bigcup_{i=1}^{\ell} f_i(K).$$

(ii) *For any probability vector $\mathbf{p} = (p_1, p_2, \ldots, p_\ell)$, there is a unique probability measure $\mu = \mu_{\mathbf{p}}$ (the* **stationary** *measure) on the attractor K such that*

$$\mu = \sum_{i=1}^{\ell} p_i \mu f_i^{-1}. \tag{2.1.3}$$

If $p_i > 0$ for all $i \le \ell$, then $\mathrm{supp}(\mu) = K$.

Recall $\mathrm{supp}(\mu)^c = \bigcup\{W \text{ open in } X : \mu(W) = 0\}$ has μ-measure zero since X is separable.

Proof (i) Let $\mathbf{Cpt}(X)$ denote the collection of non-empty compact subsets of X with the Hausdorff metric

$$d_H(C,K) = \inf\{\varepsilon : K \subset C^\varepsilon, C \subset K^\varepsilon\}, \tag{2.1.4}$$

where $A^\varepsilon = \{x : d(x,A) < \varepsilon\}$ is the ε-neighborhood of $A \subseteq X$. Then, $\mathbf{Cpt}(X)$ is a complete metric space by Blaschke's Selection Theorem (Appendix A).

Define a self-map F of $(\mathbf{Cpt}(X), d_H)$ by

$$F(C) = \bigcup_{i=1}^{\ell} f_i(C).$$

For any two sets $C, K \in \mathbf{Cpt}(X)$ we have

$$d_H(F(C), F(K)) \le \max_{1 \le i \le \ell} d_H(f_i(C), f_i(K)) \le r_{\max} d_H(C, K),$$

where $r_{\max} = \max_{1 \le i \le \ell} r_i < 1$. Thus F is a contraction of $(\mathbf{Cpt}(X), d_H)$, so by the Banach Fixed-Point Theorem (Appendix A) there is a unique $K \in \mathbf{Cpt}(X)$ such that $F(K) = K$.

(ii) We prove in Appendix A that the space $\mathbf{P}(K)$ of Borel probability measures on the compact attractor K equipped with the **dual Lipschitz metric**

$$L(v, v') = \sup_{\mathrm{Lip}(g) \leq 1} \left| \int g \, dv - \int g \, dv' \right|$$

is a compact metric space. Here

$$\mathrm{Lip}(g) = \sup_{x \neq y} \frac{|g(x) - g(y)|}{d(x, y)}$$

denotes the Lipschitz norm of g.

Consider the mapping $F_{\mathbf{p}}$ defined on $\mathbf{P}(K)$ by

$$F_{\mathbf{p}}(v) = \sum_{i=1}^{\ell} p_i v f_i^{-1}.$$

For any function $g \colon K \to \mathbb{R}$ with $\mathrm{Lip}(g) \leq 1$, we have for all $x, y \in K$

$$\sum_{i=1}^{\ell} p_i |g(f_i(x)) - g(f_i(y))| \leq \sum_{i=1}^{\ell} p_i d(f_i(x), f_i(y)) \leq \sum_{i=1}^{\ell} p_i r_i d(x, y).$$

Therefore, $\mathrm{Lip}(\sum_{i=1}^{\ell} p_i g \circ f_i) \leq \sum_{i=1}^{\ell} p_i r_i \leq r_{\max}$. Hence, for any two probability measures $v, v' \in \mathbf{P}(K)$:

$$\left| \int_K g \, dF_{\mathbf{p}}(v) - \int_K g \, dF_{\mathbf{p}}(v') \right|$$

$$= \left| \int \sum_{i=1}^{\ell} p_i g \circ f_i \, dv - \int \sum_{i=1}^{\ell} p_i g \circ f_i \, dv' \right|$$

$$\leq \mathrm{Lip} \left(\sum_{i=1}^{\ell} p_i g \circ f_i \right) L(v, v') \leq r_{\max} L(v, v).$$

Thus, $F_{\mathbf{p}}$ is a contracting self-map of $(\mathbf{P}(K), L(v, v'))$. The Banach Fixed Point Theorem ensures the existence of a unique fixed point $\mu \in \mathbf{P}(K)$.

If $p_i > 0$ for all i, then any probability measure of bounded support $v \in \mathbf{P}(X)$ such that $v = \sum_{i=1}^{\ell} p_i v f_i^{-1}$ satisfies

$$\mathrm{supp}(v) = \bigcup_{i=1}^{\ell} f_i(\mathrm{supp}(v)).$$

Since $\mathrm{supp}(v)$ is closed it must coincide with the attractor K. $\qquad\square$

For any infinite sequence

$$\xi = \{i_j\}_{j=1}^{\infty} \in \{1, 2, \ldots, \ell\}^{\mathbb{N}},$$

the decreasing sequence of compact sets in (X, d),

$$K_{\xi(n)} = f_{i_1} \circ f_{i_2} \circ \cdots \circ f_{i_n}(K)$$

has diameters tending to 0, and so converges to a point

$$\Phi(\xi) := \bigcap_{n=1}^{\infty} K_{\xi(n)}.$$

(Note that the order in the composition $f_{i_1} \circ f_{i_2} \circ \cdots \circ f_{i_n}$ in the definition of $K_{\xi(n)}$ matters, and the reverse order would not ensure that $(K_{\xi(n)})$ is a decreasing sequence of sets.) Thus

$$\Phi \colon \{1, 2, \ldots, \ell\}^{\mathbb{N}} \longrightarrow K$$

defines a map. Give $\{1, 2, \ldots, \ell\}^{\mathbb{N}}$ the product topology and the metric

$$d(\eta, \zeta) = e^{-|\eta \wedge \zeta|},$$

where $|\eta \wedge \zeta|$ denotes the length of the longest initial segment on which the two sequences agree. Then, the mapping Φ is continuous, indeed it is Hölder continuous:

$$d(\Phi(\eta), \Phi(\zeta)) \leq \operatorname{diam}(K)(r_{\max})^{|\eta \wedge \zeta|} = \operatorname{diam}(K)d(\eta, \zeta)^{\log 1/r_{\max}}.$$

Now we check that Φ is onto. It is easily seen that the image $\operatorname{Im}(\Phi)$ is a compact set satisfying

$$F(\operatorname{Im}(\Phi)) = \operatorname{Im}(\Phi).$$

Uniqueness of the attractor yields $\operatorname{Im}(\Phi) = K$. The map Φ need not be injective.

The push-forward $\nu_{\mathbf{p}}\Phi^{-1}$ of the product measure

$$\nu_{\mathbf{p}} = (p_1, p_2, \ldots, p_\ell)^{\mathbb{N}}$$

on $\{1, 2, \ldots, \ell\}^{\mathbb{N}}$ coincides with the measure μ constructed in part (ii), since it is easily verified that

$$F_{\mathbf{p}}(\nu_{\mathbf{p}}\Phi^{-1}) = \nu_{\mathbf{p}}\Phi^{-1}.$$

Probabilistically, μ is the unique stationary probability measure for the Markov process on X in which the state $f_j(x)$ follows the state x with probability p_j.

Definition 2.1.2 A mapping $f : X \to X$ is a **similitude** if

$$\exists r > 0 \text{ such that } \forall x, y \in X \quad d(f(x), f(y)) = r d(x, y).$$

When $r < 1$, the ratio r is called a **contraction ratio**. If the contracting self-maps $f_1, \ldots, f_\ell$ are all similitudes, then the attractor K is called a **self-similar set**.

Sometimes this term is used loosely when the maps f_j are not similitudes but are affine or conformal; we shall avoid this usage and refer instead to self-affine sets, etc. The examples preceding Theorem 2.1.1 are all self-similar sets. Self-affine sets are considered in Chapter 4.

Let $f_1, \ldots, f_\ell$ be similitudes, i.e., $d(f_j(x), f_j(y)) = r_j d(x, y)$, with $r_j < 1$. To guess the dimension of the attractor K assume first that the sets $\{f_j(K)\}_{j=1}^\ell$ are disjoint and

$$0 < \mathscr{H}^\alpha(K) < \infty. \tag{2.1.5}$$

By the definition of α-dimensional Hausdorff measure,

$$\mathscr{H}^\alpha(f_j(K)) = r_j^\alpha \mathscr{H}^\alpha(K). \tag{2.1.6}$$

By assumption, $\mathscr{H}^\alpha(K) = \sum_{j=1}^\ell \mathscr{H}^\alpha(f_j(K))$; it follows that

$$1 = \sum_{j=1}^\ell r_j^\alpha. \tag{2.1.7}$$

For $f_1, \ldots, f_\ell$ (not necessarily satisfying the above assumptions) the unique $\alpha > 0$ satisfying (2.1.7) is called the **similarity dimension**.

For $\sigma = (i_1, \ldots, i_n)$, write f_σ for the composition

$$f_{i_1} \circ f_{i_2} \circ \cdots \circ f_{i_n}$$

and denote

$$K_\sigma = f_\sigma(K).$$

Also, write $r_\sigma = r_{i_1} \cdot r_{i_2} \cdot \cdots \cdot r_{i_n}$. Set $r_\emptyset = 1$. Write $r_{\max}$ for $\max_{1 \le j \le \ell} r_j$, and similarly define $r_{\min}$.

The **length** n of σ is denoted by $|\sigma|$. If $(p_1, \ldots, p_\ell)$ is a vector of probabilities, write $p_\sigma = p_{i_1} \cdot p_{i_2} \cdot \cdots \cdot p_{i_n}$.

Strings σ, τ are **incomparable** if each is not a prefix of the other. Even without assuming (2.1.5) it is easy to bound the dimension of a self-similar set from above. Take $\beta > \alpha$ and notice that (recall $|K| = \mathrm{diam}(K)$)

$$\mathscr{H}_\infty^\beta(K) \le \sum_{|\sigma|=n} |K_\sigma|^\beta = \sum_{|\sigma|=n} r_\sigma^\beta |K|^\beta = \left(\sum_{j=1}^\ell r_j^\beta \right)^n |K|^\beta.$$

Since $\sum_{j=1}^{\ell} r_j^{\beta} < 1$, by letting $n \to \infty$ we obtain that $\mathcal{H}_{\infty}^{\beta}(K) = 0$ for any $\beta > \alpha$. Hence $\dim(K) \leq \alpha$.

Proposition 2.1.3 *Let $\{f_1, \ldots, f_\ell\}$ be a family of contracting similitudes with contraction ratios $\{r_1, \ldots, r_\ell\}$ and attractor K. Let α be the similarity dimension. Then*

(i) $\mathcal{H}^{\alpha}(K) = \mathcal{H}_{\infty}^{\alpha}(K) < \infty$.

(ii) $\mathcal{H}^{\alpha}(E) = \mathcal{H}_{\infty}^{\alpha}(E)$ *for any $\mathcal{H}^{\alpha}$-measurable subset E of K.*

(iii) $\mathcal{H}^{\alpha}(f_i(K) \cap f_j(K)) = 0$ *for $i \neq j$; more generally $\mathcal{H}^{\alpha}(K_\sigma \cap K_\tau) = 0$ for any two incomparable finite strings σ, $\tau \in \bigcup_{n=1}^{\infty} \{1, \ldots, \ell\}^n$.*

Part (iii) holds even if $f_i = f_j$ for some $i \neq j$. In this case, $\mathcal{H}^{\alpha}(K) = 0$ and, in fact, $\dim(K) < \alpha$.

Proof (i) For any set we have $\mathcal{H}^{\alpha}(K) \geq \mathcal{H}_{\infty}^{\alpha}(K)$, so we only have to show the opposite inequality. Let $\{E_i\}_{i \geq 1}$ be a cover of K such that

$$\sum_{i \geq 1} |E_i|^{\alpha} \leq \mathcal{H}_{\infty}^{\alpha}(K) + \varepsilon.$$

Choose n large enough so that

$$r_{\max}^n \cdot \sup_{i \geq 1} |E_i| < \varepsilon.$$

Then $\{f_\sigma(E_i) : |\sigma| = n, i \geq 1\}$ is a cover of $K = \bigcup_{|\sigma|=n} f_\sigma(K)$ by sets of diameter $< \varepsilon$ that satisfies

$$\sum_{i \geq 1} \sum_{|\sigma|=n} |f_\sigma(E_i)|^{\alpha} = \sum_{|\sigma|=n} r_\sigma^{\alpha} \sum_{i \geq 1} |E_i|^{\alpha}.$$

Since

$$\sum_{|\sigma|=n} r_\sigma^{\alpha} = \left(\sum_{j=1}^{\ell} r_j^{\alpha} \right)^n = 1,$$

this cover shows that

$$\mathcal{H}_{\varepsilon}^{\alpha}(K) \leq \mathcal{H}_{\infty}^{\alpha}(K) + \varepsilon.$$

Finally, let $\varepsilon \downarrow 0$.

(ii) Let $E \subset K$ be a $\mathcal{H}^{\alpha}$-measurable set. Then

$$\begin{aligned}
\mathcal{H}_{\infty}^{\alpha}(K) &\leq \mathcal{H}_{\infty}^{\alpha}(E) + \mathcal{H}_{\infty}^{\alpha}(K \backslash E) \\
&\leq \mathcal{H}^{\alpha}(E) + \mathcal{H}^{\alpha}(K \backslash E) \\
&= \mathcal{H}^{\alpha}(K) = \mathcal{H}_{\infty}^{\alpha}(K).
\end{aligned}$$

This implies $\mathcal{H}_{\infty}^{\alpha}(E) = \mathcal{H}^{\alpha}(E)$.

(iii) Note that

$$\mathcal{H}^\alpha(K) = \mathcal{H}^\alpha \left(\bigcup_{j=1}^\ell f_j(K) \right) \leq \sum_{j=1}^\ell \mathcal{H}^\alpha(f_j(K)) = \sum_{j=1}^\ell r_j^\alpha \mathcal{H}^\alpha(K) = \mathcal{H}^\alpha(K).$$

The equality implies that

$$\mathcal{H}^\alpha(f_i(K) \cap f_j(K)) = 0 \text{ for } i \neq j.$$

Similarly, for strings $\sigma \neq \tau$ of the same length, $\mathcal{H}^\alpha(f_\sigma(K) \cap f_\tau(K)) = 0$, which yields the assertion. $\qquad\square$

2.2 The open set condition is sufficient

Definition 2.2.1 A family of maps $\{f_1, f_2, \ldots, f_\ell\}$ of the metric space X satisfies the **open set condition (OSC)** if there is a bounded open non-empty set $V \subset X$ such that

$$f_j(V) \subset V \text{ for } 1 \leq j \leq \ell,$$

and

$$f_i(V) \cap f_j(V) = \emptyset \text{ for } i \neq j.$$

For the similitudes (2.1.2) defining the Sierpiński gasket K, the open set condition is satisfied with V the interior of the convex hull of K. The same procedure works for the similitudes defining the Sierpiński carpet and the top third of the von Koch snowflake. However, in some cases the open set V is necessarily more complicated (e.g., not simply connected).

We note that the open set condition is a property of a family $\{f_1, f_2, \ldots, f_\ell\}$, not of its attractor. For example, $K = [0, 1]$ is an attractor for both the families $F_1 = (\frac{1}{2}x, \frac{1}{2}x + \frac{1}{2})$ and $F_2 = (\frac{3}{4}x, \frac{3}{4}x + \frac{1}{4})$; the first family satisfies the open set condition, while the second does not.

Theorem 2.2.2 (Moran (1946), Hutchinson (1981)) *Let $f_1, \ldots, f_\ell$ be contracting similitudes of Euclidean space $\mathbb{R}^d$ and let K be the corresponding attractor that satisfies (2.1.1). Let α be the similarity dimension determined by $f_1, \ldots, f_\ell$. If $\{f_i\}_{i=1}^\ell$ satisfy the open set condition, then*

(i) $0 < \mathcal{H}^\alpha(K) < \infty$.
(ii) $\dim(K) = \alpha = \dim_{\mathcal{M}}(K)$.

Since we already know $\mathscr{H}^{\alpha}(K) = \mathscr{H}^{\alpha}_{\infty}(K) \le |K|^{\alpha}$, for part (i), only the lower bound $\mathscr{H}^{\alpha}(K) > 0$ must be proved. This is done via the Mass Distribution Principle, and the lemmas below.

Definition 2.2.3 A set of finite strings $\Pi \subset \bigcup_{n=1}^{\infty} \{1, 2, \ldots, \ell\}^n$ is a **cut-set** if every infinite sequence in $\{1, 2, \ldots, \ell\}^{\mathbb{N}}$ has a prefix in Π. The set of strings Π is a **minimal cut-set** if no element of Π is a prefix of another.

To motivate the terminology, just think of the space of finite sequences $\bigcup_n \{1, 2, \ldots, \ell\}^n$ as an infinite ℓ-ary tree; two sequences are connected by an edge if one is obtained from the other by concatenating one symbol. Then, a cut-set separates the root from the boundary of the tree (the boundary corresponds to the space of infinite sequences). Every minimal cut set is finite (see the proof of Lemma 3.1.1).

Lemma 2.2.4 *Let Π be a minimal cut-set in $\bigcup_n \{1, 2, \ldots, \ell\}^n$ and let $(p_1, p_2, \ldots, p_\ell)$ be a probability vector. Then*

(i) $\sum_{\sigma \in \Pi} p_\sigma = 1$.

(ii) *If μ is a measure satisfying $\mu = \sum_{i=1}^{\ell} p_i \mu f_i^{-1}$, then*

$$\mu = \sum_{\sigma \in \Pi} p_\sigma \mu f_\sigma^{-1}.$$

Proof (i) This is obvious by thinking of the product measure on $\{1, 2, \ldots, \ell\}^{\mathbb{N}}$ where each coordinate has distribution $(p_1, p_2, \ldots, p_\ell)$, since a minimal cut-set defines naturally a cover of $\{1, 2, \ldots, \ell\}^{\mathbb{N}}$ by disjoint cylinder sets. (This also shows that any minimal cut-set is finite.) Alternatively, the same induction as in (ii) works.

(ii) As noted above, Π is finite. The proof proceeds by induction on the cardinality of Π. Indeed, if the concatenation τj is a string of maximal length in Π, then all the strings $\tau 1, \tau 2, \ldots, \tau \ell$ must be in Π, and

$$\sum_{i=1}^{\ell} p_{\tau i} \mu f_{\tau i}^{-1} = p_\tau \cdot \left(\sum_{i=1}^{\ell} p_i \mu f_i^{-1} \right) \circ f_\tau^{-1} = p_\tau \mu f_\tau^{-1}.$$

Consequently, $\Pi' = \Pi \cup \{\tau\} \setminus \{\tau 1, \tau 2, \ldots, \tau \ell\}$ is a smaller minimal cut-set, and

$$\sum_{\sigma \in \Pi} p_\sigma \mu f_\sigma^{-1} = \sum_{\sigma \in \Pi'} p_\sigma \mu f_\sigma^{-1} = \mu,$$

which completes the induction step. $\square$

Lemma 2.2.5 *Let $W_1, W_2, \ldots, W_N$ be disjoint sets in $\mathbb{R}^d$ that intersect a fixed open ball of radius ρ. Assume each W_i contains a ball of radius $a\rho$ and is contained in a ball of radius $b\rho$. Then $N \le \left(\frac{1+2b}{a} \right)^d$.*

Proof The union $\bigcup_{j=1}^{N} W_j$ contains N disjoint balls of radius $a\rho$ and is contained in a fixed ball of radius $(1+2b)\rho$. Now compare volumes. $\square$

Proof of Theorem 2.2.2 (i) Since $(r_1^\alpha, r_2^\alpha, \ldots, r_\ell^\alpha)$ is a probability vector, the attractor K supports a probability measure μ such that

$$\mu = \sum_{j=1}^{\ell} r_j^\alpha \mu f_j^{-1}.$$

If V is the open set in the open set condition, then its closure must satisfy $\bar{V} \supset \bigcup_{i=1}^{\ell} f_i(\bar{V})$. Iterating this, we see that $\bar{V} \supset K$.

Given a ball B_ρ of radius $0 < \rho < 1$, consider the minimal cut-set

$$\Pi_\rho = \{\sigma : r_\sigma \le \rho < r_{\sigma'}\},$$

where σ' is obtained from σ by erasing the last coordinate. The set V contains some open ball of radius a; the sets $\{f_\sigma(V) : \sigma \in \Pi_\rho\}$ are disjoint and each contain a ball of radius $a\rho \cdot r_{\min}$. By Lemma 2.2.5,

$$\#\{\sigma \in \Pi_\rho : f_\sigma(V) \cap B_\rho \ne \emptyset\} \le \left(\frac{1+2|V|}{ar_{\min}}\right)^d = C.$$

Therefore, by Lemma 2.2.4,

$$\mu(B_\rho) = \sum_{\sigma \in \Pi_\rho} r_\sigma^\alpha \mu f_\sigma^{-1}(B_\rho) \le \sum_{\sigma \in \Pi_\rho} \rho^\alpha \mathbf{1}_{\{B_\rho \text{ intersects } f_\sigma(\bar{V})\}} \le C\rho^\alpha.$$

So, by the Mass Distribution Principle, $\mathcal{H}^\alpha(K) > 0$.

(ii) Let ψ_ρ be the cover $\{f_\sigma(K) : \sigma \in \Pi_\rho\}$. Every set in this collection has diameter less than $\rho|K|$, so by expanding each to a ball of diameter $\rho|K|$ we see that

$$N(K, \rho|K|) \le \#(\Pi_\rho).$$

Furthermore,

$$1 = \sum_{\sigma \in \Pi_\rho} r_\sigma^\alpha \ge (r_{\min}\rho)^\alpha \#(\Pi_\rho).$$

Therefore $N(K, \rho|K|) \le (r_{\min}\rho)^{-\alpha}$. But

$$\overline{\dim}_{\mathcal{M}}(K) = \limsup_{\rho \to 0} \frac{\log N(K, \rho|K|)}{\log 1/\rho},$$

so $\overline{\dim}_{\mathcal{M}}(K) \le \alpha$. Combining this with our previous result gives

$$\alpha = \dim(K) \le \underline{\dim}_{\mathcal{M}}(K) \le \overline{\dim}_{\mathcal{M}}(K) \le \alpha,$$

hence equality. $\square$

We note that the open set condition is needed in Theorem 2.2.2. For example, if we take the maps of $\mathbb{R}$

$$f_1(x) = \frac{2}{3}x, \quad f_2(x) = 1 - \frac{2}{3}x,$$

then the similarity dimension satisfies $2(2/3)^\alpha = 1$, i.e., $\alpha > 1$, which is impossible for the Hausdorff dimension of a subset of the line. We shall see in Section 9.6 that Theorem 2.2.2 always fails if the open set condition fails.

2.3 Homogeneous sets

In this section we will consider sets that, in some sense, look the same at all scales. In particular, we will present criteria that ensure that the Hausdorff and Minkowski dimensions agree.

Using b-adic intervals we define

$$N_n(K) = \# \left\{ j \in \{1, \ldots, b^n\} : [\frac{j-1}{b^n}, \frac{j}{b^n}] \cap K \neq \emptyset \right\},$$

for $K \subset [0,1]$. It is straightforward to verify that

$$\overline{\dim}_{\mathscr{M}}(K) = \limsup_{n \to \infty} \frac{\log N_n(K)}{n \log b}.$$

For an integer $b > 0$ define the b-to-1 map T_b mapping $[0,1]$ to itself by

$$T_b(x) = bx \mod 1.$$

For example, the middle thirds Cantor set is invariant under the map T_3. More generally, if $D \subset \{0, \ldots, b-1\}$ and

$$K = \left\{ \sum_{n=1}^{\infty} a_n b^{-n} : a_n \in D \right\},$$

then K is compact and invariant under T_b. We now assume that $K \subset [0,1]$ is a compact set such that $T_b K = K$.

We claim then that $N_{n+m}(K) \leq N_n(K)N_m(K)$. To see this, suppose the interval $I = [\frac{j-1}{b^m}, \frac{j}{b^m}]$ hits K and that inside this interval there are M intervals of the form $I_k = [\frac{k-1}{b^{n+m}}, \frac{k}{b^{n+m}}]$ that hit K. Multiplying by b^m and reducing mod 1 (i.e., applying T_b^m) the interval I is mapped onto $[0,1]$ and each of the intervals I_k is mapped to an interval of length b^{-n} that intersects K. Thus, $M \leq N_n(K)$ and so $N_{n+m}(K) \leq N_n(K)N_m(K)$.

It is a general fact about sequences that if $a_n > 0$ and $a_{n+m} \leq a_n + a_m$, then

$\lim_{n\to\infty} \frac{a_n}{n}$ exists and equals $\inf_n \frac{a_n}{n}$ (see Exercise 1.24). Thus, for compact T_b invariant sets, we deduce

$$\dim_{\mathscr{M}}(K) = \lim_{n\to\infty} \frac{\log N_n(K)}{n\log b}$$

exists. Next we show that the Minkowski dimension of such a set agrees with its Hausdorff dimension (Furstenberg, 1967).

Lemma 2.3.1 (Furstenberg's Lemma) *If $K \subset [0,1]$ is compact and $T_b K = K$, then $\dim(K) = \dim_{\mathscr{M}}(K)$.*

Proof Since we always have $\dim(K) \leq \overline{\dim}_{\mathscr{M}}(K)$ we have to prove the other direction, i.e., for any covering of K by b-adic intervals of possibly different sizes, there is an equally efficient covering by b-adic intervals all of the same size. In order to do this, it is convenient to introduce some notation and associate K to a subset of a sequence space.

Let $\Omega = \{0, \ldots, b-1\}$; let $\Omega^{\mathbb{N}}$ be the sequence space with the product topology. There is a natural continuous mapping ψ from $\Omega^{\mathbb{N}}$ to $[0,1]$ via the b-ary representations of real numbers $x = \sum_{n=1}^{\infty} x_n b^{-n}$. Using this we see that the map T_b can be written as

$$T_b(x) = \sum_{n=1}^{\infty} x_{n+1} b^{-n},$$

so that the induced map on Ω^{∞} is the left shift map T.

We also define Ω^* as the space of finite sequences in Ω. It is a semi-group under the operation of concatenation

$$(a_1, \ldots, a_n)(b_1, \ldots, b_m) = (a_1, \ldots, a_n, b_1, \ldots, b_m).$$

Recall that the **length** of an element $\sigma \in \Omega^*$ is the number of entries and is denoted by $|\sigma|$.

We label the b-adic intervals in $[0,1]$ by elements of Ω^* as follows. For the interval $[\frac{j-1}{b^n}, \frac{j}{b^n}]$ we write the base b expansion of $j-1$, as

$$j-1 = \sigma_{n-1} b^{n-1} + \cdots + \sigma_1 b + \sigma_0,$$

where $0 \leq \sigma_i < b$, and label this interval by the sequence

$$\sigma = (\sigma_{n-1}, \ldots, \sigma_0) \in \Omega^*,$$

of length $|\sigma| = n$ (by abuse of notation we will often not distinguish between b-adic intervals and their labelings).

Let $\mathscr{C}$ be a cover of K by open intervals. Because K is compact we may take $\mathscr{C}$ finite. Each interval in $\mathscr{C}$ can be covered by no more than $b+1$ b-adic intervals of smaller length. Let $\mathscr{C}_b$ be a cover for K formed by all these b-adic intervals. We can write this collection as $\mathscr{C}_b = \{I_\sigma : \sigma \in \Pi\}$ where Π is a cut-set in the b-ary tree and $|I_\sigma| = b^{|\sigma|}$ for $\sigma \in \Pi$. Since Π is finite,

$$L(\Pi) = \max_{\sigma \in \Pi} |\sigma|$$

is well defined, and $b^{-L(\Pi)}$ measures the shortest interval used in $\mathscr{C}_b$. Let S be all the elements of Ω^* that are prefixes of some element in $\psi^{-1}K$, so S is shift invariant (recall that $\psi : \Omega^{\mathbb{N}} \to [0,1]$ and $K \subset [0,1]$ is T_b invariant; T_b invariant subsets of $[0,1]$ correspond to shift invariant subsets of $\Omega^{\mathbb{N}}$).

Let S_Π consist of elements of S with lengths bigger than $L(\Pi)$. Note that any element of S_Π must have an initial segment belonging to Π. Therefore, any $\sigma \in S_\Pi$ can be written as $\tau_1 \sigma_1$ with $\tau_1 \in \Pi$. But because S is T_b invariant, we must have $\sigma_1 \in S$. If $|\sigma_1| \leq L(\Pi)$, we stop, otherwise an initial segment of σ_1 must belong to Π as well. By induction we can write any $\sigma \in S$ as

$$\sigma = \tau_1 \tau_2 \cdots \tau_r \sigma',$$

where $\tau_j \in \Pi$ and $|\sigma'| < L(\Pi)$. There are at most $b^{L(\Pi)}$ distinct possible values for σ'.

Suppose that for some α we have

$$\sum_{\tau \in \Pi} b^{-\alpha|\tau|} = q < 1.$$

Then

$$\sum_{\tau_1,\ldots,\tau_r \in \Pi} b^{-\alpha|\tau_1 \cdots \tau_r|} = q^r < 1,$$

and

$$\sum_{r=1}^{\infty} \sum_{\tau_1,\ldots,\tau_r \in \Pi} b^{-\alpha|\tau_1 \cdots \tau_r|} = \frac{q}{1-q}.$$

Piling up all the above we get

$$\sum_{\sigma \in S} b^{-\alpha|\sigma|} < \frac{q \cdot b^{L(\Pi)}}{1-q} < \infty,$$

but since any $\sigma \in S$ corresponds to some $\frac{j-1}{b^n}$ for some n and $j \in \{1,\ldots,b^n\}$, we find that

$$\sum_{n=1}^{\infty} N_n(K) b^{-\alpha n} = \sum_{n=1}^{\infty} \sum_{\sigma \in S, |\sigma|=n} b^{-\alpha|\sigma|} = \sum_{\sigma \in S} b^{-\alpha|\sigma|} < \infty.$$

Therefore, for all n large enough, $N_n(K)b^{-\alpha n} < 1$, i.e.,

$$\frac{\log N_n(K)}{n \log b} < \alpha,$$

and letting $n \to \infty$ we obtain $\overline{\dim}_{\mathscr{M}}(K) \leq \alpha$.

By definition, for any $\alpha > \dim(K)$ we have a covering of K that corresponds to a cut-set Π with the desired condition $\sum_{\Pi} b^{-\alpha|\tau|} < 1$, so we deduce $\dim_{\mathscr{M}}(K) \leq \dim(K)$. Recall that we always have the opposite inequality (see (1.2.3))

$$\dim(K) \leq \overline{\dim}_{\mathscr{M}}(K),$$

hence we get equality, proving the lemma. $\qquad\square$

Example 2.3.2 We have already seen the example

$$K = \left\{ \sum_{n=1}^{\infty} a_n b^{-n} : a_n \in D \right\},$$

where $D \subset \{0, \dots, b-1\}$, satisfies the hypothesis of Furstenberg's Lemma, and it is easy to check that $N_n(K) = |D|^n$, so we can deduce

$$\dim(K) = \dim_{\mathscr{M}}(K) = \frac{\log |D|}{\log b}.$$

Example 2.3.3 Another collection of compact sets that are T_b invariant are the **shifts of finite type** defined in Example 1.3.3: let $A = (a_{ij}), 0 \leq i, j < b$ be a $b \times b$ matrix of 0s and 1s, and define

$$X_A = \left\{ \sum_{n=1}^{\infty} x_n b^{-n} : A_{x_n x_{n+1}} = 1 \text{ for all } n \geq 1 \right\}.$$

This condition on the $\{x_n\}$ is clearly shift invariant and is the intersection of countably many "closed" conditions, so the set X_A is compact and T_b invariant. Since we computed the Minkowski dimension of this set in Chapter 1 we can now deduce

$$\dim(X_A) = \dim_{\mathscr{M}}(X_A) = \frac{\log \rho(A)}{\log b} = \log_b \left(\lim_{n \to \infty} \|A^n\|^{1/n} \right).$$

2.4 Microsets

A microset of K is a way of quantifying what we see as we "zoom in" closer and closer to K. Recall that if X is a compact metric space, then $\mathbf{Cpt}(X)$, the set of compact subsets of X with the Hausdorff metric d_H, is a compact metric space itself (see Theorem A.2.2 in Appendix A).

A map defined on $X = [0, 1]$ of the form

$$g(x) = \lambda x + a, \quad |\lambda| > 1$$

is called an **expanding similarity**.

Definition 2.4.1   A compact set $\tilde{K} \subset [0, 1]$ is called a **microset** of K if it is a limit point in the Hausdorff metric of $S_n(K) \cap [0, 1]$ for some sequence of expanding similarities $S_n(x) = \lambda_n x + a_n$, with $|\lambda_n| \nearrow \infty$.

Note that a microset is approximately seen at arbitrarily small scales of K, but is not necessarily a subset of K (or necessarily seen at all scales or at all locations). For example, the middle thirds Cantor set **C** is a microset of itself (take $S_n(x) = 3^n x$). On the other hand, one example of a microset of the middle thirds Cantor set **C** is the set $\{ \frac{x+1}{3} : x \in \mathbf{C} \}$ (take $S_n(x) = 3^n x + \frac{1}{3}$). .

Definition 2.4.2   A compact set K is called **Furstenberg regular** if for all microsets $\tilde{K}$ of K we have $\dim(\tilde{K}) \leq \dim(K)$.

It is fairly easy to see that $F = \{ \frac{1}{2}, \frac{1}{3}, \ldots \} \cup \{0\}$ has microset $\tilde{K} = [0, 1]$, so that it is not Furstenberg regular. In particular, this gives an example of a zero-dimensional set with a microset of dimension 1.

Definition 2.4.3   A microset $\tilde{K}$ is called a b-**microset** of K when it is a limit using expanding similarities of the form

$$S_n(x) = b^{l_n} x - a_n$$

where $0 \leq a_n < b^{l_n}$ is an integer and $l_n \to \infty$ (i.e., S_n corresponds to "blowing up" a b-adic interval to $[0, 1]$.)

We leave it as an exercise to verify that in the definition of Furstenberg regular, it suffices to require only $\dim(\tilde{K}) \leq \dim(K)$ for all b-microsets. Using this fact, we note that any compact T_b invariant set (as described in the previous section) is Furstenberg regular since then any b-microset must be a subset of K. The converse fails, see Exercise 2.20.

It is useful to keep in mind the **tree description** of a set $K \subset [0, 1]$. Recall that the collection of all b-adic intervals in $[0, 1]$ can be viewed as the vertices of a tree **T** where edges connect each interval I to the b-subintervals of length $|I|/b$. Moreover we can visualize this tree drawn in the plane with the root $([0, 1])$ at the top and the edges ordered left to right in the same order as the corresponding intervals. See Figure 2.4.1.

The boundary $\partial \mathbf{T}$ of a tree **T** is the set of maximal paths from its root.

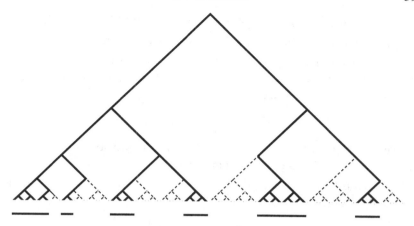

Figure 2.4.1 A set $K \subset [0, 1]$ and the corresponding tree.

There is a natural topology on $\partial \mathbf{T}$: for each vertex v take the set of infinite paths through v; this gives a basis for open sets. A sequence of rooted trees $\{T_n\}$ is said to **converge to a tree** T if for any k there is an m so that $n \geq m$ implies the kth generation truncations of T and T_n agree. With this definition of convergence, the set of rooted trees of bounded degree becomes a compact space. See Section 3.1 for a further discussion of trees and sets.

In the previous section we proved that compact sets invariant under T_b satisfy $\dim(K) = \dim_{\mathscr{M}}(K)$. We have also noted that such sets are Furstenberg regular. Now we show that this property suffices for equality of dimensions:

Lemma 2.4.4 (Extended Furstenberg Lemma)   *Any* $K \subset [0, 1]$ *has a microset* $\tilde{K}$ *with* $\dim(\tilde{K}) \geq \overline{\dim}_{\mathscr{M}}(K)$.

Corollary 2.4.5   *For any Furstenberg regular set,* $\dim(K) = \dim_{\mathscr{M}}(K)$.

Proof   The deduction of the corollary is clear, so we only prove the lemma. To do that, it is enough to show that for any $n > 1$ and $\alpha < \overline{\dim}_{\mathscr{M}}(K)$, there is a magnification of K

$$K_\alpha^{(n)} = S_n(K) \cap [0, 1],$$

where $\{S_n\}$ is a sequence of "b-adic" expanding similarities, and a probability measure $\mu_\alpha^{(n)}$ supported on $K_\alpha^{(n)}$ with the property that

$$\mu_\alpha^{(n)}\left(\left[\frac{j-1}{b^r}, \frac{j}{b^r}\right]\right) \leq b^{-r\alpha},$$

for every $r = 0, 1, \ldots, n$ and $j = 1, 2, \ldots, b^r$. If this is so, take a sequence $\{\alpha_k\}$

tending to $\beta = \overline{\dim}_{\mathcal{M}}(K)$ and get a sequence of sets $\{K_{\alpha_n}^{(n)}\}$ and measures $\{\mu_{\alpha_n}^{(n)}\}$. Taking a weak limit of a subsequence of $\{\mu_{\alpha_n}^{(n)}\}$ gives a measure μ_β supported on a set $\tilde{K}_\beta$. Passing to a subsequence we may assume $\{K_{\alpha_n}^{(n)}\}$ converges in the Hausdorff metric to a set $\tilde{K}$ that contains $\tilde{K}_\beta$ (see Exercise 2.23). Then $\tilde{K}$ is a microset of K and moreover

$$\mu_\beta\left(\left[\frac{j-1}{b^r}, \frac{j}{b^r}\right]\right) \le b^{-r\beta},$$

for every $r = 0, 1, \ldots$ and $j = 1, 2, \ldots, b^r$. The Mass Distribution Principle then implies $\dim(\tilde{K}) \ge \dim(\tilde{K}_\beta) \ge \beta = \overline{\dim}_{\mathcal{M}}(K)$.

The proof is hence reduced to showing the existence of $K_\alpha^{(n)}$ and $\mu_\alpha^{(n)}$, for a fixed n. By definition

$$\alpha < \overline{\dim}_{\mathcal{M}}(K) = \limsup_{\ell \to \infty} \frac{\log N_\ell(K)}{\ell \log b},$$

so there exists an $\varepsilon > 0$ and arbitrarily large integers M such that

$$N_M(K) > b^{M(\alpha+\varepsilon)}. \tag{2.4.1}$$

Let M be such a large integer (to be fixed at the end of the proof) and define a probability measure ν_M on K by putting equal mass on each of the $N_M(K)$ b-adic intervals of size b^{-M}.

Claim: Fix n. If M is sufficiently large and satisfies (2.4.1), then there exists a b-adic interval I so that ν_M satisfies

$$\frac{\nu_M(J)}{\nu_M(I)} \le \frac{|J|^\alpha}{|I|^\alpha}$$

for every b-adic subinterval $J \subset I$ with $|J| = |I|/b^r$ and $r \le n$.

This claim suffices to prove the lemma: we associate to such a b-adic interval I the expanding similarity $S_n = b^{l_n}x - a_n$ that sends I to $[0, 1]$, define $\mu_\alpha^{(n)}$ as the push-forward under S_n of ν_M, and set $K_\alpha^n = S_n(K) \cap [0, 1]$.

If the claim fails for M, then, in particular, it fails for $I = [0, 1]$ so there exists a b-adic subinterval I_1 of generation $r \le n$ so that

$$\nu_M(I_1) > |I_1|^\alpha.$$

Since the claim also fails for I_1, there is a b-adic subinterval $I_2 \subset I_1$ such that

$$\frac{\nu_M(I_2)}{\nu_M(I_1)} > \frac{|I_2|^\alpha}{|I_1|^\alpha}.$$

Continuing in this way we obtain a sequence $I_1 \supset I_2 \supset \cdots$, so that

$$\frac{\nu_M(I_{j+1})}{\nu_M(I_j)} > \frac{|I_{j+1}|^\alpha}{|I_j|^\alpha}.$$

Multiplying out and canceling, we get

$$v_M(I_j) > |I_j|^M.$$

Let N be the maximal index such that $|I_N| \geq b^{-M}$ and

$$v_M(I_N) > |I_N|^\alpha.$$

Our sequence of intervals skips at most n generations at a time, so I_N belongs to some level between $M - n$ and M. Hence, I_N contains at most b^n intervals of generation M. Thus, by the construction of v_M,

$$v_M(I_N) \leq \frac{b^n}{N_M(K)}.$$

Since $N_M(K) > b^{M(\alpha+\varepsilon)}$ we get

$$b^{-M\alpha} \leq v_M(I_N) \leq b^n b^{-M(\alpha+\varepsilon)}.$$

If M is large enough, this gives a contradiction, so the claim must hold. □

2.5 Poincaré sets *

Let $H \subset \mathbb{N}$ and consider the set

$$X_H = \left\{ x = \sum_{n=1}^\infty x_n 2^{-n} : x_n x_{n+h} = 0 \text{ for } h \in H \right\}.$$

This set is compact and is clearly T_2 invariant so by Furstenberg's Lemma its Hausdorff and Minkowski dimensions agree.

For example if $H = \{1\}$, then this says that in x's binary expansion, a 1 is always followed by a 0. This is the same as the shift of finite type set X_A discussed in Example 2.3.3 corresponding to the 2×2 matrix

$$A = \begin{pmatrix} 1 & 1 \\ 1 & 0 \end{pmatrix}.$$

The spectral radius of this matrix is $(1 + \sqrt{5})/2$, so the dimension of X_A is

$$\log_2 \rho(A) = \frac{\log \rho(A)}{\log 2} = \frac{\log(1 + \sqrt{5}) - \log 2}{\log 2}.$$

For a general finite $H \subset \{1, \dots, N\}$, the values of the N binary coefficients $x_{(k+1)N+1}, \dots, x_{(k+2)N}$ do not depend on the binary coefficients $x_1, \dots, x_{kN}$, if the values of $x_{kN+1}, \dots, x_{(k+1)N}$ are given. Thus, X_H can be written as a shift of finite type with a $2^N \times 2^N$ matrix A, that has value 1 in the intersection of the row corresponding to the block $x_1, \ldots, x_N$ and column corresponding to $x_{N+1}, \ldots, x_{2N}$ if $x_n x_{n+h} = 0$ for all $n \in [1,n]$ and $h \in H$ (and is 0 otherwise). The dimension of X_H can be computed from the spectral radius of this matrix:

$$\dim(X_H) = \frac{\log \rho(A)}{N \log 2}.$$

Note that for $H' \subset H$ we have $X_H \subset X_{H'}$. For infinite sets H one could try to compute $\dim(X_{H'})$ by using the spectral radius for all finite $H' \subset H$ and taking a limit, but this does not provide any useful formula. Rather than trying to compute the dimension of X_H for general infinite sets H, we will concentrate on a more modest question: For which infinite sets H do we get $\dim(X_H) = 0$? The answer turns out to depend on some interesting combinatorial properties of the set H.

Recall that for $S \subset \mathbb{N}$ the upper density of S is

$$\overline{d}(S) = \limsup_{N \to \infty} \frac{\#(S \cap \{1, \ldots, N\})}{N}.$$

Definition 2.5.1 $H \subset \mathbb{N}$ is called a **Poincaré set** if for every $S \subset \mathbb{N}$ with $\overline{d}(S) > 0$ we have $(S - S) \cap H \neq \emptyset$ (i.e., there is an $h \in H$ so that $(S+h) \cap S \neq \emptyset$).

This may seem unwieldy at first, but it turns out to be a natural concept. The original definition looks different, but is completely equivalent and may be easier to motivate.

Claim 2.5.2 *$H \subset \mathbb{N}$ is a Poincaré set if and only if for any probability measure μ on a measure space $(X, \mathscr{B})$, for any measure preserving map $T : X \to X$ and any $A \subset X$ with $\mu(A) > 0$ we have $\mu(A \cap T^{-h}(A)) > 0$ for some $h \in H$.*

See Furstenberg (1981) or Bertrand-Mathis (1986) for the proof. It may not be clear from the description in the claim that many Poincaré sets exist, but it is easy to check the definition in some cases. For example, if $\overline{d}(S) > 0$ then S certainly contains at least three distinct numbers a, b, c and if $a - b, b - c$ are both odd then $a - c$ is even. Thus, the even numbers are a Poincaré set. The same argument shows $k\mathbb{N}$ is Poincaré for any integer k. On the other hand, $S = 2\mathbb{N}$ has a positive density and only even differences, so the odd numbers do not form a Poincaré set. We notice that the square of an integer is always either 0 or $1 \mod 3$, so $n^2 + 1 \mod 3$ must be either 1 or 2. Thus, taking $S = 3\mathbb{N}$ shows $H = \{n^2 + 1 : n \in \mathbb{N}\}$ is not Poincaré. However, $H = \{n^2 : n \in \mathbb{N}\}$ is Poincaré (see Exercise 2.29).

Definition 2.5.3 Suppose $H = \{h_j\} \subset \mathbb{N}$ with $h_1 < h_2 < \cdots$. Then H is **lacunary** if there exists a $q > 1$ such that $h_{j+1} \geq q h_j$ for all j.

Lemma 2.5.4   *A lacunary set is never Poincaré.*

This is not unreasonable since Poincaré sets should be "large" and lacunary sets are very "thin" (but note that $\{n^2\}$ is both Poincaré and fairly thin). The proof depends on the fact that given a lacunary set H we have the following Diophantine approximation property: there is a positive integer d, $\alpha \in \mathbb{R}^d$, and an $\varepsilon > 0$ so that $\operatorname{dist}(h\alpha, \mathbb{Z}^d) \geq \varepsilon$ for every $h \in H$. This is left as an exercise (Exercise 2.28). If we assume this property, then the proof goes as follows.

Proof   Cover $[0,1]^d$ by a finite number of cubes $\{Q_j\}$ of diameter $\leq \varepsilon$ and let

$$S_j = \left\{ s \in \mathbb{N} : s\alpha \bmod 1 \in Q_j \right\}.$$

At least one of these sets, say S_k, has positive density in $\mathbb{N}$ and if $a, b \in S_k$, then $\operatorname{dist}((a-b)\alpha, \mathbb{Z}^d) < \varepsilon$, so by the Diophantine approximation property above, we have $(S_k - S_k) \cap H = \emptyset$. Thus, H is not Poincaré. $\qquad\qquad\square$

Theorem 2.5.5 (Furstenberg)   *With X_H defined as above, $\dim(X_H) = 0$ if and only if H is Poincaré.*

Proof   First assume H is Poincaré. It is enough to show that for

$$x = \sum_{n=1}^{\infty} x_n 2^{-n} \in X_H,$$

we have

$$\frac{1}{n} \sum_{k=1}^{n} x_k \to 0, \qquad\qquad\qquad (2.5.1)$$

for then we apply Example 1.5.5 with $p = 0$ to get $\dim(X_H) = 0$.

Let $S = \{n \in \mathbb{N} : x_n = 1\}$. For any two elements $n, n+k \in S$, we have $x_n = x_{n+k} = 1$, and since $x \in X_H$, it follows that $k \notin H$. Thus $(S - S) \cap H = \emptyset$. Since H is Poincaré, $\overline{d}(S) = 0$, which implies (2.5.1), and hence $\dim(X_H) = 0$.

Conversely, suppose H is not Poincaré and suppose $S \subset \mathbb{N}$ has positive upper density and $(S - S) \cap H = \emptyset$. Define a real number x by

$$x = \sum_{n \in S} 2^{-n}.$$

Let $\{x_n\}$ be the binary expansion of this x (i.e., the characteristic function of S) and define

$$A = \left\{ y = \sum_{n=1}^{\infty} y_n 2^{-n} : y_n \leq x_n \ \forall n \right\}.$$

Note

$$\overline{\dim}_{\mathscr{M}}(A) = \limsup_{n\to\infty} \frac{\log 2^{\#S\cap[1,n])}}{n\log 2} = \overline{d}(S) > 0.$$

If $n \notin S$, then $x_n = 0$. If $n \in S$, then for any $h \in H$ we have $n + h \notin S$, which implies $x_{n+h} = 0$. Thus, in either case $x_n x_{n+h} = 0$. This implies $A \subset X_H$, and we saw above that $\overline{\dim}_{\mathscr{M}}(A) > 0$. Since X_H is T_2 invariant, Furstenberg's Lemma implies

$$\dim(X_H) = \overline{\dim}_{\mathscr{M}}(X_H) \geq \overline{\dim}_{\mathscr{M}}(A) > 0. \qquad \square$$

How can we check whether a set is Poincaré? One sufficient (but not necessary) condition that is easier to check in practice is the following.

Definition 2.5.6 A set $H \subset \mathbb{N}$ is called a **van der Corput** set (**VDC** set) if every positive measure μ supported on the circle $\mathbb{R}/\mathbb{Z} = [0,1)$, such that $\hat{\mu}(h) = 0$ for all $h \in H$, satisfies $\mu(\{0\}) = 0$.

Here $\hat{\mu}$ denotes the Fourier transform of μ

$$\hat{\mu}(k) = \int_0^1 e^{-2\pi ik\theta}\, d\mu(\theta).$$

The examples we considered earlier in the section are VDC exactly when they are Poincaré. For instance, let us show

Lemma 2.5.7 $7\mathbb{N}$ *is a van der Corput set.*

Proof Given a positive measure μ on $[0,1]$ assume $\hat{\mu}(7n) = 0$ for all n. Then

$$0 = \frac{1}{n}\sum_{k=1}^n \hat{\mu}(7k) = \int_0^1 \frac{1}{n}\sum_{k=1}^n e^{-2\pi i7k\theta}\, d\mu(\theta) \to \int_0^1 f(\theta)\, d\mu(\theta),$$

as $n \to \infty$, where

$$f(\theta) = \lim_{n\to\infty}\frac{1}{n}\sum_{k=1}^n e^{-2\pi i7k\theta} = \begin{cases} 0, & \theta \notin \mathbb{N}/7 \\ 1, & \text{otherwise.} \end{cases}$$

Thus

$$0 = \int_0^1 f(\theta)\, d\mu(\theta) = \mu(\{0\}) + \mu\left(\left\{\frac{1}{7}\right\}\right) + \cdots + \mu\left(\left\{\frac{6}{7}\right\}\right).$$

Since μ is positive each of the above is zero; in particular $\mu(\{0\}) = 0$. $\qquad \square$

Lemma 2.5.8 $H = 7\mathbb{N} + 1$ *is not a van der Corput set.*

Proof Let μ have equidistributed mass on $\{\frac{k}{7}\}$ for $k = \{0,\ldots,6\}$. Then $\mu(\{0\}) \neq 0$, but

$$\hat{\mu}(7n+1) = \int_0^1 e^{-2\pi i(7n+1)x}\,d\mu(x) = \frac{1}{7}\sum_{n=0}^{6} e^{-2\pi in/7} = 0. \qquad \square$$

Lemma 2.5.9 *If $S \subset \mathbb{N}$ is infinite, then $(S-S) \cap \mathbb{N}$ is a van der Corput set.*

Proof Suppose $s_1,\ldots s_N$ are elements of S and let μ be a positive measure such that $\hat{\mu}(s_i - s_j) = 0$ for all $s_i, s_j \in S$, $i \neq j$. Without loss of generality, assume that μ is a probability measure. Let $e_m(x) = e^{2\pi imx}$. Then

$$N^2\mu(\{0\}) \leq \int_0^1 \left| \sum_{k=1}^{N} e_{s_k}(x) \right|^2 d\mu(x)$$

$$= \int_0^1 \sum_{1 \leq i,j \leq N} e_{s_i - s_j}(x)\,d\mu(x) = N,$$

which implies that

$$\mu(\{0\}) \leq \frac{1}{N} \to 0. \qquad \square$$

A set $H \subset \mathbb{N}$ is called **thick** if it contains arbitrarily long intervals. It is easy to see that any thick set contains an infinite difference set: let $s_1 = 1$; if $s_1 < \cdots < s_n$ are already defined, then let s_{n+1} be the right endpoint of an interval in H of length at least s_n. Then all the differences $s_{n+1} - s_j$, $1 \leq j \leq n$ are in H. Thus, the previous lemma implies:

Corollary 2.5.10 *Any thick set is a van der Corput set.*

Theorem 2.5.11 *If H is a van der Corput set, then H is Poincaré.*

Bourgain (1987) constructed a Poincaré sequence that is not van der Corput.

Proof For $S \subset \mathbb{N}$ with a positive upper density let $S_N = S \cap [0,N]$. Let

$$d\mu_N = \frac{1}{\#(S_N)} \left| \sum_{s \in S_N} e_s(x) \right|^2 dx,$$

where $e_s(x) = e^{2\pi isx}$ as above. Fix $\delta > 0$ and choose N so that $\delta \geq (10N)^{-1}$.

Let $I_N = [\frac{-1}{10N}, \frac{1}{10N}]$. Then

$$\mu_N(-\delta, \delta) \geq \mu_N(I_N)$$

$$= \int_{I_N} \frac{1}{\#(S_N)} \left| \sum_{s \in S_N} e_s(x) \right|^2 dx$$

$$\geq \int_{I_N} \frac{1}{\#S_N} \left(\sum_{s \in S_N} \mathrm{Re}(e_s(x)) \right)^2 dx.$$

Observe that for $x \in I_N$ and $s \leq N$,

$$\mathrm{Re}(e_s(x)) = \cos(2\pi s x) \geq \cos\left(2\pi \frac{N}{10N}\right) = \cos\left(\frac{\pi}{5}\right).$$

Therefore,

$$\mu_N(-\delta, \delta) \geq \frac{1}{\#(S_N)} \#(S_N)^2 \inf_{s \in S_N, x \in I_N} \left(\mathrm{Re}(e_s(x))\right)^2 \int_{I_N} dx$$

$$\geq \frac{2\cos(\frac{\pi}{5})^2}{10N} \#(S_N) = C \frac{\#(S_N)}{N}.$$

Find a subsequence $\{S_{N_k}\}$ such that $\frac{\#(S_{N_k})}{N_k} \to \overline{d}(S)$, and let μ be a weak limit point of μ_{N_k}. Thus,

$$\mu([-\delta, \delta]) \geq \overline{d}(S) > 0$$

for all $\delta > 0$, and so μ has a positive atom at 0 (i.e., $\mu(\{0\}) > 0$).

Now, suppose H is van der Corput. Then the Fourier transform of μ cannot be zero everywhere on H, since μ has an atom at 0. If we can show that $\hat{\mu}$ is supported in $S - S$, then we deduce that $H \cap (S - S) \neq \emptyset$. Thus, H is Poincaré. Therefore we need only show that for $v \notin S - S$, we have $\hat{\mu}(v) = 0$. For $v \notin S - S$

$$\hat{\mu}_N(v) = \int_0^1 \overline{e}_v(x) \frac{1}{\#(S_N)} \left| \sum_{s \in S_N} e_s(x) \right|^2 dx$$

$$= \frac{1}{\#S_N)} \int_0^1 \overline{e}_v(x) \left(\sum_{s_1 \in S_N} e_{s_1}(x) \right) \left(\sum_{s_2 \in S_N} \overline{e}_{s_2}(x) \right) dx$$

$$= \frac{1}{\#(S_N)} \sum_{s_1, s_2 \in S_N} \int_0^1 e_{s_1 - s_2} \overline{e}_v(x)(x) \, dx = 0.$$

Since μ is a weak limit point of $\{\mu_{N_k}\}$, we deduce $\hat{\mu}(v) = 0$. □

2.6 Alternative definitions of Minkowski dimension

There are several equivalent definitions for Minkowski dimension. For example, let $N_{\text{pack}}(K, \varepsilon)$ be the maximum possible number of balls in an ε-packing of K (i.e., a disjoint collection of open balls of radius ε) and replace $N(K, \varepsilon)$ by $N_{\text{pack}}(K, \varepsilon)$ in the definition of Minkowski dimension. This gives the same result. To see this, note that if $\{B_{\varepsilon}(x_1), \ldots, B_{\varepsilon}(x_n)\}$, where $n = N_{\text{pack}}(K, \varepsilon)$, is a maximal packing and x is any point of K then there is x_j so that $|x - x_j| < 2\varepsilon$ (otherwise we could insert an ε-ball round x into the packing, contradicting maximality). Thus taking a ball of radius 2ε at each x_j gives a covering by sets of diameter at most 4ε, and so

$$N(K, 4\varepsilon) \leq N_{\text{pack}}(K, \varepsilon). \tag{2.6.1}$$

On the other hand, if $B_{\varepsilon}(y)$ belongs to a minimal ε-covering then it contains at most one center of a maximal ε-packing, so

$$N_{\text{pack}}(K, \varepsilon) \leq N(K, \varepsilon). \tag{2.6.2}$$

The equivalence of the two definitions follows.

In Euclidean space $N_{\text{pack}}(K, \varepsilon) = N_{\text{sep}}(K, 2\varepsilon)$ (N_{sep} was defined in Section 1.1) since two balls of radius ε are disjoint if and only if their centers are at least distance 2ε apart. However, in general metric spaces we only have $N_{\text{pack}}(K, 2\varepsilon) \leq N_{\text{sep}}(K, 2\varepsilon) \leq N_{\text{pack}}(K, \varepsilon)$, since balls of radius ε might be disjoint if the centers are less than 2ε apart, but not if they are less than ε apart.

Minkowski's original definition was different from the one given in Chapter 1. For a bounded $K \subset \mathbb{R}^d$ he considered a "thickening" of the set K

$$K^{\varepsilon} = \{x : \text{dist}(x, K) < \varepsilon\}$$

and defined

$$\overline{\dim}_{\mathscr{M}}(K) = d + \limsup_{\varepsilon \to 0} \frac{\log \mathscr{L}_d(K^{\varepsilon})}{\log 1/\varepsilon} = \limsup_{\varepsilon \to 0} \frac{\log(\mathscr{L}_d(K^{\varepsilon})/\mathscr{L}_d(B_{\varepsilon}))}{\log 1/\varepsilon},$$

where $\mathscr{L}_d$ denotes d-dimensional Lebesgue measure and B_r denotes a ball of radius r. The equivalence of this and the former definitions can be verified by checking the two trivial inequalities

$$\mathscr{L}_d(K^{\varepsilon}) \leq N(K, \varepsilon) \mathscr{L}_d(B_{\varepsilon}),$$

$$\mathscr{L}_d(K^{\varepsilon}) \geq N_{\text{sep}}(K, \varepsilon) \mathscr{L}_d(B_{\varepsilon/2}),$$

where $N(K, \varepsilon)$ and $N_{\text{sep}}(K, \varepsilon)$ are as defined in Section 1.1.

For $n \in \mathbb{Z}$, we let $\mathscr{D}_n$ denote the collection of nth generation closed dyadic intervals

$$Q = [j2^{-n}, (j+1)2^{-n}],$$

and let $\mathscr{D}$ be the union of $\mathscr{D}_n$ over all integers n. A dyadic cube in $\mathbb{R}^d$ is any product of dyadic intervals that all have the same length. The side length of such a square is denoted $\ell(Q) = 2^{-n}$ and its diameter is denoted $|Q| = \sqrt{d}\ell(Q)$. Each dyadic cube is contained in a unique dyadic cube $Q^\uparrow$ with $|Q^\uparrow| = 2|Q|$; we call $Q^\uparrow$ the parent of Q.

Suppose $\Omega \subset \mathbb{R}^d$ is open. Every point of Ω is contained in a dyadic cube such that $Q \subseteq \Omega$ and $|Q| \leq \mathrm{dist}(Q, \partial\Omega)$. Thus every point is contained in a maximal such cube. By maximality, we have $\mathrm{dist}(Q^\uparrow, \partial\Omega) \leq |Q^\uparrow|$ and therefore $\mathrm{dist}(Q, \partial\Omega) \leq |Q^\uparrow| + |Q| = 3|Q|$. Thus the collection of such cubes forms a **Whitney decomposition** with $\lambda = 3$, i.e., a collection of cubes $\{Q_j\}$ in Ω that are disjoint except along their boundaries, whose union covers Ω and that satisfy

$$\frac{1}{\lambda} \mathrm{dist}(Q_j, \partial\Omega) \leq |Q_j| \leq \lambda \, \mathrm{dist}(Q_j, \partial\Omega),$$

for some finite λ.

For any compact set $K \subset \mathbb{R}^d$ we can define an **exponent of convergence**

$$\alpha = \alpha(K) = \inf\left\{\alpha : \sum_{Q \in \mathscr{W}} |Q|^\alpha < \infty\right\}, \tag{2.6.3}$$

where the sum is taken over all cubes in some Whitney decomposition $\mathscr{W}$ of $\Omega = \mathbb{R}^d \setminus K$ that are within distance 1 of K (we have to drop the "far away" cubes or the series might never converge). It is easy to check that α is independent of the choice of Whitney decomposition (see Exercise 2.33).

Lemma 2.6.1 *For any compact set K in $\mathbb{R}^d$, we have $\alpha \leq \overline{\dim}_{\mathscr{M}}(K)$. If K also has zero Lebesgue measure then $\alpha(K) = \overline{\dim}_{\mathscr{M}}(K)$.*

Proof Let $D = \overline{\dim}_{\mathscr{M}}(K)$. We start with the easy assertion, $\alpha(K) \leq D$. Choose $\varepsilon > 0$ and for each $n \in \mathbb{N}$, let $\mathscr{Q}_n$ be a covering of K by $O(2^{n(D+\varepsilon)})$ dyadic cubes of side length 2^{-n}. Let $\mathscr{W}$ denote the collection of maximal dyadic cubes Q that satisfy

$$|Q| \leq \mathrm{dist}(Q, \partial\Omega) \leq 1,$$

and let $\mathscr{W}_n \subset \mathscr{W}$ be the cubes with side $\ell(Q) = 2^{-n}$. For each $Q \in \mathscr{W}_n$, choose a point $x \in K$ with $\mathrm{dist}(x, Q) \leq 3|Q|$ and let $S(x, Q) \in \mathscr{Q}_n$ be a cube containing x. Since $|S(x, Q)| = |Q|$ and $\mathrm{dist}(Q, S(x, Q)) \leq 3|Q|$, each $S \in \mathscr{Q}_n$ can only be associated to at most 9^d of the Qs in $\mathscr{W}_n$. Hence $\#(\mathscr{W}_n) = O(2^{n(D+\varepsilon)})$, and thus

$$\sum_{Q \in \mathscr{W}} |Q_j|^{D+2\varepsilon} = O\left(\sum_{n=0}^{\infty} \#(\mathscr{W}_n) 2^{-n(D+2\varepsilon)} \right)$$

$$= O\left(\sum_{n=0}^{\infty} 2^{-n\varepsilon} \right)$$

$$< \infty,$$

which proves $\alpha(K) \leq D + 2\varepsilon$. Taking $\varepsilon \to 0$ gives $\alpha(K) \leq D$.

Next we assume K has zero Lebesgue measure in $\mathbb{R}^d$ and we will prove $\alpha(K) \geq D$. Let $\varepsilon > 0$. We have

$$N(K, 2^{-n}) \geq 2^{n(D-\varepsilon)},$$

for infinitely many n. Suppose this occurs for n, and let $\mathscr{S} = \{S_k\}$ be a covering of K with dyadic cubes of side 2^{-n}. Let $\mathscr{U}_n$ be the set of cubes in the dyadic Whitney decomposition of $\Omega = K^c$ with side lengths $< 2^{-n}$. For each $S_k \in \mathscr{S}$ let $\mathscr{U}_{n,k} \subset \mathscr{U}_n$ be the subcollection of cubes that intersect S_k. Because of the nesting property of dyadic cubes, every cube in $\mathscr{U}_{n,k}$ is contained in S_k. Since the volume of K is zero, this gives

$$|S_k|^d = \sum_{Q \in \mathscr{U}_{n,k}} |Q|^d.$$

(The right side is $d^{d/2}$ times the Lebesgue measure of $S_k \setminus K$, and the left side is $d^{d/2}$ times the measure of S_k; these are equal by assumption.) Note that since $-d + D - 2\varepsilon < 0$, we get

$$\sum_{Q \in \mathscr{U}_{n,k}} |Q|^{D-2\varepsilon} = \sum_{Q \in \mathscr{U}_{n,k}} |Q|^d |Q|^{-d+D-2\varepsilon}$$

$$\geq |S_k|^{-d+D-2\varepsilon} \sum_{Q \in \mathscr{U}_{n,k}} |Q|^d$$

$$= |S_k|^{D-2\varepsilon}.$$

Hence, when we sum over the entire Whitney decomposition,

$$\sum_{Q \in \mathscr{U}_0} |Q|^{D-2\varepsilon} \geq \sum_{S_k \in \mathscr{S}} \sum_{Q \in \mathscr{U}_{n,k}} |Q|^{D-2\varepsilon}$$

$$\geq \sum_{S_k \in \mathscr{S}} |S_k|^{D-2\varepsilon}$$

$$\geq N(K, 2^{-n}) \cdot 2^{-n(D-2\varepsilon)}$$

$$\geq 2^{n\varepsilon}.$$

Taking $n \to \infty$ shows $\alpha(K) \geq D - 2\varepsilon$ and taking $\varepsilon \to 0$ gives $\alpha(K) \geq D$.   □

For a compact set K in $\mathbb{R}$, it is not necessary to pass to a Whitney decomposition; the set already has the form $K = I \setminus \bigcup_k I_k$ where I is a closed interval and $\{I_k\}$ are disjoint open intervals. If $\{Q_j\}$ is a Whitney decomposition of an interval J, then it is easy to check that $\sum \ell(Q_j)^s$ is comparable in size to $|J|^s$ for any $s \in (0, 1]$ (the constant will depend on s). Thus

$$\alpha(K) = \inf\left\{ s : \sum_j |I_j|^s < \infty \right\},$$

i.e., if $K \subset \mathbb{R}$, it suffices to sum the diameters of the complementary components of K.

This is not generally true in $\mathbb{R}^d$, although it is true if the complementary components of K all have "nice shapes", e.g., are balls or cubes. If $D \subset \mathbb{R}^2$ is a disk then it is easy to check that

$$\sum_k |Q_k|^s \asymp \frac{1}{s} |D|^s,$$

for any $s > 1$, where $\{Q_k\}$ is a Whitney decomposition of D. Thus if K is a compact set obtained by removing disjoint open disks $\{D_k\}_1^\infty$ from a closed disk D, and K has zero area, then

$$\overline{\dim}_{\mathcal{M}}(K) = \alpha(K) = \inf\left\{ s : \sum_k |D_k|^s < \infty \right\}.$$

The rightmost term is called the **disk packing exponent** of K.

One specific example is the Apollonian packing. This is obtained by starting with the "circular arc triangle" bounded by three pairwise tangent disks and removing a fourth disk tangent to them all. This forms three circular arc triangles and again we remove the disk tangent to all three sides. In the limit we obtain a fractal known as the Apollonian packing complement. See Figure 2.6.1

The packings of any two circular arc triangles with pairwise tangent sides are related by a Möbius transformation, and hence have the same dimension. Moreover, any microset of an Apollonian packing is the restriction of an Apollonian packing to a square (including the degenerate case of line). This packing may not be a subset of the one we started with, but it is a Möbius image of a subset, so it has the same Hausdorff dimension. Therefore the Hausdorff dimension of any microset is at most the dimension of the original set. The proof of Lemma 2.4.4 readily extends from 1 to 2 dimensions, so we deduce that for the Apollonian packing, Hausdorff and Minkowski dimensions agree and hence both agree with the disk packing exponent.

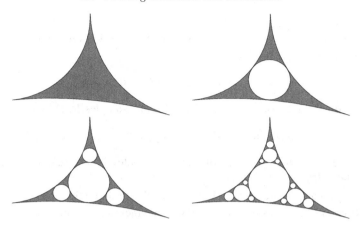

Figure 2.6.1 First four generations of an Apollonian packing.

2.7 Packing measures and dimension

Tricot (1982) introduced packing dimension, which is dual to Hausdorff dimension in several senses and comes with an associated measure.

For any increasing function $\varphi\colon [0,\infty) \to \mathbf{R}$ such that $\varphi(0) = 0$ and any set E in a metric space, define first the **packing pre-measure** (in gauge φ) by

$$\widetilde{\mathscr{P}^{\varphi}}(E) = \lim_{\varepsilon \downarrow 0} \left(\sup \sum_{j=1}^{\infty} \varphi(2r_j) \right),$$

where the supremum is over all collections of disjoint open balls $\{B(x_j, r_j)\}_{j=1}^{\infty}$ with centers in E and radii $r_j < \varepsilon$. This premeasure is finitely subadditive, but not countably subadditive, see Exercise 2.40. Then define the **packing measure** in gauge φ:

$$\mathscr{P}^{\varphi}(E) = \inf \left\{ \sum_{i=1}^{\infty} \widetilde{\mathscr{P}^{\varphi}}(E_i) : E \subset \bigcup_{i=1}^{\infty} E_i \right\}. \tag{2.7.1}$$

It is easy to check that $\mathscr{P}^{\varphi}$ is a metric outer measure, hence all Borel sets are $\mathscr{P}^{\varphi}$-measurable, see Theorem 1.2.4. When $\varphi(t) = t^{\theta}$ we write $\mathscr{P}^{\theta}$ for $\mathscr{P}^{\varphi}$ ($\mathscr{P}^{\theta}$ is called **θ-dimensional packing measure**).

Finally, define the **packing dimension** of E:

$$\dim_{\mathrm{p}}(E) = \inf \left\{ \theta : \mathscr{P}^{\theta}(E) = 0 \right\}. \tag{2.7.2}$$

We always have (this follows from Proposition 2.7.1 below)

$$\dim(E) \le \dim_{\mathrm{p}}(E) \le \overline{\dim}_{\mathscr{M}}(E). \tag{2.7.3}$$

The set $K = \{0\} \cup \{1, \frac{1}{2}, \frac{1}{3}, \frac{1}{4}, \dots\}$ is of packing dimension 0, since the packing dimension of any countable set is 0. The Minkowski dimension of K was computed in Example 1.1.4, and we get

$$\dim(K) = 0 = \dim_p(K) < 1/2 = \dim_{\mathscr{M}}(K).$$

On the other hand, for the set A_S discussed in Examples 1.3.2 and 1.4.2,

$$\dim(A_S) = \underline{d}(S) \le \overline{d}(S) = \dim_p(A_S) = \overline{\dim}_{\mathscr{M}}(A_S).$$

(Deduce the last equality from Corollary 2.8.2 below.) It is easy to construct sets S where $\underline{d}(S) < \overline{d}(S)$.

Packing measures are studied in detail in Taylor and Tricot (1985) and in Saint Raymond and Tricot (1988); here we only mention the general properties we need.

Proposition 2.7.1 *The packing dimension of any set A in a metric space may be expressed in terms of upper Minkowski dimensions:*

$$\dim_p(A) = \inf \left\{ \sup_{j \ge 1} \overline{\dim}_{\mathscr{M}}(A_j) : A \subset \bigcup_{j=1}^{\infty} A_j \right\}, \qquad (2.7.4)$$

where the infimum is over all countable covers of A.

Proof To prove "$\ge$", we first show that for any $E \subset A$ and any $\alpha > 0$ we have

$$\widetilde{\mathscr{P}}^{\alpha}(E) < \infty \Rightarrow \overline{\dim}_{\mathscr{M}}(E) \le \alpha. \qquad (2.7.5)$$

In the definition of the packing pre-measure, one choice included in the supremum is the collection of balls in the definition of $N_{\text{pack}}(E, \varepsilon)$. Hence

$$N_{\text{pack}}(E, \varepsilon) \varepsilon^{\alpha} \le 2 \widetilde{\mathscr{P}}^{\alpha}(E) < \infty$$

if ε is sufficiently small. As noted in the first paragraph of Section 2.6, this implies $\overline{\dim}_{\mathscr{M}}(E) \le \alpha$. If $\dim_p(A) < \alpha$, then $A = \bigcup_{i=1}^{\infty} A_i$ with $\widetilde{\mathscr{P}}^{\alpha}(A_i) < \infty$ for all i, so (2.7.5) implies that $\overline{\dim}_{\mathscr{M}}(A_i) \le \alpha$ for all i. Hence the right-hand side of (2.7.4) is at most α, proving the desired inequality.

To prove "$\le$" in (2.7.4), we claim that for any $E \subset A$ and any $\alpha > 0$ we have

$$\widetilde{\mathscr{P}}^{\alpha}(E) > 0 \quad \Rightarrow \quad \overline{\dim}_{\mathscr{M}}(E) \ge \alpha. \qquad (2.7.6)$$

Assume that $\widetilde{\mathscr{P}}^{\alpha}(E) > 0$. Then, there is an $s > 0$ such that for any $\varepsilon > 0$ we can find a collection of disjoint balls $\{B_j\}_j = \{B(x_j, r_j)\}_j$ with centers in E, radii smaller than ε, and $\sum_{j=1}^{\infty} r_j^{\alpha} > s$. Fix an $\varepsilon > 0$ and let $\{B_j\}$ be a corresponding family of balls. Let m_0 be the largest integer such that $2^{-m_0} > \varepsilon$. For each

$m \geq m_0$ let N_m be the number of balls in the family $\{B_j\}_j$ that have radius in the interval $[2^{-m-1}, 2^{-m})$. Then

$$\sum_{m \geq m_0} N_m 2^{(-m-1)\alpha} \geq \sum_j (\frac{r_j}{2})^\alpha > 2^{-\alpha} s.$$

Clearly, $N_{\text{pack}}(E, 2^{-m-2}) \geq N_m$ and thus

$$\sum_{m \geq m_0} N_{\text{pack}}(E, 2^{-m-2}) 2^{(-m-2)\alpha} > 4^{-\alpha} s,$$

for any $m_0 \geq 0$. Now take an arbitrary $\beta < \alpha$, define

$$a_m = N_{\text{pack}}(E, 2^{-m-2}) 2^{(-m-2)\beta},$$

and note that

$$\sum_{m \geq m_0} a_m 2^{(m+2)(\beta-\alpha)} = \sum_{m \geq m_0} N_{\text{pack}}(E, 2^{-m-2}) 2^{(-m-2)\alpha} > 4^{-\alpha} s.$$

If the sequence $\{a_m\}_m$ is bounded, then the series on the left tends to zero as m_0 increases to infinity, a contradiction. Therefore $\{a_m\}$ is unbounded, i.e.,

$$\limsup_{m \to \infty} N_{\text{pack}}(E, 2^{-m-1}) 2^{(-m-1)\beta} = \infty,$$

which implies $\overline{\dim}_{\mathcal{M}}(E) \geq \beta$ (again by the first paragraph of Section 2.6). Since $\beta < \alpha$ was arbitrary we have $\overline{\dim}_{\mathcal{M}}(E) \geq \alpha$. This is (2.7.6).

Now assume that the right-hand side of (2.7.4) is strictly smaller than α. This means that we can find a covering $A = \bigcup_{j=1}^\infty A_j$ with $\overline{\dim}_{\mathcal{M}}(A_j) < \alpha$ for all j. But then (2.7.6) implies that $\widetilde{\mathscr{P}}^\alpha(A_j) = 0$ for all j. Thus $\mathscr{P}^\alpha(A) = 0$ and so $\dim_p(A) \leq \alpha$. $\qquad\square$

The infimum in (2.7.4) may also be taken only over countable covers of A by closed sets since $\overline{\dim}_{\mathcal{M}}(A_j) = \overline{\dim}_{\mathcal{M}}(\overline{A_j})$.

Note that for packing dimension, unlike Minkowski dimension, the following property holds for any cover of A by subsets A_j, $j = 1, 2, \ldots$:

$$\dim_p A = \sup_j \dim_p A_j.$$

Another useful fact is that the packing pre-measure (in any gauge) assigns the same value to a set and to its closure. Thus, in (2.7.1), we can restrict attention to covers by closed sets.

2.8 When do packing and Minkowski dimension agree?

The following result says that if a set has large upper Minkowski dimension "everywhere", then it has large packing dimension.

Lemma 2.8.1   *Let A be a separable metric space.*

(i) *If A is complete and if the upper Minkowski dimension of every non-empty open set V in A satisfies* $\overline{\dim}_{\mathcal{M}}(V) \geq \alpha$, *then* $\dim_{\mathrm{p}}(A) \geq \alpha$.

(ii) *If* $\dim_{\mathrm{p}}(A) > \alpha$, *then there is a closed non-empty subset* $\tilde{A}$ *of A, such that* $\dim_{\mathrm{p}}(\tilde{A} \cap V) > \alpha$ *for any open set V that intersects* $\tilde{A}$.

Proof   (i) Let $A = \bigcup_{j=1}^{\infty} A_j$, where the A_j are closed. By Baire's Category Theorem there exists an open set V and an index j such that $V \subset A_j$. For this V and j we have: $\overline{\dim}_{\mathcal{M}}(A_j) \geq \overline{\dim}_{\mathcal{M}}(V) \geq \alpha$. By (2.7.4) this implies that $\dim_{\mathrm{p}}(A) \geq \alpha$.

(ii) Let W be a countable basis to the open sets in A. Define

$$\tilde{A} = A \setminus \bigcup \{ J \in W : \overline{\dim}_{\mathcal{M}}(J) \leq \alpha \}.$$

Then, any countable cover of $\tilde{A}$ together with the sets removed on the right yields a countable cover of A, giving

$$\max(\dim_{\mathrm{p}} \tilde{A}, \alpha) \geq \dim_{\mathrm{p}} A > \alpha.$$

Since $\tilde{A} \subset A$, we conclude that $\dim_{\mathrm{p}} \tilde{A} = \dim_{\mathrm{p}} A > \alpha$. If for some V open, $V \cap \tilde{A} \neq \emptyset$ and $\dim_{\mathrm{p}}(\tilde{A} \cap V) \leq \alpha$ then V contains some element J of W such that $\tilde{A} \cap J \neq \emptyset$ and

$$\dim_{\mathrm{p}}(J) \leq \max(\dim_{\mathrm{p}}(A \setminus \tilde{A}), \dim_{\mathrm{p}}(\tilde{A} \cap J))$$
$$\leq \max(\alpha, \dim_{\mathrm{p}}(\tilde{A} \cap V)) = \alpha,$$

contradicting the construction of $\tilde{A}$.   □

Corollary 2.8.2 (Tricot (1982), Falconer (1990))   *Let K be a compact set in a metric space that satisfies*

$$\overline{\dim}_{\mathcal{M}}(K \cap V) = \overline{\dim}_{\mathcal{M}}(K)$$

for every open set V that intersects K. Then

$$\dim_{\mathrm{p}}(K) = \overline{\dim}_{\mathcal{M}}(K).$$

Proof   The inequality $\dim_{\mathrm{p}}(K) \geq \overline{\dim}_{\mathcal{M}}(K)$ follows directly from part (i) of Lemma 2.8.1. The inequality $\dim_{\mathrm{p}}(K) \leq \overline{\dim}_{\mathcal{M}}(K)$ is just the second inequality in (2.7.3).   □

Next, let $\{f_j\}_{j=1}^{\ell}$ be contracting self-maps of a complete metric space. In Section 2.1 we followed Hutchinson (1981) and showed there is a unique compact set K (the attractor) that satisfies

$$K = \bigcup_{j=1}^{\ell} f_j(K). \tag{2.8.1}$$

Definition 2.8.3 A gauge function φ is called **doubling** if

$$\sup_{x>0} \frac{\varphi(2x)}{\varphi(x)} < \infty.$$

Theorem 2.8.4 *Assume $\{f_1,\dots,f_{\ell}\}$ are contracting self bi-Lipschitz maps of a complete metric space, i.e.,*

$$\varepsilon_j d(x,y) \le d(f_j(x),f_j(y)) \le r_j d(x,y)$$

for all $1 \le j \le \ell$ and any x,y, where

$$0 < \varepsilon_j \le r_j < 1.$$

Denote by K the compact attractor satisfying (2.8.1). Then

(i) $\dim_{\mathrm{p}}(K) = \overline{\dim}_{\mathcal{M}}(K)$.
(ii) *For any doubling gauge function φ such that K is σ-finite for $\mathscr{P}^{\varphi}$ we have $\tilde{\mathscr{P}}^{\varphi}(K) < \infty$.*

Proof (i) Let V be an open set that intersects K and let $x \in K \cap V$. For any $m \ge 1$ there are $j_1, j_2, \dots, j_m$ such that $x \in f_{j_1} \circ f_{j_2} \circ \cdots \circ f_{j_m}(K) \subset K$. Since $r_j < 1$, for m large enough we have $f_{j_1} \circ f_{j_2} \circ \cdots \circ f_{j_m}(K) \subset V$. The set on the left-hand side is bi-Lipschitz equivalent to K. Consequently

$$\overline{\dim}_{\mathcal{M}}(K \cap V) = \overline{\dim}_{\mathcal{M}}(K)$$

and Corollary 2.8.2 applies.

(ii) By the hypothesis, K can be covered by sets $\{E_i\}_{i=1}^{\infty}$ with $\mathscr{P}^{\varphi}(E_i) < \infty$ for all i. Thus, $E_i \subset \bigcup_{j=1}^{\infty} E_{ij}$ where $\tilde{\mathscr{P}}^{\varphi}(E_{ij}) < \infty$; without loss of generality E_{ij} are closed sets. By Baire's Theorem (see e.g. Folland (1999)) there are indices i, j and an open set V intersecting K such that $E_{ij} \supset K \cap V$. As detailed in part (i), $K \cap V$ contains a bi-Lipschitz image of K and since φ is doubling $\tilde{\mathscr{P}}^{\varphi}(K) < \infty$. □

When $K \subset \mathbb{R}^d$, the doubling condition in part (ii) can be dropped (Exercise 2.42). If $\dim(K) = \overline{\dim}_{\mathcal{M}}(K)$, the result is trivial since the packing dimension is trapped between these numbers. Non-trivial examples will be given in Chapter 4 when we consider the self-affine sets.

2.9 Notes

The fact that upper Minkowski dimension for a subset of $\mathbb{R}$ can be computed as a critical exponent for the complementary intervals has been rediscovered many times in the literature, but seems to have been first used in Besicovitch and Taylor (1954). The extension to $\mathbb{R}^d$ involves the idea of a Whitney decomposition. The connection with Whitney decompositions in higher dimensions is useful in conformal dynamics where we can sometimes associate Whitney squares to elements of an orbit. See Bishop (1996), Bishop (1997), Bishop and Jones (1997), Stratmann (2004) for applications involving Kleinian groups and Avila and Lyubich (2008), Graczyk and Smirnov (2009) for applications in polynomial dynamics. There are many more definitions of upper Minkowski dimension than we have discussed; twelve are given in Tricot (1981).

In the proof of Proposition 2.1.3, part (ii) is from Bandt and Graf (1992).

Although the open set condition is sufficient for the equality of Hausdorff and similarity dimensions, it is far from necessary, and it is believed that there should always be equality unless there is some "obvious" obstruction. Examples of obvious obstructions are exact overlaps (i.e., an equality $f_\sigma = f_\tau$ for different words σ, τ), and the similarity dimension being strictly larger than the dimension of the ambient space. It is conjectured that in $\mathbb{R}$ these are the only possible mechanisms for lack of equality; some important cases of this conjecture were established in Hochman (2014). In higher dimensions the situation is more difficult, but Hausdorff and similarity dimensions are still expected to typically agree, even in absence of the OSC, and this has been proven in certain cases by Solomyak and Xu (2003), Jordan and Pollicott (2006) and Hochman (2015); see also Lindenstrauss and Varju (2014) for a related result on the absolute continuity of self-similar measures. On the other hand, Schief (1994) shows that (in any dimension) the OSC for $\{f_i\}_{i=1}^{\ell}$ is equivalent to $0 < \mathcal{H}^\alpha(K) < +\infty$, where K and α are the attractor and similarity dimension of $\{f_i\}_{i=1}^{\ell}$. Schief's theorem will be proven in Chapter 8.

The extended Furstenberg Lemma appears in Furstenberg (1970), with an ergodic theoretic proof; the simpler proof we give here is due to François Ledrappier and the second author (unpublished). Another exposition of this proof is in Lyons and Peres (2016).

Van der Corput sets are named for Johannes G. van der Corput who proved that $\mathbb{N}$ is such a set in van der Corput (1931). An alternative definition is that H is van der Corput if given a sequence $\{s_n\}$ of real numbers so that $\{s_{n+h} - s_n\}$ is equidistributed mod 1 for all $h \in H$, then $\{s_n\}$ itself must be equidistributed mod 1. See Kamae and Mendès France (1978). More on equidistribution can be found in Kuipers and Niederreiter (1974) and Montgomery (2001)

The fact that the dimension of the Apollonian packing agrees with the disk counting exponent was first proved by Boyd (1973), with an alternate proof by Tricot (1984). The computation of this dimension is harder; Boyd (1973) gave a rigorous range $[1.300, 1.315]$ and McMullen (1998) gave the estimate ≈ 1.305688. It is conjectured that the Apollonian packing gives the smallest dimension among all planar sets (except a circle) whose complementary components are disks; this infimum is known to be > 1 by a result of Larman (1967). See also Oh (2014) for a survey of more recent results related to the Apollonian packing and its connection to number theory.

A Kleinian group G is a discrete group of Möbius transformations acting on the Riemann sphere; this can also be considered as a group of isometries of the hyperbolic 3-ball. The limit set of G is the accumulation set of any orbit in the hyperbolic ball and the Poincaré exponent is

$$\delta(G) = \inf\{s : \sum_{g \in G} \exp(-s\rho(0, g(0))) < \infty\},$$

where ρ denotes the hyperbolic distance between points. The group G is called geometrically finite if there is a finite sided fundamental region for the group action in the hyperbolic ball (this implies the group is finitely generated, but is stronger). Sullivan (1984) showed that Hausdorff dimension of the limit set of any geometrically finite Kleinian group G equals the Poincaré exponent of G (for more general groups this exponent equals the dimension of a certain subset of the limit set, known as the radial limit set, see Bishop and Jones (1997)). The Apollonian packing is a special case of such a limit set and Sullivan told the first author that understanding this example played an important role in motivating the results in Sullivan (1984). In this paper Sullivan also re-invents packing measure in order to give a metrical description of the conformal densities on the limit set (this is the "new Hausdorff measure" in the title of his paper) and only learned of its previous invention much later. Hausdorff measure describes the density when the group is co-compact or has only rank 2 parabolic subgroups; packing measure describes it when there are rank 1 parabolic subgroups but no rank 2 subgroups. Thus Kleinian limits sets are one place that packing measure arises "naturally". The general case (groups with both rank 1 and rank 2 parabolic subgroups) was done by Ala-Mattila (2011) using a variation of the usual covering/packing constructions.

Theorem 2.7.1 is from Tricot (1982), Proposition 2; see also Falconer (1990), Proposition 3.8. Part (i) of Lemma 2.8.1 is due to Tricot (1982) (see also Falconer (1990)); Part (ii) for trees can be found in (Benjamini and Peres, 1994, Prop. 4.2(b)); the general version given is in Falconer and Howroyd (1996) and in Mattila and Mauldin (1997). A very similar argument was used by Urbański

(1997) to show that packing and upper Minkowski dimension of the Julia set of any rational map agree. Rippon and Stallard (2005) extended this to meromorphic functions and used it to prove that Julia sets of certain transcendental entire functions have packing dimension 2.

2.10 Exercises

Exercise 2.1 Write down explicit linear maps giving the Sierpiński gasket as the attractor. Do the same for the top third of the von Koch snowflake.

Exercise 2.2 The standard hexagonal tiling of the plane is not self-similar, but can be modified to obtain a self-similar tiling. Replacing each hexagon by the union of seven smaller hexagons (each of area $1/7$ that of the original) yields a new tiling of the plane by 18-sided polygons. See Figure 2.10.1. Applying the above operation to each of the seven smaller hexagons yields a 54-sided polygon that tiles. Repeating this operation (properly scaled) ad infinitum, we get a sequence of polygonal tilings of the plane, that converge in the Hausdorff metric to a tiling of the plane by translates of a compact connected set G_0 called the "Gosper island". Show the dimension of the boundary of the Gosper island is $2\ln 3/\ln 7 \approx 1.12915$.

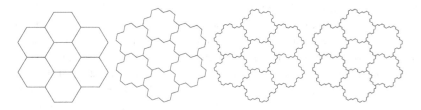

Figure 2.10.1 Gosper islands tile the plane and each island is a union of seven copies of itself.

Exercise 2.3 Show that each Gosper island is a union of seven scaled copies of itself.

Exercise 2.4 Using the notation of Theorem 2.1.1, show that K is the closure of the countable set of fixed points of all maps of the form $\{f_{j_1} \circ \cdots \circ f_{j_n}\}$.

Exercise 2.5 Suppose E is a self-similar set in $\mathbb{R}^n$. Is the projection of E onto a k-dimensional subspace necessarily a self-similar set in $\mathbb{R}^k$?

Exercise 2.6 Suppose E is a self-similar set in $\mathbb{R}^n$. Is the intersection of E with a k-dimensional subspace necessarily a self-similar set in $\mathbb{R}^k$?

Exercise 2.7 Suppose

$$f_1(x) = \frac{1}{3}x, \quad f_2(x) = \frac{2}{3} + \frac{1}{3}x, \quad f_3(x) = \frac{1}{6} + \frac{1}{3}x.$$

What is the dimension of the attractor of (f_1, f_2, f_3)?

Exercise 2.8 Suppose $K \subset \mathbb{R}^n$ is the self-similar set associated to the simili-tudes $\{f_1, \ldots, f_n\}$ that satisfy the open set condition. Prove that the set associ-ated to $\{f_1, \ldots, f_{n-1}\}$ is a subset of strictly smaller dimension.

Exercise 2.9 Suppose $K \subset \mathbb{R}^d$ is compact and that there are constants $a, r_0 > 0$ so that for any $z \in K$ and $0 < r < r_0$ there is a map $\psi : K \to K \cap D(z, r)$ such that $|\psi(x) - \psi(y)| \geq a \cdot r \cdot |x - y|$ (i.e., there is a rescaled, bi-Lipschitz copy of K inside any small neighborhood of itself). Prove that $\dim_{\mathcal{M}}(K)$ exists and $\dim(K) = \dim_{\mathcal{M}}(K)$. (See Theorem 4 of Falconer (1989a).)

Exercise 2.10 Show that a self-similar set satisfies the conditions of Exercise 2.9. Thus for any self-similar set K, the Minkowski dimension exists and equals the Hausdorff dimension.

• **Exercise 2.11** Suppose K_1 and K_2 are two self-similar sets that are Jordan arcs and suppose that $\dim(K_1) = \dim(K_2)$. Then, there is a bi-Lipschitz map-ping of one to the other. (**Bi-Lipschitz** means that there is a $C < \infty$ such that $C^{-1}|x - y| \leq |f(x) - f(y)| \leq C|x - y|$.)

Exercise 2.12 If K is a self-similar Jordan arc of dimension α, prove there is a one-to-one mapping $f : [0, 1] \to K$ of Hölder class $\beta = 1/\alpha$. (**Hölder class** β means there is a $C < \infty$ such that $|f(x) - f(y)| \leq C|x - y|^\beta$ for all x and y.)

Exercise 2.13 Suppose $f_n(x) = \frac{1}{n} + \frac{1}{n^2}x$. What is the dimension of the attrac-tor of $\{f_n\}_{n=2}^\infty$?

• **Exercise 2.14** Let $E_1, \ldots, E_N$ be compact subsets of the unit ball $B(0, 1)$ in $\mathbb{R}^d$ such that $d_H(E_i, E_j) \geq \delta > 0$ for $i \neq j$ (Hausdorff distance). Show that N has uniform bound depending only on the dimension d. In other words, the space of such sets with the Hausdorff metric is totally bounded.

Exercise 2.15 Let $\{x_n\}$ be the base 5 expansion of x. Find

$$\dim(\{x : x_n x_{n+2} = 3 \bmod 5 \text{ for all } n\}).$$

Exercise 2.16 Let $\{x_n\}$ be the base 5 expansion of x. Find

$$\dim(\{x : (x_n + x_{n-1})^2 + x_{n-2} = 2 \bmod 5 \text{ for all } n\}).$$

Exercise 2.17 Verify that in the definition of Furstenberg regular, it suffices to only require $\dim(\tilde{K}) \leq \dim(K)$ for all b-microsets.

Exercise 2.18 Is $K = \{0\} \cup \{1, \frac{1}{2}, \frac{1}{3}, \dots\}$ homogeneous? What are all its microsets?

• **Exercise 2.19** Show that there is a compact set $K \subset [0,1]$ that has every compact subset of $[0,1]$ as a microset.

Exercise 2.20 Let $\{x_n\}$ be the base 5 expansion of x. Let

$$X = \{x : x_n^2 = n^2 \bmod 5 \text{ for all } n\}.$$

Show X is Furstenberg regular, but is not invariant under $T_5 : x \to 5x \bmod 1$.

Exercise 2.21 Let $\{x_n\}$ denote the binary expansion of x. Find

$$\dim\left(\left\{x : \lim_{n \to \infty} \frac{1}{n} \sum_{k=1}^{n} x_k \text{ does not exist}\right\}\right).$$

Exercise 2.22 Given $K \subset [0,1]$, let $M(K)$ be the set of all microsets of K. Show M is connected with respect to the Hausdorff metric.

Exercise 2.23 Let (μ_n) be a sequence of probability measures on a compact space, that converges weakly to a probability measure μ. Denote the support of μ_n by K_n and the support of μ by K. Prove that the sequence of sets (K_n) has a subsequence that converges in the Hausdorff metric to a superset of K.

Exercise 2.24 Is $\{\lfloor e^{\sqrt{n}} \rfloor\}$ Poincaré?

• **Exercise 2.25** (Weyl's Equidistribution Theorem) Prove that if α is irrational then $\{n\alpha \bmod 1\}$ is dense in $[0,1]$.

Exercise 2.26 Suppose $p(x) = \alpha_n x^n + \cdots + \alpha_1 x$ has at least one irrational coefficient. Prove that $p(n) \bmod 1$ is dense in $[0,1]$.

• **Exercise 2.27** Given a lacunary sequence $\{h_n\}$, show that there is an $\alpha \in \mathbb{R}$, and an $\varepsilon > 0$ so that $\mathrm{dist}(h_n\alpha, \mathbb{Z}) \geq \varepsilon$ for every n. This is easier if we assume $h_{n+1}/h_n \geq q > 2$ for all n.

• **Exercise 2.28** Given a lacunary sequence $\{h_n\}$ show that there is $d \in \mathbb{N}$, an $\alpha \in \mathbb{R}^d$ and an $\varepsilon > 0$ so that $\mathrm{dist}(h_n\alpha, \mathbb{Z}^d) \geq \varepsilon$ for every n.

Exercise 2.29 Show that $H = \{n^2\}$ is a van der Corput set. (Hint: use the fact that $n^2\alpha$ is equidistributed for any irrational α.)

Exercise 2.30 Construct a compact set K such that the index $\alpha(K)$ in (2.6.3) is not equal to its upper Minkowski dimension.

Exercise 2.31 Suppose $K \subset \mathbb{R}^2$ is compact and $x \in K$. We say that x is a **point of approximation** (or a **conical limit point**) of K if there is a $C < \infty$ such that for every $\varepsilon > 0$ there is a $y \notin K$ with $|x - y| \leq \varepsilon$ and $\text{dist}(y, K) \geq |x - y|/C$. Denote the conical limit points of K by K_c. Show that the index (2.6.3) of K is always an upper bound for $\dim(K_c)$.

• **Exercise 2.32** Construct a set K so that $\dim(K_c) < \dim(K)$.

Exercise 2.33 Prove that the index (2.6.3) is well defined; that is, its value does not depend on the choice of Whitney decomposition.

Exercise 2.34 Is it true that $\dim(K_c)$ equals the index $\alpha(K)$ in (2.6.3) for any compact set K?

Exercise 2.35 Suppose $K \subset \mathbb{R}^2$ is compact and has zero volume. Let W_n be the number of dyadic Whitney cubes in $\Omega = \mathbb{R}^2 \setminus K$ that have side length $w(Q) = 2^{-n}$. Show that the lower Minkowski dimension is

$$\liminf_{n \to \infty} \frac{\log W_n}{n \log 2}.$$

Exercise 2.36 Suppose $K = B \setminus \bigcup_k B_k$ where $B \subset \mathbb{R}^d$ is a closed ball, and $\{B_k\}$ are disjoint open balls in B. Show that $\overline{\dim}_{\mathcal{M}}(K) \geq \inf\{s : \sum_k |B_k|^s < \infty\}$ with equality if K has zero d-measure.

Exercise 2.37 More generally if B and $\{B_k\}$ are all M-bi-Lipschitz images of d-balls in $\mathbb{R}^d$, $B_k \subset B$ for all k and $\{B_k\}$ are pairwise disjoint, prove that we have $\overline{\dim}_{\mathcal{M}}(K) \geq \inf\{s : \sum_k |B_k|^s < \infty\}$ with equality if $K = B \setminus \bigcup_k B_k$ has zero d-measure. A map is M-bi-Lipschitz if it satisfies

$$|z - y|/M \leq |f(x) - f(y)| \leq M|x - y|.$$

Exercise 2.38 Verify the claim in the text that $\dim(K) \leq \dim_p(K)$ for any K (Equation (2.7.3)).

Exercise 2.39 Suppose $E = [0, 1] \setminus \bigcup_j I_j$ where $\{I_j\}_{j=1}^{\infty}$ are disjoint open intervals of length r_j. Suppose $\sum_j r_j = 1$ and $\sum_j r_j^2 = 1/2$. Show $\dim(E) \leq 1/2$. Find an example to show this is sharp.

Exercise 2.40 For $\varphi(t) = t^{1/2}$ and the set $A = \{1/n : n \in \mathbb{N}\} \cup \{0\}$ show that $\tilde{\mathscr{P}}^{\varphi}(A) = \infty$ so $\tilde{\mathscr{P}}^{\varphi}$ is not countably subadditive.

Exercise 2.41 Construct a set K so that

$$\dim(K) < \min\{\dim_p(K), \underline{\dim}_{\mathcal{M}}(K)\}.$$

Exercise 2.42 Show that the doubling condition in Theorem 2.8.4(ii) can be dropped if $K \subset \mathbb{R}^d$.

Exercise 2.43 Suppose K is a compact set so that any open subset can be mapped to all of K by a Lipschitz map. Show $\dim_p(K) = \overline{\dim}_{\mathcal{M}}(K)$. Show the graph of the Weierstrass nowhere differentiable function on $[0, 2\pi]$ has this property.

Exercise 2.44 A map $f : K \to K$ is **topologically exact** if any open subset of K is mapped onto K by some iterate of f. If f is Lipschitz and topologically exact, then $\dim_p(K) = \overline{\dim}_{\mathcal{M}}(K)$. (Julia sets of rational maps have this property, see, for example, Section 14.2 of Milnor (2006).)

Exercise 2.45 A set K is **bi-Lipschitz homogeneous** if for any $x, y \in k$ there is a bi-Lipschitz map $f : K \to k$ with $f(x) = y$. Show that a compact bi-Lipschitz homogeneous set satisfies $\dim_p(K) = \overline{\dim}_{\mathcal{M}}(K)$. ($\dim(K) < \dim_p(K)$ is possible for such sets.)

3

Frostman's theory and capacity

In Chapter 1 we showed that the existence of certain measures on a set implied a lower bound for the Hausdorff dimension of the set, e.g., the Mass Distribution Principle and Billingsley's Lemma. In this chapter we prove the converse direction: a lower bound on the Hausdorff dimension of a set implies the existence of a measure on the set that is not too concentrated, and we use this powerful idea to derive a variety of results including the behavior of dimension under products, projections and slices.

3.1 Frostman's Lemma

Although the definition of Hausdorff dimension involves only coverings of a set, we have seen that computing the dimension of a set usually involves constructing a measure on the set. In particular, the Mass Distribution Principle says that if a set K supports a positive measure μ that satisfies

$$\mu(B) \leq C|B|^{\alpha}$$

for every ball B (recall $|B| = \operatorname{diam}(B)$), then $\mathcal{H}^{\alpha}_{\infty}(K) \geq \frac{1}{C}\mu(K)$. In this section we will prove a very sharp converse of this.

Lemma 3.1.1 (Frostman's Lemma)   *Let φ be a gauge function. Let $K \subset \mathbb{R}^d$ be a compact set with positive Hausdorff content, $\mathcal{H}^{\varphi}_{\infty}(K) > 0$. Then there is a positive Borel measure μ on K satisfying*

$$\mu(B) \leq C_d \varphi(|B|), \tag{3.1.1}$$

for all balls B and

$$\mu(K) \geq \mathcal{H}^{\varphi}_{\infty}(K).$$

Here C_d is a positive constant depending only on d.

This lemma was proved in Frostman (1935). A positive measure satisfying (3.1.1) for all balls B is called **Frostman measure**.

The assumption that K is compact can be relaxed; the result holds for all Borel sets, even the larger class of analytic sets (continuous images of Borel sets). This generalization is given in Appendix B. Using this result, many of the theorems proven in this chapter for closed sets using Frostman's theorem can easily be extended to much more general sets.

To prove Lemma 3.1.1 we will use the MaxFlow–MinCut Theorem. This is a result from graph theory about general networks, but we will only need it in the case of trees.

Let Γ be a rooted tree of depth $n \leq \infty$ with vertex set V and edge set E. We will assume Γ to be **locally finite**, that is, to have finite vertex degrees (but not necessarily bounded degrees). We let σ_0 denote the root vertex and let $|\sigma|$ denote the depth of σ, i.e., its distance from σ_0. For each vertex σ (except the root) we let σ' denote the unique vertex adjacent to σ and closer to the root. We sometimes call σ' the "parent" of σ. To each edge $e \in E$ we associate a positive number $C(e)$, called the **conductance** of the edge. If we think of the tree as a network of pipes, the conductance of any edge represents the maximum amount of material that can flow through it. A **flow** on Γ is a non-negative function f on Γ that satisfies

$$f(\sigma'\sigma) = \sum_{\tau:\tau'=\sigma} f(\sigma\tau),$$

for all vertices σ of depth from 1 to n. A **legal flow** must satisfy

$$f(\sigma'\sigma) \leq C(\sigma'\sigma)$$

for all vertices σ. The **norm** of a flow f is the sum

$$\|f\| = \sum_{|\sigma|=1} f(\sigma_0\sigma).$$

For a tree Γ we define the **boundary** $\partial\Gamma$ as the set of all maximal paths from the root. Moreover, we define a **cut-set** as a set of edges Π that intersects all maximal paths from the root. For example, taking all the edges from the $(k-1)$st to kth level vertices for some $0 < k \leq n$ forms a cut-set. A **minimal cut-set** is a cut-set that has no proper subsets that are cut-sets (i.e., the removal of any edge allows a path from the root to some nth level vertex). Given a cut-set Π we define Γ_Π to be the connected component of $\Gamma \setminus \Pi$ containing the root σ_0.

Lemma 3.1.2 *Suppose Γ is a locally finite tree. If Γ is infinite, then it contains an infinite path from the root.*

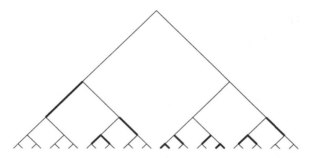

Figure 3.1.1 A minimal cut-set in a binary tree.

Proof Indeed, removing the root leaves a finite number of subtrees rooted at vertices that were adjacent to the root of Γ, and at least one of these subtrees must be infinite. Choosing this subtree and repeating the argument constructs an infinite path in Γ. $\square$

Lemma 3.1.3 *Every minimal cut-set of Γ is finite.*

Proof Since any cut-set Π of Γ hits every infinite path, Γ_Π contains no infinite paths and hence is finite by the previous lemma. Thus every minimal cut-set of Γ is finite. $\square$

We can make the boundary of a tree into a metric space by defining the distance between two rays that agree for exactly k edges to be 2^{-k}. The set of rays passing through a given edge of the tree T is an open ball on ∂T and these balls are a basis for the topology. A cut-set is simply a cover of ∂T by these open balls and the previous lemma says any such cover has a finite subcover. More generally,

Lemma 3.1.4 *With this metric, the boundary ∂T of a locally finite tree is compact.*

Proof Any open set is a union of balls as above, and hence any open cover of ∂T induces a cover of ∂T by such balls, i.e., a cut-set. Every cut-set contains a finite cut-set, so choosing one open set from our original covering that contains each element of the cut-set gives a finite subcover. Thus every open cover contains a finite subcover, which is the definition of compactness. $\square$

Given any cut-set Π, we can bound the norm of a flow by

$$\|f\| \leq \sum_{e \in \Pi} f(e).$$

Moreover, if Π is a minimal cut-set, then one can write

$$\|f\| = \sum_{|\sigma|=1} f(\sigma_0 \sigma)$$

$$= \sum_{|\sigma|=1} f(\sigma_0 \sigma) + \sum_{\sigma \in \Gamma_\Pi \setminus \{\sigma_0\}} \left[-f(\sigma'\sigma) + \sum_{\tau'=\sigma} f(\sigma\tau) \right]$$

$$= \sum_{e \in \Pi} f(e).$$

If f is a legal flow, then the right-hand side is bounded by

$$C(\Pi) \equiv \sum_{e \in \Pi} C(e).$$

This number, $C(\Pi)$, is called the *cut* corresponding to the cut-set Π. It is now clear that in a finite tree,

$$\sup_{\text{legal flows}} \|f\| \leq \min_{\text{cut-sets}} C(\Pi).$$

The MaxFlow–MinCut Theorem says that we actually have equality.

Theorem 3.1.5 (Ford and Fulkerson, 1962) *For a finite rooted tree of depth n,*

$$\max_{\text{legal flows}} \|f\| = \min_{\text{cut-sets}} C(\Pi).$$

Proof The set of legal flows is a compact subset of $\mathbb{R}^{|E|}$, where E is the set of edges of Γ. Thus, there exists a legal flow f^* attaining the maximum possible norm. Let Π_0 be the set of all saturated edges for f^*, i.e., $\Pi_0 = \{e \in E : f^*(e) = C(e)\}$. If there is a path from the root to level n that avoids Π_0, then we can add ε to f^* on each of the edges in the path and obtain another legal flow, contradicting f^*'s maximality. Therefore, there is no such path and we deduce Π_0 is a cut-set. Now let $\Pi \subset \Pi_0$ be a minimal cut-set. But then

$$\|f^*\| = \sum_{e \in \Pi} f^*(e) = \sum_{e \in \Pi} C(e) = C(\Pi),$$

which proves

$$\max_{\text{legal flows}} \|f\| = \|f^*\| \geq \min_{\text{cut-sets}} C(\Pi).$$

Since the opposite inequality is already known, we have proven the result. □

Corollary 3.1.6 *If Γ is an infinite tree,*

$$\max_{\text{legal flows}} \|f\| = \inf_{\text{cut-sets}} C(\Pi).$$

Proof We first note that a maximal flow still exists. This is because for each n there is a flow f_n^* that is maximal with respect to the first n levels of Γ and a corresponding minimal cut-set Π_n so that $\|f_n^*\| = C(\Pi_n)$. Now take a subsequential pointwise limit f^* by a Cantor diagonal argument to obtain a flow that is good for all levels. Clearly

$$\|f^*\| = \lim_{j \to \infty} \|f_{n_j}^*\| = \lim_{j \to \infty} C(\Pi_{n_j}),$$

so that $\|f^*\| = \inf_\Pi C(\Pi)$. $\qquad\qquad\qquad\qquad\qquad\qquad\qquad\qquad\square$

Proof of Lemma 3.1.1 Fix an integer $b > 1$ (say $b = 2$), and construct the b-adic tree corresponding to the compact set K (vertices correspond to b-adic cubes that hit K and so that each b-adic cube is connected to its "parent"). Define a conductance by assigning to each edge $\sigma'\sigma$, $|\sigma| = n$, the number $C(\sigma'\sigma) = \varphi(\sqrt{d}b^{-n})$.

Let f be the maximal flow given by the theorem above. We now show how to define a suitable measure on the space of infinite paths. Define

$$\tilde{\mu}(\{ \text{ all paths through } \sigma'\sigma\}) = f(\sigma'\sigma).$$

It is easily checked that the collection $\mathscr{C}$ of sets of the form

$$\{ \text{ all paths through } \sigma'\sigma\}$$

together with the empty set is a semi-algebra; this means that if $S, T \in \mathscr{C}$, then $S \cap T \in \mathscr{C}$ and S^c is a finite disjoint union of sets in $\mathscr{C}$. Because the flow through any vertex is preserved, $\tilde{\mu}$ is finitely additive. Compactness (Lemma 3.1.4) then implies countable additivity. Thus, using Theorem A1.3 of Durrett (1996), we can extend $\tilde{\mu}$ to a measure μ on the σ-algebra generated by $\mathscr{C}$. We can interpret μ as a Borel measure on K satisfying

$$\mu(I_\sigma) = f(\sigma'\sigma),$$

where I_σ is the cube associated with the vertex σ. Since K is closed, any $x \in K^c$ is in one of the sub-cubes removed during the construction. Hence, μ is supported on K. Since the flow is legal we have

$$\mu(I_\sigma) \leq \varphi(|I_\sigma|) \equiv \varphi(\sigma)$$

for every b-adic cube I_σ. Any cube J can be covered by C_d b-adic cubes of smaller size, and so since φ is increasing

$$\mu(J) \leq C_d \varphi(|J|).$$

Let Γ be a b-ary tree in the tree description of the set K (see Section 2.4). Now each b-adic cover of K corresponds to a cut-set of the tree Γ and vice versa, so

$$\inf_{\Pi} C(\Pi) = \inf_{\Pi} \sum_{e \in \Pi} C(e) = \inf_{\Pi} \sum_{k=1}^{\infty} \sum_{e \in \Pi, |e|=k} \varphi(\sqrt{d} b^{-k}) = \tilde{\mathcal{H}}_{\infty}^{\varphi}(K) \geq \mathcal{H}_{\infty}^{\varphi}(K),$$

where $\tilde{\mathcal{H}}_{\infty}^{\varphi}$ is the b-adic grid Hausdorff content of K. By the MaxFlow–MinCut Theorem,

$$\mu(K) = \|f\| = \inf_{\Pi} C(\Pi) \geq \mathcal{H}_{\infty}^{\varphi}(K),$$

and so μ satisfies all the required conditions. $\square$

3.2 The dimension of product sets

Theorem 3.2.1 (Marstrand's Product Theorem)   *If $A \subset \mathbb{R}^d$, $B \subset \mathbb{R}^n$ are compact, then*

$$\dim(A) + \dim(B) \leq \dim(A \times B) \leq \dim(A) + \dim_{\mathrm{p}}(B). \qquad (3.2.1)$$

Note that for the upper bound on $\dim(A \times B)$ we do not need the compactness of either A or B.

Proof   We start with the left-hand side. Suppose $\alpha < \dim(A)$ and $\beta < \dim(B)$. By Frostman's Lemma there are Frostman measures μ_A, μ_B, supported on A and B respectively, such that for all D, F,

$$\mu_A(D) \leq C_d |D|^{\alpha},$$

$$\mu_B(F) \leq C_n |F|^{\beta}.$$

To apply the Mass Distribution Principle to $\mu_A \times \mu_B$ it suffices to consider sets of the form $D \times F$ where $|D| = |F|$. Then

$$\mu_A \times \mu_B(D \times F) \leq C_d C_n |D|^{\alpha + \beta}.$$

The Mass Distribution Principle implies that $\dim(A \times B) \geq \alpha + \beta$, so by taking $\alpha \to \dim(A)$, $\beta \to \dim(B)$ we get the left-hand inequality.

Now for the right-hand inequality. First we prove

$$\dim(A \times B) \leq \dim(A) + \overline{\dim}_{\mathcal{M}}(B). \qquad (3.2.2)$$

Choose $\alpha > \dim(A)$ and $\beta > \overline{\dim}_{\mathcal{M}}(B)$. Then, we can find ε_0 such that $N(B, \varepsilon)$ $\leq \varepsilon^{-\beta}$ for any $\varepsilon < \varepsilon_0$. For any $\delta > 0$ we can also find a cover $\{A_j\}$ of A satisfying $|A_j| \leq \varepsilon_0$ for all j and $\sum_j |A_j|^{\alpha} < \delta$.

For each j cover B by a collection $\{B_{jk}\}_{k=1}^{L_j}$ of balls with $|B_{jk}| = |A_j|$. We can do this with $L_j \leq |A_j|^{-\beta}$. The family $\{A_j \times B_{jk}\}_{j,k}$ covers $A \times B$ and

$$\sum_j \sum_{k=1}^{L_j} |A_j \times B_{jk}|^{\alpha+\beta} \leq \sqrt{2} \sum_j L_j |A_j|^{\alpha+\beta}$$

$$\leq \sqrt{2} \sum_j |A_j|^{\alpha}$$

$$< \sqrt{2}\delta$$

for any $\delta > 0$. Thus the $\alpha + \beta$ Hausdorff content of $A \times B$ is zero, and hence $\dim(A \times B) \leq \alpha + \beta$. Taking limits as $\alpha \to \dim(A)$ and $\beta \to \overline{\dim}_{\mathscr{M}}(B)$ gives (3.2.2).

To finish the proof we use the formula (2.7.4) to cover B by closed sets $\{B_k\}$ such that $\sup_n \overline{\dim}_{\mathscr{M}}(B_k) \leq \dim_p(B) + \varepsilon$. Thus (using Exercise 1.6)

$$\dim(A \times B) \leq \sup_k \dim(A \times B_k) \leq \dim(A) + \dim_p(B) + \varepsilon.$$

Taking $\varepsilon \to 0$ gives the theorem. $\square$

Example 3.2.2   In Chapter 1 we computed the dimension of the Cantor middle thirds set **C** and of the product $\mathbf{C} \times \mathbf{C}$ (see Examples 1.1.3 and 1.3.4), and found

$$\dim(\mathbf{C} \times \mathbf{C}) = \log_3 4 = 2\log_3 2 = 2\dim(\mathbf{C}).$$

Thus equality holds in Marstrand's Product Theorem (as required since **C** is self-similar and hence its packing, Minkowski and Hausdorff dimensions agree).

Example 3.2.3   Recall Examples 1.3.2 and 1.4.2: suppose $S \subset \mathbb{N}$, and define $A_S = \{x = \sum_{k=1}^{\infty} x_k 2^{-k}\}$ where

$$x_k \in \begin{cases} \{0,1\}, & k \in S \\ \{0\}, & k \notin S. \end{cases}$$

We showed that

$$\overline{\dim}_{\mathscr{M}}(A_S) = \limsup_{N \to \infty} \frac{\#(S \cap \{1,\ldots,N\})}{N},$$

(note that by Lemma 2.8.1, $\dim_p(A_S) = \overline{\dim}_{\mathscr{M}}(A_S)$ also) and

$$\dim(A_S) = \underline{\dim}_{\mathscr{M}}(A_S) = \liminf_{N \to \infty} \frac{\#(S \cap \{1,\ldots,N\})}{N}.$$

Now let

$$S = \bigcup_{j=1}^{\infty} [(2j-1)!, (2j)!), \qquad S^c = \bigcup_{j=1}^{\infty} [(2j)!, (2j+1)!).$$

In this case it is easy to check that

$$\liminf_{N \to \infty} \frac{\#(S \cap \{1, \ldots, N\})}{N} = \liminf_{N \to \infty} \frac{\#(S^c \cap \{1, \ldots, N\})}{N} = 0,$$

so that $\dim(A_S) = \dim(A_{S^c}) = 0$. On the other hand, if

$$E = A_S \times A_{S^c} = \{(x, y) : x \in A_S, y \in A_{S^c}\},$$

then we claim $\dim(E) \geq 1$. To see this, note that

$$[0, 1] \subset A_S + A_{S^c} = \{x + y : x \in A_S, y \in A_{S^c}\},$$

since we can partition the binary expansion of any real number in $[0, 1]$ appropriately. Thus the map

$$(x, y) \to x + y$$

defines a Lipschitz map of E onto $[0, 1]$. Since a Lipschitz map cannot increase Hausdorff dimension, we deduce that $\dim(E) \geq 1$. This gives an example of strict inequality on the left side of (3.2.1) in Marstrand's Product Theorem.

Example 3.2.4   For $A = [0, 1]$ and $B = A_S$ as in the previous example, we have

$$\dim_p(B) = \dim_p(A_S) = \overline{\dim}_{\mathcal{M}}(A_S) = \overline{d}(S) = 1 > 0 = \underline{d}(S) = \dim(B),$$

and hence

$$1 \leq \dim(B \times [0, 1]) \leq \dim B + \dim_p[0, 1] = 1 < \dim[0, 1] + \dim_p B = 2.$$

This gives an example of a strict inequality in the right-hand side of (3.2.1).

3.3 Generalized Marstrand Slicing Theorem

We now return to a topic we first discussed in Chapter 1; intersecting a set with subspaces of $\mathbb{R}^n$, i.e., slicing the set. Our earlier result (Theorem 1.6.1) said that if $A \subset \mathbb{R}^2$, then for almost every $x \in \mathbb{R}$ the intersection of A with the vertical line L_x passing through x has dimension $\leq \dim(A) - 1$ if $\dim(A) \geq 1$. The generalization we consider here is based on the following question. Suppose $F \subset \mathbb{R}^2$ with $\dim(F) \geq \alpha$; given a set $A \subset [0, 1]$ with $\dim(A) \geq \alpha$, can we find an $x \in A$ with

$$\dim(F_x) \leq \dim(F) - \alpha$$

where $F_x = \{y : (x, y) \in F\}$? To see that we cannot expect to make the right-hand side any smaller, consider the following example: take sets A, B so that $\dim(B) = \dim_p(B)$. Then Theorem 3.2.1 (Marstrand's Product Theorem) says that for $F = A \times B$, we have $\dim(F_x) = \dim(B) = \dim(F) - \dim(A)$ for all $x \in A$.

Theorem 3.3.1 (Generalized Marstrand Slicing Theorem) *Suppose μ is a probability measure on $A \subset [0, 1]$ that satisfies $\mu(I) \leq C|I|^\alpha$, for all intervals $I \subset [0, 1]$. If $F \subset [0, 1]^2$ has dimension $\geq \alpha$, then*

$$\dim(F_x) \leq \dim(F) - \alpha$$

for μ-almost every x in A.

Proof The idea is to take $\gamma > \dim(F)$ and show that

$$\int_A \mathcal{H}^{\gamma - \alpha}(F_x) \, d\mu(x) < \infty,$$

and hence $\dim(F_x) \leq \gamma - \alpha$ at μ-almost every x in A.

For $\varepsilon > 0$, cover F by squares $\{Q_j\} = \{A_j \times B_j\}$ with $|Q_j| < \varepsilon$ and such that $\sum_j |Q_j|^\gamma < \varepsilon$. Let

$$f(x, y) = \sum_j |Q_j|^{\gamma - 1 - \alpha} \mathbf{1}_{Q_j}(x, y).$$

Then, with μ as in the hypothesis,

$$\int_A \int_0^1 f(x, y) \, dy \, d\mu(x) = \sum_j |Q_j|^{\gamma - 1 - \alpha} |B_j| \mu(A_j),$$

with $\mu(A_j) \leq C|A_j|^\alpha \leq C|Q_j|^\alpha$, and $|B_j| \leq |Q_j|$. So

$$\int_A \int_0^1 f(x, y) \, dy \, d\mu(x) \leq C \sum_j |Q_j|^\gamma < C\varepsilon.$$

Now define

$$Q_j^x = \begin{cases} B_j, & x \in A_j \\ \emptyset, & x \notin A_j \end{cases}.$$

The $\{Q_j^x\}$ form a cover of F_x. By Fubini's Theorem,

$$\begin{aligned}
\int_A \int_0^1 f(x, y) \, dy \, d\mu(x) &= \int_A \left(\int_0^1 f(x, y) \, dy \right) d\mu(x) \\
&\geq \int_A \sum_j |Q_j^x|^{\gamma - \alpha} \, d\mu(x) \\
&\geq \int_A \mathcal{H}_\varepsilon^{\gamma - \alpha}(F_x) \, d\mu(x).
\end{aligned}$$

Thus

$$0 \leq \int_0^1 \mathcal{H}_\varepsilon^{\gamma-\alpha}(F_x)\,d\mu(x) \leq C\varepsilon.$$

Now let $\varepsilon \searrow 0$ and we get

$$0 = \int_0^1 \mathcal{H}^{\gamma-\alpha}(F_x)\,d\mu(x),$$

so $\mathcal{H}^{\gamma-\alpha}(F_x) = 0$ μ-almost everywhere in A. Taking $\gamma = \frac{1}{n} + \dim F$ and letting $n \to \infty$ gives the result. $\square$

Corollary 3.3.2   *If E is a compact subset of $\{x : \dim(F_x) > \dim(F) - \alpha\}$, where $F \subseteq [0,1]^2$ with dimension $\geq \alpha$, then $\dim(E) \leq \alpha$.*

Proof   If $\dim(E) > \alpha$, then there is a Frostman measure μ on E with $\mu(E) = 1$ and $\mu(I) \leq C|I|^\alpha$ for all cubes. The theorem says that $\dim(F_x) \leq \dim(F) - \alpha$ for μ-almost every $x \in E$. This contradiction gives the result. $\square$

Example 3.3.3   In Example 1.6.4 we saw that if G is the Sierpiński gasket and G_x is the vertical slice at x then $G_x = A_S$ for $S = \{n : x_n = 0\}$, where $\{x_n\}$ is the binary expansion of x. Thus

$$\dim(G_x) = \liminf_{n \to \infty} \frac{1}{N} \sum_{n=1}^N (1 - x_n),$$

and

$$\dim(\{x : \dim(G_x) \geq p\}) = \dim\left(\left\{x : \limsup_{N \to \infty} \frac{1}{N} \sum_{n=1}^N x_n \leq 1 - p\right\}\right)$$
$$= \dim(\widetilde{A}_{1-p}),$$

where $\widetilde{A}_p$ is the set defined in Example 1.5.5. It follows from the argument there that the dimension of this set is

$$\dim(\widetilde{A}_{1-p}) = \begin{cases} h_2(p), & p \geq 1/2 \\ 1, & p \leq 1/2. \end{cases}$$

Thus

$$\dim(\{x : \dim(G_x) \geq \dim(G) - \alpha\}) = \dim(\{x : \dim(G_x) \geq \log_2 3 - \alpha\})$$
$$= \begin{cases} 0, & \alpha \leq \log_2 3 - 1 \\ h_2(\log_2 3 - \alpha), & \log_2 3 - 1 \leq \alpha \leq \log_2 3 - 1/2 \\ 1, & \alpha \geq \log_2 3 - 1/2. \end{cases}$$

Since $\log_2 3 - \frac{1}{2} \approx 1.08496 > 1$, the third case is consistent with the corollary. A little more work shows

$$h_2(p) \le \log_2 3 - p,$$

on $[1/2, 1]$, hence taking $p = \log_2 3 - \alpha$,

$$h_2(\log_2 3 - \alpha) \le \alpha,$$

on $[\log_2 3 - 1, \log_2 3 - 1/2]$, again as required by the corollary.

3.4 Capacity and dimension

Motivated by Frostman's Lemma we make the following definition.

Definition 3.4.1 For a Borel measure μ on $\mathbb{R}^d$ and an $\alpha > 0$ we define the **α-dimensional energy** of μ to be

$$\mathscr{E}_\alpha(\mu) = \int_{\mathbb{R}^d} \int_{\mathbb{R}^d} \frac{d\mu(x)d\mu(y)}{|x-y|^\alpha},$$

and for a set $K \subset \mathbb{R}^d$ we define the **α-dimensional capacity** of K to be

$$\left[\inf_\mu \mathscr{E}_\alpha(\mu) \right]^{-1},$$

where the infimum is over all Borel probability measures supported on K. If $\mathscr{E}_\alpha(\mu) = \infty$ for all such μ, then we say $\mathrm{Cap}_\alpha(K) = 0$.

Pólya and Szegő defined the critical α for the vanishing of $\mathrm{Cap}_\alpha(K)$ to be the **capacitary dimension** of K, but this name is no longer used because of the following result.

Theorem 3.4.2 (Frostman (1935)) *Suppose $K \subset \mathbb{R}^d$ is compact.*

(i) *If $\mathscr{H}^\alpha(K) > 0$ then $\mathrm{Cap}_\beta(K) > 0$ for all $\beta < \alpha$.*
(ii) *If $\mathrm{Cap}_\alpha(K) > 0$ then $\mathscr{H}^\alpha(K) = \infty$.*

In particular, $\dim(K) = \inf\{\alpha : \mathrm{Cap}_\alpha(K) = 0\}$.

Using the inner regularity of Borel measures in $\mathbb{R}^d$ (see Folland (1999)) one can deduce that $\mathrm{Cap}_\alpha(A) > 0$ for a Borel set A implies that $\mathrm{Cap}_\alpha(\Lambda) > 0$ for some compact $\Lambda \subset A$. This implies that the set K is not σ-finite for $\mathscr{H}^\alpha$ whenever $\mathrm{Cap}_\alpha(K) > 0$ (use Exercise 3.14).

Proof   We start with (i). Suppose $\mathscr{H}^\alpha(K) > 0$. Then by Frostman's Lemma there is a positive Borel probability measure μ supported on K so that

$$\mu(B) \leq C|B|^\alpha,$$

for all balls B. Take a $\beta < \alpha$ and fix a point $x \in K$. If $|K| \leq 2^r$, then

$$\int_K \frac{d\mu(y)}{|x-y|^\beta} = \sum_{n=-r}^\infty \int_{2^{-n-1} \leq |y-x| \leq 2^{-n}} \frac{d\mu(y)}{|x-y|^\beta}$$

$$\leq \sum_{n=-r}^\infty \mu(B(x,2^{-n}))2^{\beta(n+1)}$$

$$\leq C \sum_{n=-r}^\infty 2^{-n\alpha}2^{n\beta} \leq M < \infty$$

where C and M are independent of x. Thus $\mathscr{E}_\beta(\mu) \leq M\|\mu\| < \infty$ and therefore $\mathrm{Cap}_\beta(K) > 0$.

Now for part (ii). Since $\mathrm{Cap}_\alpha(K) > 0$, there is a positive Borel measure μ on K with $\mathscr{E}_\alpha(\mu) < \infty$. Hence for sufficiently large M, the set

$$K_1 = \{x \in K : \int_K \frac{d\mu(y)}{|x-y|^\alpha} \leq M\}$$

has positive μ measure. One can write this integral as a sum and use summation by parts to get,

$$\int_K \frac{d\mu(y)}{|x-y|^\alpha} \geq \sum_{n=-r}^\infty \mu(\{2^{-n-1} \leq |y-x| \leq 2^{-n}\})2^{n\alpha}$$

$$= \sum_{n=-r}^\infty \left(\mu(B(x,2^{-n})) - \mu(B(x,2^{-n-1}))\right)2^{n\alpha}$$

$$= C_1 + \sum_{n=-r}^\infty \left((2^{n\alpha} - 2^{(n-1)\alpha})\mu(B(x,2^{-n}))\right)$$

$$= C_1 + C_2 \sum_{n=-r}^\infty 2^{n\alpha}\mu(B(x,2^{-n})).$$

If $x \in K_1$, then the integral, and thus the sum, is finite, so

$$\lim_{n\to\infty} \frac{\mu(B(x,2^{-n}))}{2^{-n\alpha}} = 0.$$

The Mass Distribution Principle implies $\mathscr{H}^\alpha(K_1) = \infty$, hence $\mathscr{H}^\alpha(K) = \infty$.
□

The capacities defined above with respect to the kernels $L(x,y) = |x-y|^{-\alpha}$ can be generalized to other kernels, just as Hausdorff measures generalize to other gauge functions besides t^{α}. One kernel that is important for the study of Brownian motion and harmonic measure in $\mathbb{R}^2$ is the logarithmic kernel,

$$L(x,y) = \log \frac{1}{|x-y|}.$$

This kernel is not positive, and special definitions and arguments are needed in this case. As before, we define the logarithmic energy of a measure as

$$\mathcal{E}(\mu) := \int_{\mathbb{R}^d} \int_{\mathbb{R}^d} \log \frac{1}{|x-y|} \, d\mu(x) d\mu(y),$$

and define

$$\gamma(K) := \inf_{\mu \in \mathrm{Pr}(K)} \mathcal{E}(\mu).$$

This is called Robin's constant. It can take any value in $(-\infty, \infty]$ and satisfies $\gamma(\lambda E) = \gamma(E) - \log \lambda$ for any $\lambda > 0$, so expanding a set makes the Robin constant smaller. It is therefore more convenient to work with

$$\mathrm{Cap}_{\log}(K) = \exp(-\gamma(K)).$$

This is a monotone set function, and if K is a disk then $\mathrm{Cap}_{\log}(K)$ equals its radius. For a good introduction to logarithmic capacity see Garnett and Marshall (2005). An equivalent definition of logarithmic capacity is described in Exercise 3.37. Logarithmic capacity has a close connection to 2-dimensional Brownian motion; a set in the plane is eventually hit by Brownian motion if and only if it has positive logarithmic capacity.

3.5 Marstrand's Projection Theorem

Let $K \subset \mathbb{R}^2$ be a compact set and let Π_θ be the orthogonal projection of $\mathbb{R}^2$ onto the line through the origin in the direction $(\cos \theta, \sin \theta)$, for $\theta \in [0, \pi)$. Since Π_θ is Lipschitz and $\Pi_\theta K \subset \mathbb{R}$,

$$\dim(\Pi_\theta K) \leq \min\{\dim(K), 1\}.$$

Marstrand's Theorem (Marstrand, 1954) states that for almost all directions equality holds. We will prove the following stronger version due to Kaufman (1968):

Theorem 3.5.1 *If* $\mathrm{Cap}_\alpha(K) > 0$ *for some* $0 < \alpha < 1$, *then for almost every* θ, *we have* $\mathrm{Cap}_\alpha(\Pi_\theta K) > 0$.

This has the following corollary.

Corollary 3.5.2 *If* $\dim K \leq 1$, *then* $\dim(\Pi_\theta K) = \dim(K)$ *for almost every* θ.

Proof Take $\alpha < \dim(K)$. By Part (i) of Theorem 3.4.2, $\mathrm{Cap}_\alpha(K) > 0$, so by Theorem 3.5.1, $\mathrm{Cap}_\alpha(\Pi_\theta K) > 0$ for almost every θ. By Part (ii) of Theorem 3.4.2, $\dim(\Pi_\theta K) \geq \alpha$ for almost every θ. The set of θ depends on α, but by intersecting over all rational $\alpha < \dim(K)$ we get a single set of full measure for which the result holds. □

Proof of Theorem 3.5.1 Let μ be a measure on K with $\mathscr{E}_\alpha(\mu) < \infty$, for some $0 < \alpha < 1$ and consider the projected measure $\mu_\theta = \Pi_\theta \mu$ on $\mathbb{R}$ defined by $\mu_\theta(A) = \mu(\Pi_\theta^{-1}(A))$ or equivalently by

$$\int_\mathbb{R} f(x) d\mu_\theta(x) = \int_{\mathbb{R}^2} f(\Pi_\theta(y)) \, d\mu(y).$$

It is enough to show that μ_θ has finite energy for a.e. θ, which will follow from the stronger fact

$$\int_0^\pi \mathscr{E}_\alpha(\mu_\theta) \, d\theta < \infty.$$

Let us write this integral explicitly (using Fubini's Theorem):

$$\begin{aligned}
\int_0^\pi \mathscr{E}_\alpha(\mu_\theta) \, d\theta &= \int_0^\pi \iint \frac{d\mu_\theta(t) d\mu_\theta(s)}{|t-s|^\alpha} \, d\theta \\
&= \int_0^\pi \int_K \int_K \frac{d\mu(x) d\mu(y)}{|\Pi_\theta(x-y)|^\alpha} \, d\theta \\
&= \int_K \int_K \int_0^\pi \frac{d\theta}{|\Pi_\theta(x-y)|^\alpha} \, d\mu(x) d\mu(y) \\
&= \int_K \int_K \frac{1}{|x-y|^\alpha} \int_0^\pi \frac{d\theta}{|\Pi_\theta(u)|^\alpha} \, d\mu(x) d\mu(y),
\end{aligned}$$

where $u = (x-y)/|x-y|$ is the unit vector in the direction $(x-y)$. The inner integral does not depend on u and converges (since $\alpha < 1$), so we can write it as a constant C_α and pull it out of the integral. Thus

$$\int_0^\pi \mathscr{E}_\alpha(\mu_\theta) \, d\theta = C_\alpha \int_K \int_K \frac{d\mu(t) d\mu(s)}{|x-y|^\alpha} = C_\alpha \mathscr{E}_\alpha(\mu) < \infty,$$

by hypothesis. □

If K has dimension 1, then Marstrand's Theorem says $\dim(\Pi_\theta K) = 1$ for almost every θ. However, even if $\mathscr{H}^1(K)$ is positive this does not guarantee that $\mathscr{L}_1(\Pi_\theta K)$ is positive, where $\mathscr{L}_1$ is the Lebesgue measure on the real line. For example, the standard $\frac{1}{4}$-Cantor set in the plane (also called the four corner Cantor set) has projections of zero length in almost all directions. The proof of this will be given as Theorem 9.5.3. In contrast to this example, we have the following result (Kaufman, 1968).

Theorem 3.5.3 *If* $\mathrm{Cap}_1(K) > 0$, *then* $\mathscr{L}_1(\Pi_\theta K) > 0$ *for almost every* θ.

The following version of Theorem 3.5.3 is due to Mattila (1990). It involves the notion of the **Favard length** of a set

$$\mathrm{Fav}(K) = \int_0^\pi \mathscr{L}_1(\Pi_\theta K)\,d\theta.$$

(We will deal with Favard length in greater detail in Chapter 10.)

Theorem 3.5.4 *If* K *is compact,* $\mathrm{Fav}(K) \geq \pi^2 \mathrm{Cap}_1(K)$.

The constant π^2 is sharp; equality is obtained when K is a disk. No such estimate is true in the opposite direction since a line segment has positive Favard length but zero 1-dimensional capacity by Theorem 3.4.2, since its $\mathscr{H}^1$ measure is finite. The statement of Mattila's Theorem does not imply Kaufman's Theorem, since positive Favard length only implies $\mathscr{L}_1(\Pi_\theta K) > 0$ for a positive measure set of directions and not almost all directions. However, as a consequence of the proof we will obtain

Lemma 3.5.5 $\int_0^\pi \mathscr{L}_1(\Pi_\theta K)^{-1}\,d\theta \leq \mathscr{E}_1(\mu)$, *for every probability measure* μ *supported on* K.

This implies Kaufman's Theorem since the integral would be infinite if $\mathscr{L}_1(\Pi_\theta K) = 0$ for a positive measure set of θ. However, the lemma says that $\mathscr{L}_1(\Pi_\theta K)^{-1} \in L^1(0,\pi)$.

We first sketch how Kaufman proved Theorem 3.5.3 and will then give complete proofs of Theorem 3.5.4 and Lemma 3.5.5. Given a measure μ on K with $\mathscr{E}_1(\mu) < \infty$, Kaufman used Fourier transforms and elementary facts about Bessel's functions to show that

$$\int_0^\pi \int_{\mathbb{R}} |\hat{\mu}_\theta(s)|^2\,ds\,d\theta < \infty.$$

Thus, $\hat{\mu}_\theta$ is in L^2 for almost every θ, and so by the Plancherel formula we have $d\mu_\theta(x) = f_\theta(x)dx$ for some L^2 function f_θ. In particular, μ_θ is absolutely continuous with respect to Lebesgue measure, so $\mathrm{supp}(\mu_\theta) \subset \Pi_\theta K$ must have positive length for almost every θ.

Before beginning the proof of Mattila's Theorem we record a lemma on the differentiation of measures that we will need.

Lemma 3.5.6 *Suppose* ν *is a finite Borel measure on* $\mathbb{R}$ *satisfying*

$$\liminf_{\delta \to 0} \frac{\nu(x-\delta, x+\delta)}{2\delta} < \infty,$$

for ν-*a.e.* x. *Then* $\nu \ll \mathscr{L}_1$, *i.e.,* ν *is absolutely continuous with respect to Lebesgue measure.*

Proof Let $A \subset \mathbb{R}$ with $\mathscr{L}_1(A) = 0$. We wish to show $\nu(A) = 0$. We define

$$A_m = \left\{ x \in A : \liminf_{\delta \to 0} \frac{\nu(x - \delta, x + \delta)}{2\delta} \leq m \right\}.$$

By the assumption on ν, given an $\varepsilon > 0$ there is an m big enough so that $\nu(A \setminus A_m) < \varepsilon$. Since $A_m \subset A$ has zero Lebesgue measure we can take an open set U containing A_m with $\mathscr{L}_1(U) \leq \varepsilon / m$. For each $x \in A_m$ take a dyadic interval J_x with $x \in J_x \subset U$ and

$$\frac{\nu(J_x)}{|J_x|} \leq 8m.$$

There is such an interval J_x because x is the center of an interval $I \subset U$ such that $\nu(I) \leq 2m|I|$. If we let J_x be the dyadic interval of maximal length contained in I and containing x, then $|J_x| \geq |I|/4$ and $\nu(J_x) \leq \nu(I)$, as desired. The family $S = \{J_x : x \in A_m\}$ is a countable cover of A_m. Because the intervals are dyadic we can take a disjoint subcover (take the maximal intervals) and thus

$$\nu(A_m) \leq \sum_{J \in S} \nu(J) \leq 8m \sum_{J \in S} |J| \leq 8m\mathscr{L}_1(U) \leq 8\varepsilon.$$

Letting $\varepsilon \to 0$ proves $\nu(A) = 0$ as desired. $\square$

Proof of Lemma 3.5.5 Let $K \subset \mathbb{R}^2$ and μ be a probability measure on K with $\mathscr{E}_1(\mu) < \infty$. Put $\mu_\theta = \Pi_\theta \mu$. Further, let

$$\phi_\theta(t) = \liminf_{\delta \to 0} \frac{\mu_\theta(t - \delta, t + \delta)}{2\delta}.$$

We claim that

$$\int_0^\pi \int_{\mathbb{R}} \phi_\theta(t) \, d\mu_\theta(t) d\theta \leq \mathscr{E}_1(\mu). \tag{3.5.1}$$

To prove this we start with Fatou's lemma, which states that

$$\liminf_n \int f_n d\mu \geq \int (\liminf f_n) \, d\mu.$$

This implies that the left-hand side of (3.5.1) is bounded above by

$$\liminf_{\delta \to 0} \frac{1}{2\delta} \int_0^\pi \int_{\mathbb{R}} \mu_\theta(t - \delta, t + \delta) d\mu_\theta(t) \, d\theta$$

$$= \liminf_{\delta \to 0} \frac{1}{2\delta} \int_0^\pi \int_{\mathbb{R}} \int_{\mathbb{R}} \mathbf{1}_{(t - \delta, t + \delta)}(s) \, d\mu_\theta(s) d\mu_\theta(t) d\theta.$$

If $s = \Pi_\theta(x)$ and $t = \Pi_\theta(y)$ then $\mathbf{1}_{(t - \delta, t + \delta)}(s) = 1$ if and only if

$$|\Pi_\theta(x - y)| = |\Pi_\theta(x) - \Pi_\theta(y)| < \delta.$$

Using this and Fubini's Theorem the integral becomes

$$= \liminf_{\delta \to 0} \frac{1}{2\delta} \int_0^\pi \int_K \int_K \mathbf{1}_{|\Pi_\theta (x-y)| < \delta}(x,y) \, d\mu(x) \, d\mu(y) \, d\theta$$

$$= \liminf_{\delta \to 0} \frac{1}{2\delta} \int_K \int_K \int_0^\pi \mathbf{1}_{|\Pi_\theta (x-y)| < \delta}(x,y) \, d\theta \, d\mu(x) \, d\mu(y).$$

It is fairly easy to check that given x and y, $\{\theta : |\Pi_\theta(x-y)| < \delta\}$ is an interval of length $2\arcsin(\delta/|x-y|)$. Thus, the integral becomes

$$= \liminf_{\delta \to 0} \frac{1}{2\delta} \int_K \int_K 2\arcsin \frac{\delta}{|x-y|} \, d\mu(x) \, d\mu(y). \qquad (3.5.2)$$

Note that since $\arcsin(0) = 0$, the definition of derivative implies

$$\lim_{\delta \to 0} \frac{1}{\delta} \arcsin \frac{\delta}{|x-y|} = \frac{1}{|x-y|},$$

and it is easy to check that

$$\sup_{\delta \to 0} \frac{1}{\delta} \arcsin \frac{\delta}{|x-y|} \le \frac{\pi}{|x-y|},$$

which is in $L^1(\mu \times \mu)$ by assumption. By the Lebesgue Dominated Convergence Theorem we can bring the limit inside the integral and (3.5.2) becomes

$$= \int_K \int_K \liminf_{\delta \to 0} \frac{1}{\delta} \arcsin \frac{\delta}{|x-y|} \, d\mu(x) \, d\mu(y)$$

$$= \int_K \int_K \frac{1}{|x-y|} \, d\mu(x) \, d\mu(y) = \mathscr{E}_1(\mu).$$

This is the claim, (3.5.1), we wished to prove.

Using the claim we now know that for a.e. θ

$$\int_{\mathbb{R}} \phi_\theta(t) \, d\mu_\theta(t) < \infty,$$

which implies that $\phi_\theta(t) < \infty$ for μ_θ-a.e. t. By the previous lemma this implies μ_θ is absolutely continuous for a.e. θ and that

$$\phi_\theta = \frac{d\mu_\theta}{dt}.$$

Using this, the claim becomes

$$\int_0^\pi \int_{\mathbb{R}} \left(\frac{d\mu_\theta}{dt} \right)^2 dt \, d\theta \le \mathscr{E}_1(\mu).$$

Now use the Cauchy–Schwarz inequality

$$1 = (\mu_\theta(\Pi_\theta K))^2 = \left(\int_{\mathbb{R}} \mathbf{1}_{\Pi_\theta K} \frac{d\mu_\theta}{dt} \, dt \right)^2 \le \mathscr{L}_1(\Pi_\theta K) \int_{\mathbb{R}} \left(\frac{d\mu_\theta}{dt} \right)^2 dt.$$

Thus

$$\mathscr{L}_1(\Pi_\theta K)^{-1} \le \int_{\mathbb{R}} \left(\frac{d\mu_\theta}{dt} \right)^2 dt.$$

Integrating $d\theta$ over $[0, \pi]$ gives

$$\int_0^\pi \mathscr{L}_1(\Pi_\theta K)^{-1} d\theta \le \mathscr{E}_1(\mu). \qquad \Box$$

Proof of Theorem 3.5.4   To finish the proof of Mattila's Theorem we use the Cauchy–Schwarz Inequality again and apply the above inequality

$$\begin{aligned}
\pi^2 &= \left(\int_0^\pi \mathscr{L}_1(\Pi_\theta K)^{1/2} \mathscr{L}_1(\Pi_\theta K)^{-1/2} d\theta \right)^2 \\
&\le \int_0^\pi \mathscr{L}_1(\Pi_\theta K) d\theta \int_0^\pi \mathscr{L}_1(\Pi_\theta K)^{-1} d\theta \\
&\le \mathrm{Fav}(K) \mathscr{E}_1(\mu).
\end{aligned}$$

Thus

$$\mathrm{Fav}(K) \ge \frac{\pi^2}{\mathscr{E}_1(\mu)}.$$

Taking the supremum over all probability measures supported on K gives the result. $\Box$

Much more can be said about projections of sets K that have finite $\mathscr{H}^1$ measure. We will prove in Section 9.5 that certain self-similar Cantor sets of dimension 1 project onto zero length in almost all directions. This is a special case of a more general result of Besicovitch (see the Notes at the end of Chapter 9).

3.6 Mapping a tree to Euclidean space preserves capacity

We have already seen that it is very useful to think of b-adic cubes in $\mathbb{R}^d$ as forming a tree and associating subsets of $\mathbb{R}^d$ with subtrees of this tree. This is a theme that will also be important later in the book. In this section we record a result that allows us to compare the capacity of a subset of $\mathbb{R}^d$ with the capacity of the associated tree.

In order to interpret theorems proved on trees as theorems about sets in Euclidean space, we employ the canonical mapping $\mathscr{R}$ from the boundary of a b^d-ary tree $\Gamma^{(b^d)}$ (every vertex has b^d children) to the cube $[0, 1]^d$. Formally, label the edges from each vertex to its children in a one-to-one manner with the vectors in $\Omega = \{0, 1, \ldots, b-1\}^d$. Then, the boundary $\partial \Gamma^{(b^d)}$ is identified with the sequence space $\Omega^{\mathbb{N}}$ and we define $\mathscr{R} : \Omega^{\mathbb{N}} \to [0,1]^d$ by

$$\mathscr{R}(\omega_1, \omega_2, \ldots) = \sum_{n=1}^{\infty} \omega_n b^{-n}. \tag{3.6.1}$$

Similarly, a vertex σ of $\Gamma^{(b^d)}$ is identified with $(\omega_1, \ldots, \omega_k) \in \Omega^k$ if $|\sigma| = k$, and we write $\mathscr{R}(\sigma)$ for the cube of side b^{-k} obtained as the image under $\mathscr{R}$ of all sequences in $\Omega^{\mathbb{N}}$ with prefix $(\omega_1, \ldots, \omega_k)$.

With the notation above, let T be a subtree of the b^d-ary tree $\Gamma^{(b^d)}$, so we may take $\partial T \subseteq \Omega^{\mathbb{N}}$. Two points ξ and η in the boundary of a tree correspond to infinite paths in the tree, and we let $\xi \wedge \eta$ denote the maximum subpath they have in common and $|\xi \wedge \eta|$ the length of this path. Moreover, if σ is a vertex of the tree, we write $\xi \geq \sigma$ to denote that σ is on the path corresponding to ξ (i.e., ξ is "below" σ in the tree). Equip $X = \partial T$ with the metric

$$\rho(\xi, \eta) = b^{-|\xi \wedge \eta|}.$$

Let μ be a Borel measure on ∂T. For increasing $f : \mathbb{N} \cup \{0\} \to (0, \infty)$, define

$$\mathscr{E}_f(\mu) = \int_{\partial T} \int_{\partial T} f(|\xi \wedge \eta|) d\mu(\xi) d\mu(\eta),$$

and

$$\mathrm{Cap}_f(T) = \left(\inf_{\mu(\partial T)=1} \mathscr{E}_f(\mu) \right)^{-1}.$$

The notions of energy and capacity are meaningful on any compact metric space (X, ρ). Given a decreasing function $g : (0, \infty) \to (0, \infty)$, define the g-energy of a Borel measure μ by

$$\mathscr{E}_g(\mu) = \int_X \int_X g(\rho(x,y)) d\mu(x) d\mu(y),$$

and the g-capacity of a set Λ by

$$\mathrm{Cap}_g(\Lambda) = \left(\inf_{\mu(\Lambda)=1} \mathscr{E}_g(\mu) \right)^{-1}.$$

Rather than using different symbols for energy and capacity when they occur on different spaces, we use the same symbols for both and depend on context (where does the measure or set involved live) to supply the correct definition to apply. For example, the following result uses energy and capacity on both the boundary of a tree and on Euclidean space (trees on the left side of the displayed equations and Euclidean space on the right side).

Theorem 3.6.1   *Given a decreasing function* $g : (0, \infty) \to (0, \infty)$ *define the kernel* $f(n) = g(b^{-n})$. *Then for any finite measure* μ *on* ∂T *we have*

$$\mathscr{E}_f(\mu) < \infty \Leftrightarrow \mathscr{E}_g(\mu \mathscr{R}^{-1}) < \infty, \tag{3.6.2}$$

and, in fact, the ratio of these two energies is bounded between positive constants depending only on the dimension d. *It follows that*

$$\mathrm{Cap}_f(T) > 0 \Leftrightarrow \mathrm{Cap}_g(\mathscr{R}(\partial T)) > 0.$$

Proof   Using summation by parts, the energy on ∂T can be rewritten:

$$\mathscr{E}_f(\mu) = \sum_{n=1}^{\infty} \int \int_{|\xi \wedge \eta| = n} f(n) \, d\mu(\xi) \, d\mu(\eta)$$

$$= \sum_{k=0}^{\infty} h(k)(\mu \times \mu) \{ |\xi \wedge \eta| \ge k \},$$

where $h(k) = f(k) - f(k-1)$ and by convention $f(-1) = 0$. Since

$$\{ |\xi \wedge \eta| \ge k \} = \bigcup_{|\sigma| = k} \{ \xi \wedge \eta \ge \sigma \} = \bigcup_{|\sigma| = k} (\{ \xi \ge \sigma \} \cap \{ \eta \ge \sigma \}),$$

we obtain

$$\mathscr{E}_f(\mu) = \sum_{k=0}^{\infty} h(k) \sum_{|\sigma| = k} \mu(\sigma)^2 = \sum_{k=0}^{\infty} h(k) S_k \tag{3.6.3}$$

where $S_k = S_k(\mu) = \sum_{|\sigma| = k} \mu(\sigma)^2$. Now we wish to adapt this calculation to the set $\mathscr{R}(\partial T)$ in the cube $[0,1]^d$. First observe that the same argument yields

$$\mathscr{E}_g(\mu \mathscr{R}^{-1}) \le \sum_{n=-r}^{\infty} g(b^{-n})(\mu \mathscr{R}^{-1} \times \mu \mathscr{R}^{-1}) \{ (x,y) : b^{-n} < |x-y| \le b^{1-n} \}$$

$$= \sum_{k=-r}^{\infty} h(k)(\mu \mathscr{R}^{-1} \times \mu \mathscr{R}^{-1}) \{ (x,y) : |x-y| \le b^{1-k} \}, \tag{3.6.4}$$

where r is a positive integer such that $\sqrt{d} \le b^{r+1}$.

For vertices σ, τ of T we write $\sigma \sim \tau$ if $\mathscr{R}(\sigma)$ and $\mathscr{R}(\tau)$ intersect (this is not an equivalence relation!). If $x, y \in \mathscr{R}(\partial T)$ satisfy $|x-y| \le b^{1-k}$, then there exist vertices σ, τ of T with $|\sigma| = |\tau| = k-1$ and $\sigma \sim \tau$ satisfying $x \in \mathscr{R}(\sigma)$ and $y \in \mathscr{R}(\tau)$. Therefore

$$(\mu \mathscr{R}^{-1} \times \mu \mathscr{R}^{-1}) \{ (x,y) : |x-y| \le b^{1-k} \} \le \sum_{|\sigma| = |\tau| = k-1} \mathbf{1}_{\{\sigma \sim \tau\}} \mu(\sigma)\mu(\tau).$$

Now use the inequality

$$\mu(\sigma)\mu(\tau) \leq \frac{\mu(\sigma)^2 + \mu(\tau)^2}{2}$$

and the key observation that

$$\#(\{\tau \in T : |\tau| = |\sigma| \text{ and } \tau \sim \sigma\}) \leq 3^d \quad \text{for all } \sigma \in T,$$

to conclude that

$$(\mu\mathscr{R}^{-1} \times \mu\mathscr{R}^{-1})\{(x,y) : |x - y| \leq b^{1-k}\} \tag{3.6.5}$$

$$\leq \frac{1}{2}\left(\sum_{\sigma}\sum_{\tau \sim \sigma}\mu(\sigma)^2 + \sum_{\tau}\sum_{\sigma \sim \tau}\mu(\tau)^2\right)$$

$$\leq \frac{1}{2}\left(\sum_{\sigma}\mu(\sigma)^2 3^d + \sum_{\tau}\mu(\tau)^2 3^d\right)$$

$$\leq 3^d S_{k-1}.$$

It is easy to compare S_{k-1} to S_k: Clearly, $|\sigma| = k - 1$ implies that

$$\mu(\sigma)^2 = \left(\sum_{\tau \geq \sigma; |\tau|=k}\mu(\tau)\right)^2 \leq b^d \sum_{\tau \geq \sigma; |\tau|=k}\mu(\tau)^2$$

and therefore

$$S_{k-1} \leq b^d S_k. \tag{3.6.6}$$

Combining this with (3.6.3), (3.6.4) and (3.6.5) yields

$$\mathscr{E}_g(\mu\mathscr{R}^{-1}) \leq (3b)^d \sum_{k=0}^{\infty} h(k)S_k = (3b)^d \mathscr{E}_f(\mu).$$

This proves the direction ($\Rightarrow$) in (3.6.2).

The other direction is immediate in dimension $d = 1$ and easy in general:

$$\mathscr{E}_g(\mu\mathscr{R}^{-1}) \geq \sum_{k=0}^{\infty} g(b^{-k})(\mu\mathscr{R}^{-1} \times \mu\mathscr{R}^{-1})\{(x,y) : b^{-k-1} < |x - y| \leq b^{-k}\}$$

$$= \sum_{n=0}^{\infty} h(n)(\mu\mathscr{R}^{-1} \times \mu\mathscr{R}^{-1})\{(x,y) : |x - y| \leq b^{-n}\}$$

$$\geq \sum_{n=0}^{\infty} h(n)S_{n+l},$$

where l is chosen to satisfy $b^l \geq d^{1/2}$ and therefore

$$\{(x,y) : |x - y| \leq b^{-n}\} \supseteq \bigcup_{|\sigma|=n+l} (\mathscr{R}(\sigma) \times \mathscr{R}(\sigma)).$$

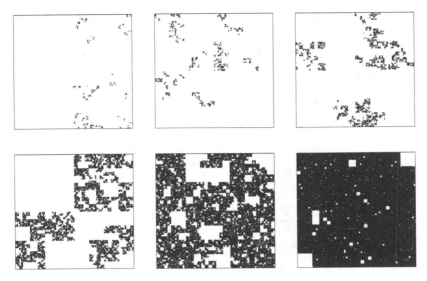

Figure 3.7.1 Random Cantor sets for $d = b = 2$ and $p = .5, .6, .7, .8, .9, .99$.

Invoking (3.6.6) we get

$$\mathscr{E}_g(\mu\mathscr{R}^{-1}) \geq b^{-dl} \sum_{n=0}^{\infty} h(n)S_n = b^{-dl}\mathscr{E}_f(\mu),$$

which completes the proof of (3.6.2). The capacity assertion of the theorem follows, since any measure ν on $\mathscr{R}(\partial T) \subseteq [0,1]^d$ can be written as $\mu\mathscr{R}^{-1}$ for an appropriate measure μ on ∂T.   □

3.7 Dimension of random Cantor sets

Fix a number $0 < p < 1$ and define a random set by dividing the unit cube $[0,1]^d$ of $\mathbb{R}^d$ into b^d b-adic cubes in the usual way. Each cube is kept with probability p, and the kept cubes are further subdivided, and again each subcube is kept with probability p. The result is a random Cantor set denoted by $\Lambda_d(b,p)$, or more precisely, a measure on the space of compact subsets of $[0,1]^d$ that we will denote μ_p in this section. If a property holds on a set of full μ_p measure, we will say it holds almost surely. A few examples of such sets are illustrated in Figure 3.7.1. These were drawn using 8 generations and the values $d = b = 2$ and $p \in \{.5, .6, .7, .8, .9, .99\}$.

Theorem 3.7.1 *If $p \leq b^{-d}$, then the random set $\Lambda_d(b,p)$ is empty μ_p-almost surely. If $p > b^{-d}$, then μ_p-almost surely, $\Lambda_d(b,p)$ is either empty or has Hausdorff and Minkowski dimension $d + \log_b p$. Moreover, the latter case occurs with positive probability.*

This was first proved in Hawkes (1981). We will actually give another proof of this later in the book, using Lyons' theorem on percolation on trees (see Section 8.3). However, the direct proof is simple and is a nice illustration of the ideas introduced in this chapter.

In terms of the corresponding trees, these random Cantor sets have a simple interpretation. We start with the usual rooted tree $\Gamma = \Gamma^{(b^d)}$ where each vertex has b^d children and we select edges with probability p and consider the connected component of the chosen edges that contains the root. Our random Cantor sets then correspond to the boundary of these random subtrees. The equivalence of dimension (see Section 1.3) and capacity under the canonical mapping from trees to sets yields that it suffices to do all our computations in the tree setting.

The first thing we want to do is compute the expected number of vertices at level n. Given a vertex we can easily compute the probability of having exactly k children as

$$q_k = p^k (1-p)^{b^d - k} \binom{b^d}{k}.$$

Thus, the expected number of vertices in the first generation is

$$m = pb^d = \sum_{k=0}^{\infty} kq_k.$$

(In our case, this sum has only finitely many terms since $q_k = 0$ for $k > b^d$, but many of our calculations are valid for branching processes given by the data $\{q_k\}$ only assuming finite mean and variance.)

Let Γ denote the full b^d-ary tree and let $Q = Q(\Gamma)$ denote the set of (finite or infinite) rooted subtrees of Γ (with the same root as Γ). We say a finite tree in Q has depth n if the maximum distance from any vertex of T to the root is n. A basis for the topology of Q is given by "cylinder sets" $Q(T)$, determined by fixing a finite rooted tree T of depth at most n by setting $Q(T)$ equal to the set of all $T' \in Q$ whose truncation to the first n levels equals T. The μ_p measure of such a cylinder set is

$$\mu_p(Q(T)) = \prod_{\sigma \in T, |\sigma| < n} p^{k(\sigma)} (1-p)^{b^d - k(\sigma)},$$

where $k(\sigma)$ is the number of children of the vertex σ and ∂T is identified with the leaves of T. Let $Z_n(T)$ denote the number of nth generation vertices of T. This is a random variable on the probability space (Q, μ_p) and, as usual, we define its expectation by

$$\mathbb{E}Z_n = \int_Q Z_n(T) \, d\mu_p(T).$$

Also in keeping with probabilistic notation, if $A, B \subset Q$, we write,

$$\mathbb{P}(A) = \mu_p(A),$$

and let

$$\mathbb{P}(A|B) = \mu_p(A \cap B)/\mu_p(B)$$

be the probability of A conditioned on B.

Lemma 3.7.2 $\mathbb{E}Z_n = m^n$.

Proof   This is easy by induction. We have already observed $\mathbb{E}Z_1 = m$ and in general,

$$\begin{aligned}
\mathbb{E}Z_n &= \sum_{k=0}^{\infty} \mathbb{E}(Z_n | Z_{n-1} = k) \mathbb{P}(Z_{n-1} = k) \\
&= \sum_k mk \mathbb{P}(Z_{n-1} = k) \\
&= m \mathbb{E}Z_{n-1} = m^n.
\end{aligned}$$ $\square$

Already, this is enough to give the correct upper estimate for the dimension of our random Cantor sets. Recall that $N(A, \varepsilon)$ is the minimum number of sets of diameter ε needed to cover A.

Lemma 3.7.3   *Suppose $\alpha \geq 0$ and A is a random set in $[0, 1]^d$ so that*

$$\mathbb{E}N(A, b^{-n}) \leq Cb^{\alpha n}.$$

Then $\overline{\dim}_{\mathcal{M}}(A) \leq \alpha$.

Proof   Fix $\varepsilon > 0$ and note that by the Monotone Convergence Theorem

$$\mathbb{E} \sum_{n=0}^{\infty} \frac{N(A, b^{-n})}{b^{n(\alpha+\varepsilon)}} = \sum_{n=0}^{\infty} \frac{\mathbb{E}N(A, b^{-n})}{b^{n(\alpha+\varepsilon)}} \leq \sum_{n=0}^{\infty} \frac{Cb^{\alpha n}}{b^{n(\alpha+\varepsilon)}} < \infty.$$

Thus $\sum_{n=0}^{\infty} \frac{N(A, b^{-n})}{b^{n(\alpha+\varepsilon)}}$ converges, and hence $N(A, b^{-n}) \leq Cb^{n(\alpha+\varepsilon)}$, for almost every A. Letting $\varepsilon \downarrow 0$ through a countable sequence gives the lemma. $\square$

When $p \leq b^{-d}$ we get $m \leq 1$ and the lemma only says that $\overline{\dim}_{\mathcal{M}}(A) = 0$. However, we shall prove later that A is almost surely empty in this case.

Combining the last two lemmas shows that $\Lambda_d(b,p)$ has dimension at most $d + \log_b p$, since

$$m^n = b^{n \log_b m} = b^{n \log_b(b^d p)} = b^{n(d + \log_b p)}.$$

To prove that this is also a lower bound (when we condition on the event that the set is non-empty) is harder, and will occupy most of the rest of this section. The main point is to show that for almost every $T \in Q(\Gamma)$, either $Z_n(T) = 0$ for large enough n or $Z_n(T) \geq Cm^n$ for some positive constant C (which may depend on T). To do this, define the random variable on Q,

$$W_n = m^{-n} Z_n.$$

The main result we need is that

Theorem 3.7.4 *Assume $\sum_k k^2 q_k < \infty$ (which is automatic for random Cantor sets). There is a function $W \in L^2(Q, \mu_p)$ so that $W_n \to W$ pointwise almost surely and in L^2. Moreover, the sets $\{T : W(T) = 0\}$ and $\bigcup_n \{T : Z_n(T) = 0\}$ are the same up to a set of μ_p measure zero.*

We will prove this later in the section. For the present, we simply use this result to compute the capacity of our sets using the kernel $f(n) = b^{\alpha n}$ on the tree. Under the assumptions of the theorem, for every vertex $\sigma \in T$ the following limit exists a.s.:

$$W(\sigma) = \lim_{n \to \infty} m^{-n} Z_n(T, \sigma),$$

where $Z_n(T, \sigma)$ denotes the number of nth level vertices of T that are descendants of σ. It is easy to check that this is a flow on T with total mass $W(T)$. Moreover, $W(\sigma)$ is a random variable on Q with the same distribution as $m^{-|\sigma|} W$ for every $\sigma \in T$. For $\sigma \notin T$ we let $W(\sigma) = 0$. Therefore $\mathbb{E} W(\sigma)^2 = m^{-2|\sigma|} \mathbb{E} W^2 \mathbb{P}(\sigma \in T) < \infty$ by Theorem 3.7.4.

According to formula (3.6.3), the f-energy of the measure determined by the flow above is given by

$$\tilde{\mathcal{E}}_f(W) = \sum_{n=0}^{\infty} \sum_{\sigma \in \Gamma_n} [f(n) - f(n-1)] W(\sigma)^2,$$

where Γ_n denotes the nth level vertices of the full tree Γ. Thus, the expected energy is

$$\mathbb{E} \tilde{\mathcal{E}}_f(W) = \sum_{n=0}^{\infty} \sum_{\sigma \in \Gamma_n} [f(n) - f(n-1)] \mathbb{E}(W(\sigma)^2)$$

$$= \sum_{n=0}^{\infty} \sum_{\sigma \in \Gamma_n} [f(n) - f(n-1)] m^{-2n} \mathbb{E}(W^2) \mathbb{P}(\sigma \in T).$$

The probability that $\sigma \in T$ is the probability that all the edges connecting it to the root are in T. Each edge is chosen independently with probability p, so this probability is p^n. Thus

$$\mathbb{E}\tilde{\mathscr{E}}_f(\nu) = \mathbb{E}(W^2) \sum_{n=0}^{\infty} \sum_{\sigma \in \Gamma_n} (b^{\alpha n} - b^{\alpha(n-1)}) m^{-2n} p^n$$

$$= \mathbb{E}(W^2)(1 - b^{-\alpha}) \sum_{n=0}^{\infty} b^{\alpha n} b^{dn} m^{-2n} p^n$$

$$= \mathbb{E}(W^2)(1 - b^{-\alpha}) \sum_{n=0}^{\infty} b^{\alpha n} b^{dn} (b^d p)^{-2n} p^n$$

$$= \mathbb{E}(W^2)(1 - b^{-\alpha}) \sum_{n=0}^{\infty} b^{n(\alpha - \log_b m)}.$$

This sum is clearly finite if $\alpha < \log_b m = d + \log_b p$. Thus, if $W(T) > 0$, the corresponding set has positive α-capacity for every $\alpha < d + \log_b p$ and hence dimension $\geq d + \log_b p$. This completes the proof of Theorem 3.7.1 except for the proof of Theorem 3.7.4.

A short proof of Theorem 3.7.4 can be given using martingales. Here we give a more elementary but slightly longer proof.

Proof of Theorem 3.7.4 Our first goal is to show

$$\mathbb{E}(|W_n - W_{n+1}|^2) = \|W_n - W_{n+1}\|_2^2 < Cm^{-n}.$$

Clearly,

$$\mathbb{E}(|W_n - W_{n+1}|^2) = m^{-2n-2} \mathbb{E}(|mZ_n - Z_{n+1}|^2)$$

so it is enough to show

$$\mathbb{E}(|Z_{n+1} - mZ_n|^2) \leq Cm^n. \tag{3.7.1}$$

To prove this, we will condition on Z_n and write $Z_{n+1} - mZ_n$ as a sum of Z_n orthogonal functions $\{X_j\}$ on the space Q of random trees. To do this we will actually condition on the first n levels of the tree, i.e., fix some finite n-level tree T with a fixed number Z_n of nth level leaves and let $Q(T)$ be the associated cylinder set in Q (i.e., all the infinite trees that agree with T in the first n levels). Let

$$\mathbb{E}_T(X) = \frac{1}{\mu_p(Q(T))} \int_{Q(T)} X \, d\mu_p$$

denote the average of X over $Q(T)$. We will order the Z_n leaves of T in some way and let

$$X_j = \sum_k (k - m) \mathbf{1}_{Q(j,k)},$$

where $\mathbf{1}_A$ denotes the indicator function of A and $Q(j,k)$ is the set of infinite trees in $Q(T)$ so that the jth leaf of the finite tree T has k children. Note that $\mu_p(Q(j,k)) = q_k \mu_p(Q(T))$. Also note that for every tree $T' \in Q(T)$, we have

$$Z_{n+1}(T') - mZ_n(T') = \sum_{j=1}^{Z_n} X_j(T').$$

Observe that

$$\mathbb{E}_T(X_j) = \sum_k (k-m)q_k = 0,$$

since $\sum_k kq_k = m$. Next we want to see that X_i and X_j are orthogonal if $i \neq j$. Choose $i < j$. The sets $Q(i,l)$ and $Q(j,k)$ are independent, so

$$\mu\Big(Q(i,l) \cap Q(j,k)\Big) = \mu(Q(i,l))\mu(Q(j,k))$$

and hence $\mathbb{E}_T(X_iX_j) = \mathbb{E}_T(X_i)\mathbb{E}_T(X_j) = 0$. Finally, we need the L^2 norm of the X_j to be bounded uniformly. Note,

$$\mathbb{E}_T(|X_j|^2) = \sum_k |k-m|^2 q_k \equiv \sigma^2,$$

which we assume is bounded.

Thus, if we take expected values over $Q(T)$ we get

$$\mathbb{E}_T(|Z_{n+1} - mZ_n|^2) = \mathbb{E}_T\left(\left|\sum_{j=1}^{Z_n} X_j\right|^2\right) = \sum_{j=1}^{Z_n} \mathbb{E}_T(|X_j|^2) = \sigma^2 Z_n(T),$$

as desired. Now we have to write Q as a finite disjoint union of sets of the form $Q(T)$. Since the average value over $Q(T)$ is $\sigma^2 Z_n(T)$ and the expectation (over all n-level trees T) of $Z_n(T)$ is m^n, we infer that

$$\mathbb{E}(|Z_{n+1} - mZ_n|^2) = \sum_T \mathbb{E}_T(|Z_{n+1} - mZ_n|^2)\mu_p(Q(T))$$

$$= \sigma^2 \sum_T Z_n(T)\mu_p(Q(T)) = \sigma^2 m^n,$$

as desired.

We have now proven that $\|W_n - W_{n-1}\|_2 \leq Cm^{-n/2}$. This easily implies that the sequence $\{W_n\}$ is Cauchy in L^2 and hence converges in the L^2 norm to a function W. Moreover, $W \geq 0$ since the same is true for every W_n. Furthermore,

$$\mathbb{E}W = \lim_n \mathbb{E}W_n = \lim_n m^{-n}\mathbb{E}Z_n = 1,$$

and so $\{T : W(T) > 0\}$ must have positive measure.

To see that W_n converges pointwise almost everywhere to W, note that using the Cauchy–Schwarz Inequality we get

$$\mathbb{E} \sum_n |W_n - W_{n-1}| = \sum_n \mathbb{E}|W_n - W_{n-1}|$$
$$\leq \sum_n \|W_n - W_{n-1}\|_2$$
$$\leq \sum_n Cm^{-n/2},$$

which converges if $m > 1$ (the case $m \leq 1$ is considered later). Since the left-hand side must be finite almost everywhere and hence the series converges absolutely for almost every T. Thus $W_n \to W$ almost everywhere, as desired.

Since $Z_n(T) = 0$ for some n implies $W(T) = 0$, the final claim of the theorem will follow if we show the two sets have the same measure. This can be done by computing each of them using a generating function argument. Let

$$f(x) = \sum_{k=0}^{\infty} q_k x^k$$

be the generating function associated to the sequence $\{q_k\}$. (As noted before, in the case of random Cantor sets this is a finite sum and f is a polynomial, but the argument that follows is more general.) Since f is a positive sum of the increasing, convex functions $\{q_k x^k\}$, it is also increasing and convex on $[0, 1]$. Furthermore, $f(1) = 1$, $f(0) = q_0$ and $f'(1) = \sum_k kq_k = m$. If $m > 1$, this means that f has a unique fixed point $x_0 < 1$ that is zero if $q_0 = 0$ and is $q_0 < x_0 < 1$ if $q_0 > 0$. The plot in Figure 3.7.2 shows some of these generating functions.

We claim that $\mathbb{P}(Z_n = 0)$ and $\mathbb{P}(W = 0)$ can be computed in terms of the generating function f. Let $q_{k,j} = \mathbb{P}(Z_{n+1} = j | Z_n = k)$. Each of the k vertices has children with distribution $\{q_i\}$ independent of the other $k-1$ vertices. Thus, the probability of choosing exactly j children overall is the coefficient of x^j in

$$(q_0 + q_1 x + q_2 x^2 + \cdots)^k,$$

or, in other words, the generating function of $\{q_{k,j}\}$ is

$$\sum_j q_{k,j} s^j = [f(s)]^k.$$

Now let $q_j^{(n)} = \mathbb{P}(Z_n = j | Z_0 = 1)$ be the n-step transition probabilities. Let F_n be the corresponding generating function associated to the sequence $(q_j^{(n)})_j$. We claim that $F_n = f^{(n)}$ where $f^{(n)} = f \circ \cdots \circ f$ denotes the n fold composition of f with itself. We prove this by induction, the case $n = 1$ being obvious.

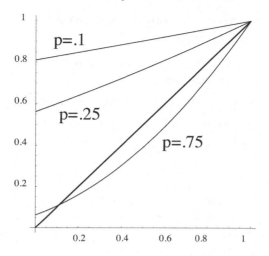

Figure 3.7.2 The generating functions f described in the text for $d = 1, b = 2$ and $p = .1, .25, .75$. Note that $f'(1) = \sum_k k q_k$ is the expected number of children.

Computing $q_j^{(n+1)}$ from $q_j^{(n)}$ gives

$$q_j^{(n+1)} = \sum_k q_k^{(n)} q_{k,j},$$

so in terms of generating functions,

$$F_{n+1}(x) = \sum_j q_j^{(n+1)} x^j = \sum_k \sum_j q_k^{(n)} q_{k,j} x^j = \sum_k q_k^{(n)} [f(x)]^k$$
$$= F_n(f(x)) = f^{(n)}(f(x)) = f^{(n+1)}(x),$$

as desired.

Thus $\mathbb{P}(Z_n = 0) = f^{(n)}(0)$. Since f is a strictly increasing function on $[0, 1]$, the iterates of 0 increase to the first fixed point of f. If $m > 1$, this is the point x_0 described above. If $m \leq 1$, then the first fixed point is 1. Thus, if $m > 1, Z_n$ remains positive with positive probability and if $m \leq 1$, it is eventually equal to zero with probability equal to 1.

To compute the probability of $\{W = 0\}$, let $r = \mathbb{P}(W = 0)$ and note that

$$r = \sum_k \mathbb{P}(W = 0 | Z_1 = k) \mathbb{P}(Z_1 = k) = \sum_k r^k q_k = f(r).$$

Thus r is a fixed point of f. If $m > 1$, then we saw above that $\mathbb{E}W = 1$, so we must have $r < 1$ and so $r = x_0 = \mathbb{P}(Z_n = 0$ for some $n)$, as desired. $\square$

Next we discuss the probability that a random set Λ hits a deterministic set

A and see that this depends crucially on the dimensions of the two sets. For simplicity we assume $A \subset [0, 1]$ and $\Lambda = \Lambda(p) = \Lambda_1(2, p)$.

Lemma 3.7.5 *If $A \subset [0, 1]$ is closed and intersects the random set $\Lambda(2^{-\alpha})$ with positive probability then $\dim(A) \geq \alpha$.*

Proof Let $b = \mathbb{P}(A \cap \Lambda(2^{-\alpha}) \neq \emptyset) > 0$. Then for any countable collection $\{I_j\}$ of dyadic subintervals of $[0, 1]$ such that $A \subset \bigcup_j I_j$ we have

$$b \leq \sum_j \mathbb{P}(I_j \cap \Lambda(2^{-\alpha}) \neq \emptyset) \leq \sum_j \mathbb{P}(I_j \cap \Lambda^{n_j}(2^{-\alpha}) \neq \emptyset).$$

Here Λ^n is the nth generation of the construction of $\Lambda(2^{-\alpha})$ and n_j is defined so I_j has length 2^{-n_j}. Thus the summand on the right equals $(2^{-\alpha})^{n_j} = |I_j|^{\alpha}$. Therefore $b \leq \sum_j |I_j|^{\alpha}$. However, if $\dim(A) = \alpha_1 < \alpha$, then there exists a collection of dyadic intervals $\{I_j\}$ so that $A \subset \bigcup_j I_j$ and so that $\sum_j |I_j|^{\alpha} < b$. $\square$

Thus $\Lambda(2^{-\alpha})$ almost surely doesn't hit A if $\dim(A) < \alpha$. In Theorem 8.5.2 we will show these sets do intersect with positive probability if $\dim(A) > \alpha$. By a random closed set A in $[0, 1]$ we mean a set chosen from any probability distribution on the space of compact subsets of $[0, 1]$.

Corollary 3.7.6 *If a random closed set $A \subset [0, 1]$ intersects the independent random set $\Lambda(2^{-\alpha})$ (which has dimension $1 - \alpha$) with positive probability then $\mathbb{P}(\dim(A) \geq \alpha) > 0$.*

Proof This follows from Lemma 3.7.5 by conditioning on A. $\square$

Lemma 3.7.7 *If a random closed set A satisfies $\mathbb{P}(A \cap K \neq \emptyset) > 0$ for all fixed closed sets K such that $\dim K \geq \beta$, then $||\dim(A)||_{\infty} \geq 1 - \beta$.*

Proof We know from Theorem 3.7.1 that $\dim(\Lambda(2^{1-\beta})) = \beta = \beta$ almost surely (if the set is non-empty) and so by hypothesis is hit by A with positive probability. Thus by Corollary 3.7.6, $\dim(A) \geq 1 - \beta$ with positive probability and hence $||\dim(A)||_{\infty} \geq 1 - \beta$. $\square$

3.8 Notes

The left-hand inequality of Theorem 3.2.1 is Marstrand's Product Theorem (Marstrand, 1954) and the right-hand side is due to Tricot (1982), refining an early result of Besicovitch and Moran (1945). Theorem 3.2.1 gives upper and lower bounds for the Hausdorff dimension of a product $A \times B$ as

$$\dim(A) + \dim(B) \leq \dim(A \times B) \leq \dim_p(A) + \dim(B).$$

We saw an example where both inequalities were sharp, but there is a more subtle question: For any A are both inequalities sharp? In other words, given A, do there exist sets B_1 and B_2 so that

$$\dim(A \times B_1) = \dim(A) + \dim(B_1),$$

$$\dim(A \times B_2) = \dim_p(A) + \dim(B_2)?$$

The first is easy since any set B_1 where $\dim(B_1) = \dim_p(B_1)$ will work, say $B_1 = [0, 1]$, or any self-similar set. The second question is harder; in Bishop and Peres (1996) we prove that it is true by constructing such a B_2. A similar result for sets in $\mathbb{R}$ was given independently by Xiao (1996). We state here the claim for compact sets. By quoting results of Joyce and Preiss (1995) this implies it is true for the much larger class of analytic sets (which contain all Borel sets as a proper subclass).

Theorem 3.8.1 *Given a compact set $A \subset \mathbb{R}^d$ and $\varepsilon > 0$ there is a compact $B \subset \mathbb{R}^d$ so that $\dim(B) \leq d - \dim_p(A) + \varepsilon$ and $\dim(A \times B) \geq d - \varepsilon$.*

Thus

$$\dim_p(A) = \sup_B (\dim(A \times B) - \dim(B)),$$

where the supremum is over all compact subsets of $\mathbb{R}^d$. It is a simple matter to modify the proof to take $\varepsilon = 0$. The proof is by construction (that will not be given here) and applies Lemma 2.8.1(ii).

As discovered by Howroyd (1995) both Frostman's Lemma 3.1.1 and part (i) of Theorem 3.4.2 also extend to compact metric spaces, but the proof is different, see Mattila (1995). The proof of part (ii) extends directly to any metric space, see Theorem 6.4.6.

Theorem 3.6.1 for $g(t) = \log(1/t)$ and $f(n) = n \log b$ was proved by Benjamini and Peres (1992).

Consider a Borel set A in the plane of dimension at α. For $\beta < \alpha \wedge 1$ (the minimum of α and 1), define the set E_β of directions where the projection of A has Hausdorff dimension at most β. Then Kaufman (1968) showed that $\dim(E_\beta) \leq \alpha$. This was improved by Falconer (1982), who showed that $\dim(E_\beta) \leq 1 + \beta - \alpha$, with the convention that the empty set has dimension $-\infty$. It was extended further in Peres and Schlag (2000), Proposition 6.1. This result, and many others about projections, are treated in detail by Mattila (2015).

Many recent papers address the problem of improving Marstrand's Projection Theorem for *specific* sets, and in particular for self-similar sets. Let K be the attractor of $\{f_i\}_{i=1}^\ell$, where the f_i are planar similitudes. In (Peres and

Shmerkin, 2009, Theorem 5) it is shown that $\dim_H(\Pi_\theta K) = \min(\dim_H K, 1)$ for *all* θ, provided the linear part of some f_i is a scaled irrational rotation. Without the irrationality assumption this is not true, but it follows from work of Hochman (2014) that $\dim_H(\Pi_\theta K) = \min(\dim_H K, 1)$ outside of a set of θ of zero packing (and therefore also Hausdorff) dimension. The methods of these papers are intrinsically about dimension but, building upon them, in (Shmerkin and Solomyak, 2014, Theorem C) it is shown that if $\dim_H K > 1$, then $\mathscr{L}_1(\Pi_\theta K) > 0$ outside of a set of zero Hausdorff dimension of (possible) exceptions θ. General methods for studying the dimensions of projections of many sets and measures of dynamical origin, including self-similar sets and measures in arbitrary dimension, were developed in Hochman and Shmerkin (2012) and Hochman (2013). The proofs of all these results rely on some version of Marstrand's Projection Theorem.

A related problem concerns arithmetic sums of Cantor sets: if $K, K' \subset \mathbb{R}$ and $u \in \mathbb{R}$, the arithmetic sum $K + uK' = \{x + uy : x \in K, y \in K'\}$ is, up to homothety, the same as $\Pi_\theta(K \times K')$ where $u = \tan(\theta)$. For these sums, similar results have been obtained. For example, (Peres and Shmerkin, 2009, Theorem 2) shows that if there are contraction ratios r, r' among the similitudes generating self-similar sets $K, K' \subset \mathbb{R}$ such that $\log r / \log r'$ is irrational, then $\dim_H(K + uK') = \min(\dim_H K + \dim_H K', 1)$ for all $u \in \mathbb{R} \setminus \{0\}$, and in particular for $u = 1.$j

The arguments of Section 3.7 prove, given a kernel f, that almost surely our non-empty random Cantor sets have positive capacity with respect to f if and only if

$$\sum_{n=1}^{\infty} m^{-n} f(n) < \infty. \tag{3.8.1}$$

The set of full measure depends on f however. Pemantle and Peres (1995) proved that there is a set of full μ_p measure (conditioned on survival) such that ∂T has positive capacity for the kernel f if and only if (3.8.1) holds (with the set independent of the kernel).

The sets $\Lambda_d(b, p)$ defined in Section 3.7 were described by Mandelbrot (1974) and analyzed by Kahane and Peyrière (1976). The exact measure of more general random sets was found by Graf et al. (1988). Chayes et al. (1988) noted that the dimension could be easily inferred from the work in Kahane and Peyrière (1976) and proved the remarkable result that $\Lambda_2(b, p)$ contains connected components for every $b > 1$ and p close enough to 1 (depending on b).

The Mass Distribution Principle (Theorem 1.2.8) gives a lower bound for the dimension of a set E in terms of a class of measures supported on the set, and Frostman's Lemma (Lemma 3.1) shows that this bound is always sharp. A lower bound is also given by the Fourier dimension, i.e., the supremum of all $s \in \mathbb{R}$ so that E supports a measure μ whose Fourier transform is bounded by $|x|^{s/2}$. In general, the two dimensions are not equal (e.g., they differ for the middle thirds Cantor set), and sets where they agree are called Salem sets, after Raphaël Salem, who constructed examples using randomization. Salem sets and a vast array of Fourier techniques for estimating dimension are discussed in Mattila (2015).

3.9 Exercises

Exercise 3.1 Construct an infinite tree with a legal flow so that no cut-set attains the infimum cut.

Exercise 3.2 A rooted tree T is called super-periodic if the subtree below any vertex $v \in T$ contains a copy of T (with the root corresponding to v). Show that if $K \subset [0,1]^d$ corresponds to a super-periodic tree via dyadic cubes, then $\dim_{\mathcal{M}}(K)$ exists and $\dim(K) = \dim_{\mathcal{M}}(K)$. (This is a simpler tree version of Exercise 2.9 in Chapter 2.)

Exercise 3.3 Give an example of sets $E, F \subset \mathbb{R}$ with

$$\dim(E) + \dim(F) < \dim(E \times F) < \dim(E) + \dim(F) + 1.$$

Exercise 3.4 We define the lower packing dimension $\underline{\dim}_p(A)$ of a set A in a metric space to be

$$\underline{\dim}_p(A) = \inf \left\{ \sup_{j \geq 1} \underline{\dim}_{\mathcal{M}}(A_j) \mid A \subset \bigcup_{j=1}^{\infty} A_j \right\}. \tag{3.9.1}$$

Show that $\dim(A) \leq \underline{\dim}_p(A)$.

Exercise 3.5 Prove that $\dim_p(E \times F) \geq \underline{\dim}_p(E) + \dim_p(F)$. This is from Bishop and Peres (1996) and improves a result of Tricot (1982).

• **Exercise 3.6** Show that strict inequality is possible in the previous problem, i.e., there is a compact E so that

$$\dim_p(E \times F) - \dim_p(F) > \underline{\dim}_p(E),$$

for every compact F. See Bishop and Peres (1996).

Exercise 3.7   Show that there are compact sets $A \subset \mathbb{R}^d$ so that

$$\dim(A \times B) = \dim_p(A) + \dim(B)$$

if and only if $\dim(B) = d - \dim_p(A)$.

• **Exercise 3.8**   Construct two closed sets $A, B \subset \mathbb{R}^2$, both of dimension 1, so that $\dim(A \cap (B + x)) = 1$ for every $x \in \mathbb{R}^2$.

• **Exercise 3.9**   Show that we can take $\varepsilon = 0$ in Theorem 3.8.1.

Exercise 3.10   Construct a set $E \subset \mathbb{R}^2$ so that for every $0 \le \alpha \le 1$ there is a direction such that $\dim(\Pi_\theta E) = \alpha$.

Exercise 3.11   Suppose a Cantor set $E = \cap E_n \subset \mathbb{R}$ is defined by inductively taking E_0 to be an interval and defining E_{n+1} by removing an open interval $J(I)$ from each component interval I of E_n. The **thickness** of E is the largest constant τ so that both components of $I \setminus J(I)$ have length at least $\tau |J(I)|$, at every stage of the construction. Show that if two Cantor sets E, F have thicknesses τ_E, τ_F that satisfy $\tau_E \cdot \tau_F \ge 1$, and their convex hulls intersect, then $E \cap F \ne \emptyset$.

   The definition and result are due to Sheldon Newhouse, see Section 3 of Newhouse (1970). For more on this topic see Solomyak (1997).

• **Exercise 3.12**   Let $\mathbf{C}$ be the middle thirds Cantor set and let $E = \mathbf{C} \times \mathbf{C}$. Show that for almost every direction θ, $\Pi_\theta E$ contains an interval. (In fact, this holds for all directions except for two.)

Exercise 3.13   Give an example to show that α-capacity is not additive.

Exercise 3.14   Show that if $\{K_n\}$ is an increasing sequence of sets, that is, $K_n \subset K_{n+1}$ for all $n \in \mathbb{N}$, then $\mathrm{Cap}_\alpha(\bigcup_n K_n) = \lim_n \mathrm{Cap}_\alpha(K_n)$.

Exercise 3.15   Prove $\mathrm{Cap}_\alpha(E \cup F) \le \mathrm{Cap}_\alpha(E) + \mathrm{Cap}_\alpha(F)$.

Exercise 3.16   Given two probability measures on E, prove

$$\mathscr{E}_g\left(\frac{1}{2}(\mu + \nu)\right) \le \frac{1}{2}\left(\mathscr{E}_g(\mu) + \mathscr{E}_g(\nu)\right).$$

• **Exercise 3.17**   Show that if T is a regular tree (every vertex has the same number of children), and f is increasing, then $\mathscr{E}_f$ is minimized by the probability measure that evenly divides mass among child vertices.

Exercise 3.18   Compute the $(\log_3 2)$-capacity of the middle thirds Cantor set.

Exercise 3.19   Prove E has σ-finite $\mathscr{H}^{\varphi}$ measure if and only if $\mathscr{H}^{\psi}(E) = 0$ for every ψ such that

$$\lim_{t \to 0} \frac{\psi(t)}{\varphi(t)} = 0.$$

This is from Besicovitch (1956).

Exercise 3.20   Construct a Jordan curve of σ-finite $\mathscr{H}^1$ measure, but which has no subarc of finite $\mathscr{H}^1$ measure.

Exercise 3.21   Show the π in Theorem 3.5.4 is sharp.

Exercise 3.22   Modify the proof of Lemma 3.7.3 to show that almost surely $\mathscr{H}^{\alpha}(A) < \infty$.

Exercise 3.23   Suppose we generate random subsets of $[0, 1]$ by choosing dyadic subintervals with probability p. Compute the probability of getting the empty set (as a function of p).

• **Exercise 3.24**   Let $Z_n(T)$ denote the number of nth generation vertices of the random tree T as in Section 3.7, and, as before, let $m = pb^d$ be the expected number first generation vertices. Use generating functions to show that for $m \neq 1$,

$$\text{Var}Z_n = \mathbb{E}(|Z_n - \mathbb{E}Z_n|^2) = \sigma^2 m^{n-1}(m^n - 1)/(m - 1),$$

where

$$\sigma^2 = \text{Var}Z_1 = \sum_k (k - m)^2 q_k.$$

If $m = 1$ show the variance of Z_n is $n\sigma^2$. (See Athreya and Ney (1972).)

Exercise 3.25   If $m < 1$, show that $\mathbb{P}(Z_n > 0)$ decays like $O(m^n)$.

Exercise 3.26   Suppose we define a random subset of $[0, 1]$ iteratively by dividing an interval I into two disjoint subintervals with lengths chosen uniformly from $[0, \frac{1}{2}|I|]$. What is the dimension of the resulting set almost surely?

Exercise 3.27   Suppose A is a random set in $\mathbb{R}^2$ generated with b-adic squares chosen with probability p (we shall call this a (b, p) random set in $\mathbb{R}^2$). What is the almost sure dimension of the vertical projection of A onto the real axis?

Exercise 3.28   If $p > 1/b$, show that the vertical projection has positive length almost surely (if it is non-empty).

Exercise 3.29   If $p > 1/b$, does the vertical projection contain an interval almost surely (if it is non-empty)?

Exercise 3.30   With A as above, what is the almost sure dimension of an intersection with a random horizontal line?

Exercise 3.31   Suppose A_1, A_2 are random sets generated from b-adic cubes in $\mathbb{R}^d$ with probabilities p_1 and p_2. Give necessary and sufficient conditions on p_1 and p_2 for $A_1 \cap A_2 \neq \emptyset$ with positive probability.

Exercise 3.32   Prove that $\Lambda_2(b, p)$ is totally disconnected almost surely if $p < 1/b$ (see Section 3.7).

Exercise 3.33   Prove the same if $p = 1/b$.

Exercise 3.34   Is there an $\varepsilon > 0$ so that the set $\Lambda_2(b, p)$ is almost surely disconnected if $p < (1 + \varepsilon)/b$?

Exercise 3.35   Consider random sets constructed by replacing p by a sequence $\{p_n\}$ (so that edges at level n are chosen independently with probability p_n). Compute the dimension in terms of $\{p_n\}$.

Exercise 3.36   With sets as in the previous exercise, characterize the sequences that give sets of positive Lebesgue measure in $\mathbb{R}^d$.

• **Exercise 3.37**   If $E \subset \mathbb{C}$ and $n \geq 2$ is an integer, we define

$$p_n(E) = \sup \prod_{j<k} |z_j - z_k|,$$

where the supremum is over all sets $\{z_k\}$ of size n in E. Show that

$$d_n(E) = p_n(E)^{\frac{2}{n(n-1)}}$$

is decreasing in n. The limit, $d(E)$, is called the transfinite diameter of E and is equal to the logarithmic capacity of E (but this is not obvious; see Appendix E of Garnett and Marshall (2005)).

4

Self-affine sets

In Chapter 2 we discussed criteria that imply that the Hausdorff and Minkowski dimensions of a set agree. In this chapter we will describe a class of self-affine sets for which the two dimensions usually differ. These sets are invariant under affine maps of the form $(x, y) \to (nx, my)$. We will compute the Minkowski, Hausdorff and packing dimensions as well as the Hausdorff measure in the critical dimensions.

4.1 Construction and Minkowski dimension

Suppose $n \geq m$ are integers and divide the unit square $[0, 1]^2$ into $n \times m$ equal closed rectangles, each with width n^{-1} and height m^{-1}. Choose some subset of these rectangles and throw away the rest. Divide the remaining rectangles into $n \times m$ subrectangles each of width n^{-2} and height m^{-2} and keep those corresponding to the same pattern used above. At the kth stage we have a collection of $n^{-k} \times m^{-k}$ rectangles. To obtain the next level we subdivide each into $n \times m$ rectangles as above, and keep the ones corresponding to our pattern. Continuing indefinitely gives a compact set that we will denote K. See Figure 1.3.5 for the construction when $n = 3$, $m = 2$.

Here are two alternate ways of describing the same construction.

1. Given an $m \times n$ matrix A of 0's and 1's, with rows labeled by 0 to $m - 1$ from bottom to top and columns by 0 to $n - 1$ from left to right, we let

$$K[A] = \left\{ (x, y) = \left(\sum_k x_k n^{-k}, \sum_k y_k m^{-k} \right) : A_{y_k, x_k} = 1 \text{ for all } k \right\}.$$

119

2. Given a subset $D \subset \{0, \ldots, n-1\} \times \{0, \ldots, m-1\}$, we let

$$K(D) = \{\sum_{k=1}^{\infty} (a_k n^{-k}, b_k m^{-k}) : (a_k, b_k) \in D \text{ for all } k\}.$$

Thus, the McMullen set described above corresponds either to the matrix

$$A_M = \begin{pmatrix} 0 & 1 & 0 \\ 1 & 0 & 1 \end{pmatrix},$$

or to the pattern

$$D_M = \{(0,0), (1,1), (2,0)\}.$$

The matrix representation is nice because it is easy to "see" the pattern in the matrix, but the subset notation is often more convenient. In what follows, sums are often indexed by the elements of D, and $\#(D)$, the number of elements in D, will be an important parameter.

A self-affine set K associated to the pattern D is obviously the attractor for the maps

$$g_i(x,y) = (x/n, y/m) + (a_i/n, b_i/m),$$

where $\{(a_i, b_i)\}$ is an enumeration of the points in D. If each of these maps were a similarity (which occurs if and only if $n = m$), then K would be a self-similar set and the Hausdorff and Minkowski dimensions would both be equal to $\log \#(D)/\log n$. In general, however, $\dim(K) < \dim_{\mathscr{M}}(K)$.

Theorem 4.1.1 *Suppose every row contains a chosen rectangle and assume that $n > m$. Then*

$$\dim_{\mathscr{M}}(K(D)) = 1 + \log_n \frac{\#(D)}{m}. \tag{4.1.1}$$

The assumption that every row contains a chosen rectangle is necessary. In general, if π denotes the projection onto the second coordinate, then $\#(\pi(D))$ is the number of occupied rows, and we get

$$\dim_{\mathscr{M}}(K(D)) = \log_m \#(\pi(D)) + \log_n \frac{\#(D)}{\#(\pi(D))}.$$

The general case is left as an exercise (Exercise 4.1).

Proof Let $r = \#(D)$ be the number of rectangles in the pattern. At stage j we have r^j rectangles of width n^{-j} and height m^{-j} (recall that $n^{-j} \ll m^{-j}$ for integers $n > m$). Let $k = \lceil \frac{\log n}{\log m} j \rceil$ (round up to next integer). Then, we can cover each rectangle by m^{k-j} squares of side m^{-k} ($\sim n^{-j}$) and m^{k-1-j} such squares are needed. This is where we use the assumption that every row has a rectangle in it; thus for any generational rectangle R the horizontal projection of $K(D) \cap R$ onto a vertical side of R is the whole side.

Therefore the total number of squares of side m^{-k} $(\sim n^{-j})$ needed to cover $K(D)$ is $r^j m^{k-j}$ and so

$$
\begin{aligned}
\dim_{\mathscr{M}}(K(D)) &= \lim_{j \to \infty} \frac{\log r^j m^{k-j}}{\log n^j} \\
&= \lim_{j \to \infty} \frac{j \log r + (k-j) \log m}{j \log n} \\
&= \frac{\log r}{\log n} + \frac{\log n}{\log n} - \frac{\log m}{\log n} \\
&= 1 + \log_n \frac{r}{m}.
\end{aligned}
$$
□

4.2 The Hausdorff dimension of self-affine sets

Theorem 4.2.1 *Suppose every row contains a chosen rectangle and assume that $n > m$. Then*

$$
\dim(K(D)) = \log_m \left(\sum_{j=1}^{m} r(j)^{\log_n m} \right), \tag{4.2.1}
$$

where $r(j)$ is the number of rectangles of the pattern lying in the jth row.

For example, the McMullen set K_M is formed with $m = 2, n = 3, r(1) = 1$ and $r(2) = 2$. Thus

$$
\dim_{\mathscr{M}}(K_M) = 1 + \log_3 \frac{3}{2} = 1.36907\ldots,
$$
$$
\dim(K_M) = \log_2(1 + 2^{\log_3 2}) = 1.34968\ldots.
$$

The difference arises because the set K_M contains long, narrow rectangles, some of which lie next to each other. Such "groups" of thin rectangles can be efficiently covered simultaneously, rather than each covered individually. Thus, allowing squares of different sizes allows much more efficient coverings. The proof of (4.1.1) was by a simple box counting argument. The proof of (4.2.1) is by Billingsley's Lemma (Lemma 1.4.1), applied to an appropriately constructed measure.

Fix integers $m < n$ and let $\alpha = \frac{\log m}{\log n} < 1$. Following McMullen (1984) we use approximate squares to calculate dimension.

Definition 4.2.2 Suppose $(x, y) \in [0, 1)^2$ have base n and base m expansions $\{x_k\}, \{y_k\}$, respectively. We define the **approximate square** of generation k at

(x,y), $Q_k(x,y)$, to be the closure of the set of points $(x',y') \in [0,1)^2$ such that the first $\lfloor \alpha k \rfloor$ digits in the base n expansions of x and x' coincide, and the first k digits in the base m expansions of y and y' coincide. We refer to the rectangle $Q_k(x,y)$ as an approximate square of generation k since its width $n^{-\lfloor \alpha k \rfloor}$ and height m^{-k} satisfy:

$$m^{-k} \leq n^{-\lfloor \alpha k \rfloor} \leq nm^{-k}$$

and hence (recall $|Q| = \mathrm{diam}(Q)$)

$$m^{-k} \leq |Q_k(\omega)| \leq (n+1)m^{-k}.$$

In the definition of Hausdorff measure, we can restrict attention to covers by such approximate squares since any set of diameter less than m^{-k} can be covered by a bounded number of approximate squares Q_k, see also the remarks following the proof of Billingsley's Lemma in Chapter 1. In what follows we write αk where more precisely we should have written $\lfloor \alpha k \rfloor$.

Proof of Theorem 4.2.1   Any probability vector $\{p(d) : d \in D\} = \mathbf{p}$ defines a probability measure $\mu_\mathbf{p}$ on $K(D)$ that is the image of the product measure $\mathbf{p}^{\mathbb{N}}$ under the representation map

$$R: D^{\mathbb{N}} \to K(D) \tag{4.2.2}$$

given by

$$\{(a_k, b_k)\}_{k=1}^\infty \to \sum_{k=1}^\infty (a_k n^{-k}, b_k m^{-k}).$$

Any such measure is supported on $K(D)$, so the dimensions of these measures all give lower bounds for the dimension of $K(D)$. (See Definition 1.4.3 of the dimension of a measure.) We shall show that the supremum of these dimensions is exactly $\dim(K(D))$. In fact, we will restrict attention to measures coming from probability vectors $\mathbf{p}$ such that

$$p(d) \text{ depends only on the second coordinate of } d, \tag{4.2.3}$$

i.e., all rectangles in the same row get the same mass.

Let (x,y) be in $K(D)$. Suppose $\{x_v\}$, $\{y_v\}$ are the n-ary and m-ary expansions of x and y. We claim that

$$\mu_\mathbf{p}(Q_k(x,y)) = \prod_{v=1}^k p(x_v, y_v) \prod_{v=\alpha k+1}^k r(y_v), \tag{4.2.4}$$

where for $d = (i,j) \in D$ we denote $r(d) = r(j)$, the number of elements in row j. To see this, note that the $n^{-k} \times m^{-k}$ rectangle defined by specifying the first k

digits in the base n expansion of x and the first k digits in the base m expansion of y has $\mu_{\mathbf{p}}$-measure $\prod_{v=1}^{k} p(x_v, y_v)$. The approximate square $Q_k(x,y)$ contains

$$r(y_{\alpha k+1}) \cdot r(y_{\alpha k+2}) \cdots r(y_k)$$

such rectangles, all with the same $\mu_{\mathbf{p}}$-measure by our assumption (4.2.3), and so (4.2.4) follows (for notational simplicity we are omitting the integer part symbol that should be applied to αk). Now take logarithms in (4.2.4),

$$\log\left(\mu_{\mathbf{p}}(Q_k(x,y))\right) = \sum_{v=1}^{k} \log p(x_v, y_v) + \sum_{v=\alpha k+1}^{k} \log r(y_v). \quad (4.2.5)$$

Since $\{(x_v, y_v)\}_{v \geq 1}$ are i.i.d. random variables with respect to $\mu_{\mathbf{p}}$, the Strong Law of Large Numbers yields for $\mu_{\mathbf{p}}$-almost every (x,y):

$$\lim_{k \to \infty} \frac{1}{k} \log\left(\mu_{\mathbf{p}}(Q_k(x,y))\right) \quad (4.2.6)$$
$$= \sum_{d \in D} p(d) \log p(d) + (1-\alpha) \sum_{d \in D} p(d) \log r(d).$$

The proof of Billingsley's Lemma extends to this setting (see also Lemma 1.4.4) and implies

$$\dim(\mu_{\mathbf{p}}) = \frac{1}{\log m} \sum_{d \in D} p(d)\left(\log \frac{1}{p(d)} + \log(r(d)^{\alpha-1})\right). \quad (4.2.7)$$

An easy and well-known calculation says that if $\{a_k\}_{k=1}^{n}$ are real numbers then the maximum of the function

$$F(\mathbf{p}) = \sum_{k=1}^{n} p_k \log \frac{1}{p_k} + \sum_{k=1}^{n} p_k a_k$$

over all probability measure $\mathbf{p}$ is attained at $p_k = e^{a_k}/\sum_l e^{a_l}, k = 1, \ldots, n$. This is known as Boltzmann's Principle. (See Exercise 4.8.) In the case at hand it says that $\dim(\mu_{\mathbf{p}})$ will be maximized if

$$p(d) = \frac{1}{Z} r(d)^{\alpha-1} \quad (4.2.8)$$

where

$$Z = \sum_{d \in D} r(d)^{\alpha-1} = \sum_{j=0}^{m-1} r(j)^{\alpha}.$$

For the rest of the proof we fix this choice of $\mathbf{p}$ and write μ for $\mu_{\mathbf{p}}$. Note that

$$\dim(\mu) = \log_m(Z), \quad (4.2.9)$$

so this is a lower bound for $\dim(K(D))$. To obtain an upper bound denote

$$S_k(x,y) = \sum_{v=1}^{k} \log r(y_v).$$

Note that $\frac{1}{k}S_k(x,y)$ is uniformly bounded. Using (4.2.8), rewrite (4.2.5) as

$$\log \mu(Q_k(x,y)) = \sum_{v=1}^{k} \log \frac{1}{Z} r(y_v)^{\alpha-1} + \left(\sum_{v=1}^{k} \log r(y_v) - \sum_{v=1}^{\alpha k} \log r(y_v) \right)$$

$$= -\sum_{v=1}^{k} \log Z + (\alpha - 1)S_k(x,y) + S_k(x,y) - S_{\alpha k}(x,y).$$

Thus

$$\log \mu(Q_k(x,y)) + k \log Z = \alpha S_k(x,y) - S_{\alpha k}(x,y).$$

Therefore,

$$\frac{1}{\alpha k} \log \mu(Q_k(x,y)) + \frac{1}{\alpha} \log Z = \frac{S_k(x,y)}{k} - \frac{S_{\alpha k}(x,y)}{\alpha k}. \qquad (4.2.10)$$

Summing the right-hand side of (4.2.10) along $k = \alpha^{-1}, \alpha^{-2}, \dots$ gives a telescoping series (strictly speaking, we have to take integer parts of $\alpha^{-1}, \alpha^{-2}, \dots$, but it differs from an honest telescoping series by a convergent series).

Since $S_k(x,y)/k$ remains bounded for all k, it is easy to see

$$\limsup_{k\to\infty} \left(\frac{S_k(x,y)}{k} - \frac{S_{\alpha k}(x,y)}{\alpha k} \right) \geq 0,$$

since otherwise the sum would tend to $-\infty$. Therefore, by (4.2.10), we have for every $(x,y) \in K(D)$

$$\limsup_{k\to\infty} \left(\log \mu(Q_k(x,y)) + k \log Z \right) \geq 0.$$

This implies

$$\liminf_{k\to\infty} \frac{\log \mu(Q_k(x,y))}{-k} \leq \log Z.$$

Since $m^{-k} \leq |Q_k(x,y)| \leq (n+1)m^{-k}$, the last inequality, along with Billingsley's Lemma, implies that

$$\dim(K(D)) \leq \log_m(Z).$$

Combining this with the lower bound given by (4.2.9), we get

$$\dim(K(D)) = \frac{\log Z}{\log m},$$

which is (4.2.1). $\square$

4.3 A dichotomy for Hausdorff measure

Definition 4.3.1   Given a pair (i, j), let $\pi(i, j) = j$ denote projection onto the second coordinate. A digit set $D \subset \{0, 1, \ldots, n-1\} \times \{0, 1, \ldots, m-1\}$ has **uniform horizontal fibers** if all non-empty rows in D have the same cardinality, i.e., for all $j \in \pi(D)$ the preimages $\pi^{-1}(j)$ have identical cardinalities (otherwise, D has **non-uniform horizontal fibers**).

In his paper, McMullen points out that if D has uniform horizontal fibers then $K(D)$ has positive, finite Hausdorff measure in its dimension, and asks what is the Hausdorff measure in other cases. See Exercise 4.3 and part (2) of Theorem 4.5.2. A short and elegant argument found by Lalley and Gatzouras (1992) shows that in these cases the Hausdorff measure of $K(D)$ in any dimension is either zero or infinity. Here we prove

Theorem 4.3.2   *For any gauge function φ, the Hausdorff measure $\mathscr{H}^{\varphi}(K(D))$ is either zero or infinity, provided D has non-uniform horizontal fibers.*

Proof   Assume that φ is a gauge function satisfying

$$0 < \mathscr{H}^{\varphi}(K(D)) < \infty. \tag{4.3.1}$$

Since the intersections of $K(D)$ with the $\#(D)$ rectangles of the first generation are translates of each other and the Hausdorff measure $\mathscr{H}^{\varphi}$ is translation invariant, its restriction to $K(D)$ must assign all these rectangles the same measure. Continuing in the same fashion inside each rectangle, it follows that the restriction of $\mathscr{H}^{\varphi}$ to $K(D)$ is a positive constant multiple of μ_u, where $u(d) = \frac{1}{\#(D)}$ for $d \in D$, i.e., u is the uniform probability vector (μ_u is one of the measures $\mu_{\mathbf{p}}$ considered in the previous proof). The expression (4.2.7) for $\dim(\mu_{\mathbf{p}})$ shows that the function $\mathbf{p} \mapsto \dim(\mu_{\mathbf{p}})$, defined on probability vectors indexed by D, is *strictly* concave and hence attains a unique maximum at the vector $\mathbf{p}$ given in (4.2.8). Since D has non-uniform horizontal fibers, this is *not* the uniform vector u and therefore

$$\dim(\mu_u) < \dim(\mu_{\mathbf{p}}) = \gamma = \dim(K(D)).$$

For $\varphi(t) = t^{\gamma}$ (the case considered in Lalley and Gatzouras (1992)) this yields the desired contradiction immediately; for a general gauge function φ one further remark is needed. Choose β such that $\dim(\mu_u) < \beta < \gamma = \dim(K(D))$. By Egorov's Theorem (e.g., Theorem 2.33 in Folland (1999)) and (4.2.6) there is a set $E \subset K(D)$ of positive μ_u-measure such that the convergence

$$\frac{-1}{k \log m} \log \mu_u(Q_k(x, y)) \to \dim(\mu_u)$$

as $k \to \infty$, is uniform on E. Therefore, for large k, the set E may be covered by $m^{\lfloor \beta k \rfloor}$ approximate squares Q_k.

Since the restriction of $\mathscr{H}^\varphi$ to $K(D)$ is a positive multiple of μ_u, necessarily $\mathscr{H}^\varphi(E) > 0$, so the coverings mentioned above force

$$\liminf_{t \downarrow 0} \frac{\varphi(t)}{t^\beta} > 0.$$

This implies that $\mathscr{H}^\varphi$ assigns infinite measure to any set of dimension strictly greater than β and in particular to $K(D)$. $\qquad\qquad\qquad\qquad\qquad\square$

The proof of Theorem 4.3.2 above gives no hint whether $\mathscr{H}^\gamma(K(D))$ is zero or infinity, but we shall prove in the next section that it is infinity.

The following theorem, due to Rogers and Taylor (1959), refines Billingsley's Lemma (Lemma 1.4.1). Whereas Billingsley's Lemma allows one to compute Hausdorff dimensions, this result allows us to estimate Hausdorff measure and can handle gauges other than power functions. The proof we give is taken from Mörters and Peres (2010).

Theorem 4.3.3 (Rogers–Taylor Theorem)   *Let μ be a Borel measure on $\mathbb{R}^d$ and let φ be a Hausdorff gauge function.*

(i) *If $\Lambda \subset \mathbb{R}^d$ is a Borel set and*

$$\limsup_{r \downarrow 0} \frac{\mu(B(x,r))}{\varphi(r)} < \alpha$$

for all $x \in \Lambda$, then $\mathscr{H}^\varphi(\Lambda) \geq \alpha^{-1} \mu(\Lambda)$.

(ii) *If $\Lambda \subset \mathbb{R}^d$ is a Borel set and*

$$\limsup_{r \downarrow 0} \frac{\mu(B(x,r))}{\varphi(r)} > \theta$$

for all $x \in \Lambda$, then $\mathscr{H}^\varphi(\Lambda) \leq \kappa_d \theta^{-1} \mu(V)$ for any open set $V \subset \mathbb{R}^d$ that contains Λ, where κ_d depends only on d.

If μ is finite on compact sets, then $\mu(\Lambda)$ is the infimum of $\mu(V)$ over all open sets $V \supset \Lambda$, see, for example, Section 2.18 in Rudin (1987) or Theorem 7.8 in Folland (1999). Hence $\mu(V)$ can be replaced by $\mu(\Lambda)$ on the right-hand side of the conclusion in (ii).

Proof   (i) We write

$$\Lambda_\varepsilon = \left\{ x \in \Lambda : \sup_{r \in (0,\varepsilon)} \frac{\mu(B(x,r))}{\varphi(r)} < \alpha \right\}$$

and note that $\mu(\Lambda_\varepsilon) \to \mu(\Lambda)$ as $\varepsilon \downarrow 0$.

Fix $\varepsilon > 0$ and consider a cover $\{A_j\}$ of Λ_ε. Suppose that A_j intersects Λ_ε and $r_j = |A_j| < \varepsilon$ for all j. Choose $x_j \in A_j \cap \Lambda_\varepsilon$ for each j. Then we have $\mu(B(x_j, r_j)) < \alpha \varphi(r_j)$ for every j, whence

$$\sum_{j \geq 1} \varphi(r_j) \geq \alpha^{-1} \sum_{j \geq 1} \mu(B(x_j, r_j)) \geq \alpha^{-1} \mu(\Lambda_\varepsilon).$$

Thus $\mathscr{H}_\varepsilon^\varphi(\Lambda) \geq \mathscr{H}_\varepsilon^\varphi(\Lambda_\varepsilon) \geq \alpha^{-1} \mu(\Lambda_\varepsilon)$. Letting $\varepsilon \downarrow 0$ proves (i).

(ii) Let $\varepsilon > 0$. For each $x \in \Lambda$, choose a positive $r_x < \varepsilon$ such that $B(x, 2r_x) \subset V$ and $\mu(B(x, r_x)) > \theta \varphi(r_x)$; then among the dyadic cubes of diameter at most r_x that intersect $B(x, r_x)$, let Q_x be a cube with $\mu(Q_x)$ maximal. (We consider here dyadic cubes of the form $\prod_{i=1}^d [a_i/2^m, (a_i + 1)/2^m)$ where a_i are integers.) In particular, $Q_x \subset V$ and $|Q_x| > r_x/2$ so the side-length of Q_x is at least $r_x/(2\sqrt{d})$. Let $N_d = 1 + 8\lceil \sqrt{d} \rceil$ and let Q_x^* be the cube with the same center z_x as Q_x, scaled by N_d (i.e., $Q_x^* = z_x + N_d(Q_x - z_x)$). Observe that Q_x^* contains $B(x, r_x)$, so $B(x, r_x)$ is covered by at most N_d^d dyadic cubes that are translates of Q_x. Therefore, for every $x \in \Lambda$, we have

$$\mu(Q_x) \geq N_d^{-d} \mu(B(x, r_x)) > N_d^{-d} \theta \varphi(r_x).$$

Let $\{Q_{x(j)} : j \geq 1\}$ be any enumeration of the maximal dyadic cubes among $\{Q_x : x \in \Lambda\}$. Since these cubes are pairwise disjoint and $Q_x \subset D(x, r_x)$ we get

$$\mu(V) \geq \sum_{j \geq 1} \mu(Q_{x(j)}) \geq N_d^{-d} \theta \sum_{j \geq 1} \varphi(r_{x(j)}).$$

The collection of cubes $\{Q_{x(j)}^* : j \geq 1\}$ forms a cover of Λ. Since each of these cubes is covered by N_d^d cubes of diameter at most $r_{x(j)}$, we infer that

$$\mathscr{H}_\varepsilon^\varphi(\Lambda) \leq N_d^d \sum_{j \geq 1} \varphi(r_{x(j)}) \leq N_d^{2d} \theta^{-1} \mu(V).$$

Letting $\varepsilon \downarrow 0$ proves (ii). $\qquad\qquad\qquad\qquad\qquad\qquad\qquad\qquad\quad \square$

4.4 The Hausdorff measure is infinite *

We saw in the last section that the Hausdorff measure of a self-affine set (in the non-uniform case) must be either 0 or ∞. Now we will prove that it must be the latter.

Theorem 4.4.1 *Assume the digit set D has non-uniform horizontal fibers and let $\gamma = \dim(K(D))$. Then*

$$\mathscr{H}^\gamma(K(D)) = \infty.$$

Furthermore, $K(D)$ is not σ-finite for $\mathcal{H}^\gamma$.

Note the contrast with the self-similar sets that have positive and finite Hausdorff measure in their dimension (see Section 2.1). More precise information can be expressed using gauge functions:

Theorem 4.4.2 *If D has non-uniform horizontal fibers and $\gamma = \dim(K(D))$, then $\mathcal{H}^\varphi(K(D)) = \infty$ for*

$$\varphi(t) = t^\gamma \exp\left(-c\frac{|\log t|}{(\log|\log t|)^2}\right)$$

provided $c > 0$ is sufficiently small. Moreover, $K(D)$ is not σ-finite with respect to $\mathcal{H}^\varphi$.

Let us start with the motivation for the proof of Theorem 4.4.1. If the horizontal and vertical expansion factors n and m were equal, the factors $r(y_v)$ would disappear from the formula (4.2.4) for the $\mu_\mathbf{p}$-measure of an approximate square and to maximize $\dim(\mu_\mathbf{p})$ we would take $\mathbf{p}$ to be uniform. In our case $m < n$ and the factors $r(y_v)$ in (4.2.4) occur only for $v > \alpha k$. Therefore, it is reasonable to perturb McMullen's measure $\mu_\mathbf{p}$ from (4.2.4) by making it more uniform in the initial generations. Thus, we obtain a measure on $K(D)$ that is slightly "smoother" then the one previously constructed, which implies $K(D)$ is "large" in some sense.

Fine-tuning this idea leads to the following. Assume D has non-uniform horizontal fibers, and let $\mathbf{p}$ be the probability vector given by (4.2.8). Let $p^{(1)} = \mathbf{p}$. Fix a small $\delta > 0$. For each $k \geq 2$ define a probability vector $p^{(k)}$ on D by

$$p^{(k)} = \left(\frac{\delta}{\log k}\right)u + \left(1 - \frac{\delta}{\log k}\right)\mathbf{p}, \tag{4.4.1}$$

where u is the uniform probability vector on D. Denote by λ_δ the image under the representation map R (see (4.2.2)) of the product measure $\prod_{k=1}^\infty p^{(k)}$ on $D^\mathbb{N}$.

Theorem 4.4.3 *With the notation above, there exist $\delta > 0$ and $c > 0$ such that the measure λ_δ satisfies*

$$\lim_{k\to\infty} \log \lambda_\delta(Q_k(x,y)) - \log \varphi(m^{-k}) = -\infty \tag{4.4.2}$$

for λ_δ-a.e. (x,y) in $K(D)$, where

$$\varphi(t) = t^\gamma \exp\left(-c\frac{|\log t|}{(\log|\log t|)^2}\right) \tag{4.4.3}$$

and

$$\gamma = \dim(K(D)).$$

Proof   In complete analogy with (4.2.5) above, we obtain

$$\log \lambda_\delta(Q_k(x,y)) = \sum_{v=1}^{k} \log p^{(v)}(x_v,y_v) + \sum_{v=\alpha k+1}^{k} \log r(y_v).$$

On the right-hand side we have sums of independent random variables, that are typically close to their expectations

$$f_\delta(k) \stackrel{\text{def}}{=} \int_{K(D)} \log \lambda_\delta(Q_k(x,y)) \, d\lambda_\delta$$

$$= \sum_{v=1}^{k} \sum_{d\in D} p^{(v)}(d) \log p^{(v)}(d) + \sum_{v=\alpha k+1}^{k} \sum_{d\in D} p^{(v)}(d) \log r(d). \quad (4.4.4)$$

More precisely, by the Law of the Iterated Logarithm (see Section 7.3.3),

$$|\log \lambda_\delta(Q_k(x,y)) - f_\delta(k)| = O\left((k \log\log k)^{1/2}\right) \quad (4.4.5)$$

for λ_δ-almost all (x,y) in $K(D)$. In fact, weaker estimates $(O(k^{1-\varepsilon}))$ would suffice for our purposes.

For any probability vector $\{q(d)\}_{d\in D}$ denote

$$J(q) = \sum_{d\in D} q(d) \log q(d).$$

Observe that for $\varepsilon > 0$:

$$J((1-\varepsilon)p + \varepsilon u) = J(p) - \varepsilon\Delta + O(\varepsilon^2),$$

where

$$\Delta = \sum_{d\in D} (p(d) - u(d)) \log p(d)$$

is strictly positive because

$$\Delta = \frac{1}{2|D|} \sum_{d\in D} \sum_{d'\in D} (p(d) - p(d'))(\log p(d) - \log p(d'))$$

is a sum of non-negative numbers and $p \neq u$.

Returning to (4.4.4), we find that

$$f_\delta(k) = \sum_{v=1}^k J(p^{(v)}) + \sum_{v=\alpha k+1}^k \sum_{d\in D} p^{(v)}(d)\log r(d)$$

$$= kJ(p) - \sum_{v=2}^k \frac{\delta}{\log v}\Delta + O\left(\sum_{v=2}^k \left(\frac{\delta}{\log v}\right)^2\right)$$

$$+ (1-\alpha)k \sum_{d\in D} p(d)\log r(d)$$

$$+ \sum_{v=\alpha k+1}^k \frac{\delta}{\log v} \sum_{d\in D} (u(d) - p(d))\log r(d) + O(1).$$

Recalling that $\log p(d) = (\alpha - 1)\log r(d) - \log Z$ we get

$$f_\delta(k) + k\log Z = -\Delta\delta \sum_{v=2}^k \frac{1}{\log v} + \frac{k}{\log^2 k}O(\delta^2)$$

$$+ \frac{\Delta\delta}{1-\alpha} \sum_{v=\alpha k+1}^k \frac{1}{\log v} + O(1).$$

Therefore,

$$f_\delta(k) + k\log Z = \frac{\Delta\delta}{1-\alpha}\left((\alpha - 1)\int_2^k \frac{dt}{\log t} + \int_{\alpha k}^k \frac{dt}{\log t}\right) \qquad (4.4.6)$$

$$+ \frac{k}{\log^2 k}O(\delta^2) + O(1).$$

A change of variables shows that

$$(\alpha - 1)\int_2^k \frac{dt}{\log t} + \int_{\alpha k}^k \frac{dt}{\log t} = \alpha\int_2^k \frac{dt}{\log t} - \int_{2\alpha^{-1}}^k \frac{\alpha ds}{\log(\alpha s)}$$

$$= \alpha\log\alpha \int_{2\alpha^{-1}}^k \frac{ds}{\log s\log(\alpha s)} + O(1)$$

$$\le \alpha\log\alpha \frac{k}{\log^2(k)} + O(1)$$

since $\log\alpha < 0$. Thus, if we choose $\delta > 0$ and $c_1 > 0$ small enough, (4.4.6) implies

$$f_\delta(k) + k\log Z \le -2c_1\frac{k}{\log^2 k} + O(1).$$

Utilizing (4.4.5) we infer that for λ_δ-a.e. (x,y),

$$\log\lambda_\delta(Q_k(x,y)) \le -k\log Z - c_1\frac{k}{\log^2 k} + O(1). \qquad (4.4.7)$$

Finally, if c in (4.4.3) is small (e.g., $2c \log m \leq c_1$), then (4.4.2) follows from (4.4.7). □

Proof of Theorem 4.4.1 and Theorem 4.4.2 Choose $\delta > 0$ and $c > 0$ such that the measure λ_δ defined in the previous proposition satisfies (4.4.2). The set

$$K_c = \left\{ (x,y) \in K(D) \mid \lim_{k \to \infty} \frac{\varphi(m^{-k})}{\lambda_\delta(Q_k(x,y))} = \infty \right\}$$

has full λ_δ-measure by (4.4.2). By Part (i) of Theorem 4.3.3, we know that $\mathcal{H}^\varphi(K(D)) \geq \mathcal{H}^\varphi(K_c) = \infty$, and this proves Theorem 4.4.2. On the other hand, since $\varphi(t) \leq t^\gamma$ for $0 < t < 1/e$, it follows that $\mathcal{H}^\gamma(K(D)) = \infty$, proving Theorem 4.4.1. Non σ-finiteness follows easily by changing the value of c. □

The observant reader may have noticed a slight discrepancy between the statement of the Rogers–Taylor Theorem (Theorem 4.3.3) and its application in the proof above. The theorem was stated in terms of shrinking balls centered at x, but was applied to a sequence of shrinking "approximate cubes" $\{Q_k\}$ containing x. The proof of part (i) of the Rogers–Taylor theorem gives a lower bound for the φ-sum of a covering of a set; to estimate Hausdorff measure up to a constant factor, it suffices to consider coverings by approximate cubes, and this is the version applied above.

4.5 Notes

The expression (4.2.10) for the μ-measure of an approximate square implies that if D has non-uniform horizontal fibers, then μ is carried by a subset of $K(D)$ of zero γ-dimensional measure. Indeed, by the Law of the Iterated Logarithm (see Theorem 7.2.1), the set

$$K_* = \left\{ (x,y) \in K(D) : \limsup_{k \to \infty} \frac{\alpha S_k(x,y) - S_{\alpha k}(x,y)}{(k \log \log k)^{\frac{1}{2}}} > 0 \right\}$$

has full μ-measure. By (4.2.10), $\limsup_{k \to \infty} \mu(Q_k(x,y)) m^{k\gamma} = \infty$ for $(x,y) \in K_*$, and Proposition 4.3.3 below yields $\mathcal{H}^\gamma(K_*) = 0$. Urbański (1990) proves refinements of this involving gauge functions.

In Chapter 2 we defined packing dimension $\dim_p$, the packing measure $\mathcal{P}_\theta$ and the packing pre-measure $\tilde{\mathcal{P}}_\theta$. We also proved there (Theorem 2.8.4) that if K is a compact set that is the attractor for strictly contracting, bi-Lipschitz mappings $\{f_1, \ldots, f_n\}$, then the packing dimension of K equals its upper Minkowski dimension. We have previously noted that a self-affine set is the attractor for a set of affine contractions so this result applies, i.e.,

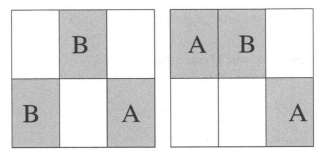

Figure 4.5.1 The "A" and "B" boxes in the finite type definition.

Lemma 4.5.1 *Let $D \subset \{0,\ldots,n-1\} \times \{0,\ldots,m-1\}$ be a pattern. Then*

$$\dim_{\mathrm{p}}(K(D)) = \dim_{\mathscr{M}}(K(D)) = \log_m \#(\pi(D)) + \log_n \frac{\#(D)}{\#(\pi(D))}.$$

The following theorem describes the packing measure at the critical dimension (Peres, 1994a).

Theorem 4.5.2 *Let $\theta = \dim_{\mathrm{p}}(K(D))$.*

(1) *If D has non-uniform horizontal fibers, then $\mathscr{P}_\theta(K(D)) = \infty$ and furthermore, $K(D)$ is not σ-finite for $\mathscr{P}_\theta$.*
(2) *If D has uniform horizontal fibers, then*

$$0 < \mathscr{P}_\theta(K(D)) < \infty.$$

There is a more complicated construction that generalizes the self-affine sets, and which is analogous to the shifts of finite type we considered earlier (see Example 1.3.3). In the earlier construction we had a single collection of squares. Here we have a set of patterns $\{D_1,\ldots,D_N\}$. Each pattern consists of a collection of level 1 squares and a labeling of each square with one of the numbers $\{1,\ldots,N\}$ (repeats allowed). We start by assigning the half-open unit square a label in $\{1,\ldots,N\}$, say J. We then divide the unit square into $n \times m$ half-open rectangles and keep those corresponding to the pattern D_J. Each of the subsquares that we keep has a label assigned to it by the pattern D_J and we subdivide the squares using the pattern corresponding to the label.

In Figure 4.5.1 we have two patterns labeled "A" and "B", respectively. Figure 4.5.2 shows the set generated after four generations by these patterns if we assume the unit square is type A.

The Minkowski dimension of these sets can be computed in terms of the spectral radius of certain matrices and the Hausdorff dimension in terms of

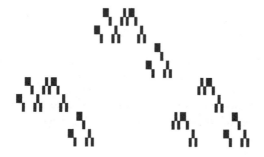

Figure 4.5.2 A Finite type self-affine Cantor set (4 generations).

the behavior of certain random products of matrices. For details the reader is referred to Kenyon and Peres (1996).

Other generalizations of the McMullen–Bedford carpets was analyzed by Lalley and Gatzouras (1992) and Barański (2007). Earlier, quite general self-affine sets were considered by Falconer (1988) who showed that for almost all choices of parameters, the resulting self-affine has equal Hausdorff and Minkowski dimensions. For surveys of self-affine sets see Peres and Solomyak (2000) and Falconer (2013).

Let $K = K(D)$, where $D \subset \{1,\dots,m\} \times \{1,\dots,n\}$ with $\log m / \log n$ irrational. It is proved in Ferguson et al. (2010) that $\dim_H(P_\theta K) = \min(\dim_H K, 1)$ for all $\theta \in (0, \pi/2) \cup (\pi/2, \pi)$; in other words, the only exceptional directions for the dimension part of Marstrand's Projection Theorem are the principal directions (which clearly are exceptional).

Theorems 4.4.1 and 4.4.2 are due to Peres (1994b). That paper also proves that if the exponent 2 on the $\log |\log t|$ is replaced by $\theta < 2$ in Theorem 4.4.2, then $\mathcal{H}^\varphi(K(D)) = 0$.

4.6 Exercises

Exercise 4.1 Show that if $\#(\pi(D)) \le m$ of the rows are occupied then the Minkowski dimension of the self-affine set in Theorem 4.1.1 is

$$\log_m \#(\pi(D)) + \log_n \frac{\#(D)}{\#(\pi(D))}.$$

Exercise 4.2 Prove that the Hausdorff and Minkowski dimensions of $K(D)$ agree if and only if D has uniform horizontal fibers.

Exercise 4.3 Show that if D has uniform horizontal fibers, then $K(D)$ has positive finite Hausdorff measure in its dimension.

Exercise 4.4 Show that Theorem 4.2.1 is still correct even if not every row contains a chosen square (we interpret $0^{\log_n m}$ as 0).

Exercise 4.5 Compute the dimension of a self-affine set of finite type if we assume the number of rectangles in corresponding rows is the same for each pattern.

Exercise 4.6 Construct a finite type self-affine set that is the graph of a continuous real-valued function. Can this be done with a regular (not finite type) set?

Exercise 4.7 Let K_M be the McMullen set. What is $\sup \dim(\tilde{K})$ as $\tilde{K}$ ranges over all microsets of K_M?

Exercise 4.8 Prove Boltzmann's Principle: If $\{a_k\}_{k=1}^n$ are real then the maximum of the function

$$F(\mathbf{p}) = \sum_{k=1}^n p_k \log \frac{1}{p_k} + \sum_{k=1}^n p_k a_k$$

over all probability measures $\mathbf{p}$ is attained at $p_k = e^{a_k} / \sum_l e^{a_l}, k = 1, \ldots, n$.

Exercise 4.9 Suppose K is compact and is $T_{a,b}$ invariant. How big can the difference $\dim_{\mathcal{M}}(K) - \dim(K)$ be?

Exercise 4.10 Fix $0 < \alpha < 1$. Use self-affine sets to construct a set K such that almost every vertical cross section has dimension 0 and almost every horizontal cross section has dimension α.

Exercise 4.11 Fix $0 < \alpha, \beta < 1$. Use self-affine sets to construct a set K such that almost every vertical cross section has dimension β and almost every horizontal cross section has dimension α.

Exercise 4.12 How small can the dimension of K be in the previous exercise? By the Slicing Theorem it must be at least $1 + \max(\alpha, \beta)$. Can it be this small?

Exercise 4.13 Suppose K is a finite type self-affine set and let $n(i, j)$ be the number of rows in pattern i that have j elements. If $n(i, j)$ is independent of i show that

$$\dim(K) = \log_m \sum_j n(i, j) j^{\log_n m}.$$

Exercise 4.14 This and the following exercises compute the dimensions of some random self-affine sets. Divide a square into nm rectangles of size $\frac{1}{n} \times \frac{1}{m}$, assuming $m = n^a$ with $a < 1$. Assume each rectangle is retained with probability $p = n^{-b}$ and repeat the process to get a limiting set K. Show K is non-empty if $mnp > 1$.

• **Exercise 4.15** Show the random affine set in the previous problem has upper Minkowski dimension $\leq 2 - b$.

• **Exercise 4.16** If $np > 1$, show the random affine set in Exercise 4.14 has Hausdorff dimension $\geq 2 - b$.

5

Graphs of continuous functions

In this chapter we consider very special sets: graphs of continuous functions. The primary example we examine is the Weierstrass nowhere differentiable function. We will give a proof of its non-differentiability and compute the Minkowski dimension of its graph.

5.1 Hölder continuous functions

Given a function f on $[a,b] \subset \mathbb{R}$, its graph is the set in the plane

$$G_f = \{(x, f(x)) : a \le x \le b\}.$$

If f is continuous, this is a closed set.

Definition 5.1.1 A function f is **Hölder of order** α on an interval I if it satisfies the estimate

$$|f(x) - f(y)| \le C|x - y|^{\alpha},$$

for some fixed $C < \infty$ and all x, y in I. If $\alpha = 1$, then f is called **Lipschitz**. More generally we say that f has **modulus of continuity** η if η is a positive continuous function on $(0, \infty)$ and

$$|f(x) - f(y)| \le C\eta(|x - y|),$$

for some $C < \infty$.

Lemma 5.1.2 *Suppose f is a real-valued function that is Hölder of order α on an interval I. Then the upper Minkowski dimension (and hence the Hausdorff dimension) of its graph is at most $2 - \alpha$.*

136

Proof Divide I into intervals of length r. Using the Hölder condition we see that the part of the graph above any such interval can be covered by $(C+1)r^{\alpha-1}$ squares of size r. Thus the whole graph can be covered by at most $C'r^{\alpha-2}$ such squares, which proves the claim. □

Lemma 5.1.3 *Suppose $K \subset \mathbb{R}^d$ and $f : \mathbb{R}^d \to \mathbb{R}^n$ is Hölder of order γ. Then*

$$\dim(f(K)) \leq \frac{1}{\gamma}\dim(K).$$

Proof Let $\alpha > \dim(K)$ and $\varepsilon > 0$ and let $\{U_j\}$ be a covering of K such that $\sum_j |U_j|^\alpha < \varepsilon$ (recall $|U| = \text{diam}(U)$). The covering $\{U_j\}$ is mapped by f to the covering $\{W_j\} = \{f(U_j)\}$ that satisfies

$$|W_j| \leq C|U_j|^\gamma.$$

Thus

$$\sum_j |W_j|^{\alpha/\gamma} \leq C\sum_j |U_j|^\alpha < C\varepsilon.$$

Since $\alpha > \dim(K)$ and $\varepsilon > 0$ are arbitrary, this proves the lemma. □

Example 5.1.4 Suppose $\mathbf{C}$ is the usual Cantor middle thirds set, and μ is the standard singular measure on $\mathbf{C}$ (see Example 1.4.5), then

$$f(x) = \int_0^x d\mu$$

is a continuous function that maps $\mathbf{C}$ to $[0,1]$. On the Cantor set itself, we can define the function by $f(x) = \sum_{n=1}^\infty a_n 2^{-n}$ if $x = 2\sum_{n=1}^\infty a_n 3^{-n}$ with $a_n \in \{0,1\}$, and then set it to be constant on each complementary interval. See Figure 5.1.1. Moreover,

$$|f(x) - f(y)| = \mu([x,y]) \leq C|x-y|^{\log_3 2},$$

so f is Hölder of order $\alpha = \log_3 2$. This shows that Lemma 5.1.3 is sharp. This function is called the Cantor singular function.

Definition 5.1.5 A function f satisfies a **lower Hölder estimate** of order α in I if there exists $C > 0$ such that for any subinterval J of I, there exist points $x, y \in J$ such that

$$|f(x) - f(y)| \geq C|J|^\alpha.$$

Lemma 5.1.6 *Suppose f is a continuous real-valued function that satisfies a lower Hölder estimate of order α in I. Then the lower Minkowski dimension of its graph is at least $2 - \alpha$.*

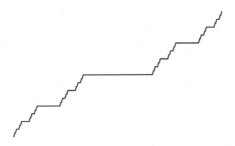

Figure 5.1.1 The Cantor singular function.

Proof   There is a $C > 0$ so that every interval of length r contains two points x, y so that

$$|f(x) - f(y)| \geq Cr^{\alpha}.$$

Therefore, using continuity, at least $C'r^{\alpha-2}$ squares of size r are needed to cover the graph. $\square$

The lower Hölder estimate does not imply by itself that the graph of f has Hausdorff dimension greater than 1 (Exercise 5.52), but this is true if we also assume f is Hölder of the same order (Theorem 5.3.3). However, even in this case the Hausdorff dimension can be strictly smaller than the Minkowski dimension (Exercise 5.42).

Example 5.1.7   The Weierstrass function

$$f_{\alpha,b}(x) = \sum_{n=1}^{\infty} b^{-n\alpha} \cos(b^n x),$$

where b is an integer larger than 1 and $0 < \alpha \leq 1$. See Figure 5.1.2.

The formula of the Weierstrass functions has terms that scale vertically by $b^{-\alpha}$ and horizontally by b, so we might expect it to behave like a self-affine set with $n = b$ and $m = b^{\alpha}$ and one element chosen from each column, so $D = n$. Then Theorem 4.1.1 predicts the Minkowski dimension of the graph is $1 + \log_n(D/m) = 2 - \alpha$, and we shall prove this later (Corollary 5.3.2). We shall see that the Weierstrass function is Hölder of order α, so $2 - \alpha$ is automatically an upper bound for the Minkowski (and hence Hausdorff) dimension.

It had been long conjectured that the Hausdorff dimension of the Weierstrass graph is also $2 - \alpha$. This was proved by Barański et al. (2014) when b is an integer and α is close enough to zero, depending on b. Shen (2015) showed this holds for all integers $b \geq 2$ and all $0 < \alpha < 1$. Shen also shows the dimension of the graph of $\sum_{n=0}^{\infty} b^{-n\alpha}\phi(b^n x)$ is $2 - \alpha$ for any 2π-periodic, non-constant C^2 function, whenever α is close enough to 1, depending on ϕ and b.

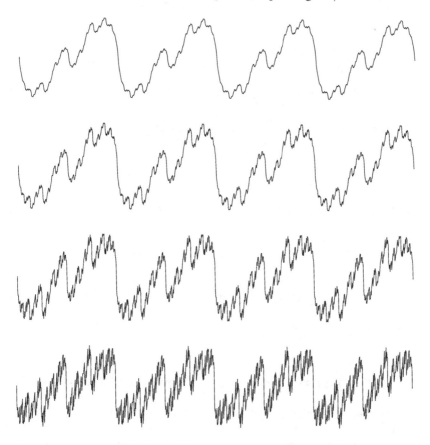

Figure 5.1.2 The graph of the Weierstrass function $f_{\alpha,b}$ for $b = 2$ and $\alpha = 1, \frac{3}{4}, \frac{1}{2}, \frac{1}{4}$ on $[0, 4\pi]$.

Lemma 5.1.8 *If $0 < \alpha < 1$ then the Weierstrass function $f_{\alpha,b}$ is Hölder of order α. If $\alpha = 1$ then $f_{\alpha,b}$ has modulus of continuity $\eta(t) = t(1 + \log_b^+ t^{-1})$.*

Proof Since $f_{\alpha,b}$ is bounded, it clearly satisfies the desired estimates when $|x - y| \geq 1$. So fix x, y with $|x - y| < 1$ and choose a positive integer n so that $b^{-n} \leq |x - y| < b^{-n+1}$. Split the series defining $f_{\alpha,b}$ at the nth term,

$$f_{\alpha,b}(x) = f_1(x) + f_2(x) = \sum_{k=1}^{n} b^{-k\alpha} \cos(b^k x) + \sum_{k=n+1}^{\infty} b^{-k\alpha} \cos(b^k x).$$

If $\alpha < 1$, then the first term f_1 has derivative bounded by

$$|f_1'(t)| \leq \sum_{k=1}^{n} b^{k(1-\alpha)} \leq C b^{n(1-\alpha)}.$$

Therefore

$$|f_1(x) - f_1(y)| \leq C b^{n(1-\alpha)} |x-y| \leq C b^{1-\alpha} |x-y|^{\alpha}.$$

If $\alpha = 1$, then instead of a geometric sum we get

$$|f_1'(t)| \leq \sum_{k=1}^{n} 1 = n \leq C + C \log_b |x-y|^{-1},$$

so

$$|f_1(x) - f_1(y)| \leq C|x-y|(1 + \log_b |x-y|^{-1}).$$

To handle the second term, f_2, we note that for $0 < \alpha \leq 1$,

$$|f_2(x)| \leq \sum_{k=n+1}^{\infty} b^{-k\alpha} \leq \frac{b^{-(n+1)\alpha}}{1 - b^{-\alpha}} \leq C b^{-n\alpha}.$$

Thus,

$$|f_2(x) - f_2(y)| \leq 2C b^{-n\alpha} \leq 2C |x-y|^{\alpha}.$$

Combining the estimates for f_1 and f_2 gives the estimate for $f_{\alpha,b}$. $\qquad\square$

5.2 The Weierstrass function is nowhere differentiable

The nowhere differentiability of $f_{\alpha,b}$ was proven by Weierstrass in the case $\alpha < 1 - \log_b(1 + \frac{3}{2}\pi)$ (Weierstrass, 1872) and by Hardy when $\alpha \leq 1$. We will give an easy proof of Hardy's result using an idea of Freud (1962) as described in Kahane (1964). Also see Izumi et al. (1965).

It will be convenient to think of the Weierstrass function in the context of more general Fourier series. Suppose f can be written in the form

$$f(x) = \sum_{n=-\infty}^{\infty} a_n e^{inx},$$

and assume (as will happen in all the cases we consider) that $\sum_n |a_n| < \infty$. Then the series converges uniformly to f. Since

$$\int_{-\pi}^{\pi} e^{ikx} e^{-inx} \, dx = 2\pi \delta_{n,k} = \begin{cases} 0, & n \neq k \\ 2\pi, & n = k \end{cases},$$

we have

$$a_n = \lim_{m \to \infty} \frac{1}{2\pi} \int_{-\pi}^{\pi} [\sum_{k=-m}^{m} a_k e^{ikx}] e^{-inx} \, dx = \frac{1}{2\pi} \int_{-\pi}^{\pi} f(x) e^{-inx} \, dx.$$

In complex notation, we can write the Weierstrass function as

$$f_{a,b}(x) = -\frac{1}{2} + \sum_{k=-\infty}^{\infty} \frac{1}{2} b^{-|k|\alpha} \exp(i \operatorname{sgn}(k) b^{|k|} x),$$

where $\operatorname{sgn}(k)$ is the sign of the integer k and $\operatorname{sgn}(0) = 0$.

We will need two other properties of Fourier series. First,

$$f(x - x_0) = \sum_{n=-\infty}^{\infty} a_n e^{in(x-x_0)} = \sum_{n=-\infty}^{\infty} (a_n e^{-inx_0}) e^{inx},$$

i.e., translating a function changes only the argument of the Fourier coefficients. Second, convolving two functions is equivalent to multiplying their Fourier coefficients. More precisely, if

$$f(x) = \sum_{n=-\infty}^{\infty} c_n e^{inx}, \quad g(x) = \sum_{n=-\infty}^{\infty} d_n e^{inx},$$

then the convolution

$$f * g(x) = \frac{1}{2\pi} \int_{-\pi}^{\pi} f(y) g(x - y) \, dy = \sum_{n=-\infty}^{\infty} c_n d_n e^{inx}.$$

We leave the verification to the reader (Exercise 5.15).

Definition 5.2.1 Recall that an infinite sequence of positive integers $\{n_k\}_{k \geq 1}$ is **lacunary** if there is a $q > 1$ so that $n_{k+1} \geq q n_k$ for all k. A Fourier series is lacunary (or satisfies the **Hadamard gap condition**) if there exists a lacunary sequence $\{n_k\}$ so that $\{|n| : a_n \neq 0\} \subset \{n_k\}$. Clearly the Weierstrass functions are examples of such series.

In a lacunary Fourier series each term oscillates on a scale for which previous terms are close to constant. This means that they roughly look like sums of independent random variables, and indeed, many results for such sums have analogs for lacunary Fourier series.

Suppose f has Fourier series

$$f(x) = \sum_{k=-\infty}^{\infty} a_k e^{in_k x}.$$

We saw above that for a general Fourier series the coefficients $\{a_k\}$ could be computed as

$$a_n = \frac{1}{2\pi} \int_{-\pi}^{\pi} f(x)e^{-inx}\,dx.$$

For lacunary Fourier series it is possible to replace e^{-inx} by a more general trigonometric polynomial (i.e., a finite sum of terms $a_n e^{inx}$). Suppose that $T_k(x) = \sum_{n=-L_k}^{L_k} c_n e^{inx}$ is a trigonometric polynomial with degree bounded by $L_k < \min(n_{k+1} - n_k, n_k - n_{k-1})$. Let $g(x) = T_k(x)e^{in_k x}$. The nth Fourier coefficient of g is c_{n-n_k}, hence all the coefficients are 0, except in the interval $[n_k - L_k, n_k + L_k]$ and the only non-zero coefficient of f in this interval is at n_k. Therefore

$$\frac{1}{2\pi} \int_{-\pi}^{\pi} f(x)T_k(x)e^{-in_k x}\,dx = \frac{1}{2\pi} \int_{-\pi}^{\pi} f(x)g(x)\,dx$$

$$= \sum_{n=-\infty}^{\infty} a_n c_{n_k - n} \qquad (5.2.1)$$

$$= a_{n_k} c_0.$$

Here we have used a well-known property of Fourier series: if

$$f(x) = \sum_{n=-\infty}^{\infty} a_n e^{inx}, \qquad g(x) = \sum_{n=-\infty}^{\infty} b_n e^{inx},$$

and we define their convolution by

$$h(x) = \frac{1}{2\pi} \int_{0}^{2\pi} f(t)g(x-t)\,dt,$$

then

$$h(x) = \sum_{n=-\infty}^{\infty} a_n b_n e^{inx}. \qquad (5.2.2)$$

This follows from a simple application of Fubini's Theorem (see, e.g., Section 8.3 of Folland (1999)). Equation (5.2.1) follows from (5.2.2) by setting $x = 0$ and using the fact that

$$g(-x) = \sum_{n=-\infty}^{\infty} b_{-n} e^{inx},$$

and in our case $b_n = c_{n-n_k}$.

Since

$$c_0 = \frac{1}{2\pi} \int_{-\pi}^{\pi} T_k(x)\,dx,$$

Equation (5.2.1) implies that if $c_0 \neq 0$, then the n_k coefficient of f is given by

$$a_{n_k} = \frac{\int_{-\pi}^{\pi} f(x)T_k(x)e^{in_k x}\,dx}{\int_{-\pi}^{\pi} T_k(x)\,dx}.$$

One can obtain estimates on the coefficients $\{a_n\}$ by estimating these integrals. The trick is to choose an appropriate T_k. To prove the Weierstrass function is nowhere differentiable it is convenient to take $T_n = F_n^2$, i.e., where F_n is the Fejér kernel

$$F_n(x) = 1 + 2\sum_{k=1}^{n}\left(1 - \frac{k}{n}\right)\cos(kx) = \frac{1}{n}\left(\frac{\sin(nx/2)}{\sin(x/2)}\right)^2.$$

(The proof of this identity is Exercise 5.17.) From the formula it is easy to check that the Fejér kernel satisfies, for absolute constants C_1, C_2,

$$F_n(x) \leq \min\left(n, \frac{C_1}{nx^2}\right), \tag{5.2.3}$$

for all $-\pi \leq x \leq \pi$ and

$$F_n(x) \geq C_2 n, \tag{5.2.4}$$

for $|x| \leq \frac{\pi}{2n}$ (see Exercises 5.18 and 5.19).

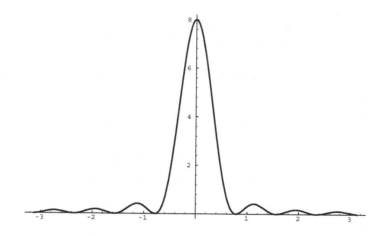

Figure 5.2.1 The graph of the Fejér kernel, $n = 8$.

Lemma 5.2.2 *Let $T_n = F_n^2$. There is a $C_3 < \infty$, so that for every n*

$$\int_{-\pi}^{\pi} |x| T_n(x)\, dx \leq \frac{C_3}{n}\int_{-\pi}^{\pi} T_n(x)\, dx.$$

Moreover, there is a constant $C_4 < \infty$ and for any $\delta > 0$ there is a $C = C(\delta) < \infty$

so that

$$\int_{\delta<|x|<\pi} T_n(x)\,dx \le \frac{C_4}{\delta^3 n^3}\int_{-\pi}^{\pi} T_n(x)\,dx.$$

Proof The estimate (5.2.4) implies

$$\int_{-\pi}^{\pi} T_n(x)\,dx \ge C_2^2 \int_0^{\pi/2n} n^2\,dx \ge \frac{\pi C_2^2}{2}n, \qquad (5.2.5)$$

and (5.2.3) implies

$$\int_{-\pi}^{\pi} |xT_n(x)|\,dx \le \int_{-\pi}^{\pi} |x|\min\left(n^2,\frac{C_1^2}{n^2 x^4}\right)dx$$

$$\le 1+2\int_{1/n}^{\pi}\frac{C_1^2}{n^2 x^3}\,dx$$

$$\le 1+C_1^2.$$

Thus

$$\int_{-\pi}^{\pi} |x|T_n(x)\,dx \le C_3 n^{-1}\int_{-\pi}^{\pi} T_n(x)\,dx,$$

for $C_3 = 2(1+C_1^2)/\pi C_2^2$, which is the first part of the lemma. To prove the second part of the lemma fix $\delta>0$. Then (5.2.3) implies

$$\int_{\delta<|x|<\pi} T_n(x)\,dx \le C_1^2 \int_{\delta<|x|<\pi}(nx^2)^{-2}\,dx$$

$$\le \frac{C_1^2}{n^2}\int_{\delta<|x|<\pi} x^{-4}\,dx$$

$$\le \frac{C_1^2}{3\delta^3 n^2}.$$

Combined with (5.2.5), this gives the desired result with $C_4 = 2C_1^2/3\pi C_2^2$. □

Theorem 5.2.3 *If $\alpha \le 1$ then $f_{\alpha,b}$ is nowhere differentiable.*

Proof Suppose $f_{\alpha,b}$ is differentiable at x_0. Then there is a function of the form $f(x) = f_{\alpha,b}(x-x_0) - c_0 - c_1 e^{ix}$, so that $f(0)=0$ and $f'(0)=0$. Thus for any $\varepsilon>0$ we can choose a $\delta>0$ so that $|f(x)| \le \varepsilon|x|$ for $x\in[-\delta,\delta]$. Moreover, all the Fourier coefficients for $|n|>1$ of f have the same modulus as those for $f_{\alpha,b}$, i.e.,

$$f(x) = \sum_{n=-\infty}^{\infty} a_n e^{inx},$$

where $a_n = |n|^{-\alpha}$ if $|n|=b^k$ and are 0 otherwise (if $|n|>1$).

Now fix k and consider the b^k coefficient. Choose n so $2n < b^k - b^{k-1} < 4n$. Then by our earlier remarks (and recalling $n_k = b^k$)

$$a_{b^k} = \frac{\int_{-\pi}^{\pi} f(x) T_n(x) e^{in_k x} \, dx}{\int_{-\pi}^{\pi} T_n(x) \, dx}. \tag{5.2.6}$$

By Lemma 5.2.2 the numerator is bounded

$$\left| \int_{-\pi}^{\pi} f(x) T_n(x) e^{in_k x} \, dx \right| \leq \int_{-\pi}^{\pi} \varepsilon |x| T_n(x) \, dx + \int_{\delta < |x| < \pi} C_1 T_n(x) \, dx$$

$$\leq C_3 \frac{\varepsilon}{n} \int_{-\pi}^{\pi} T_n(x) \, dx + \frac{C_4}{\delta^3 n^3} \int_{-\pi}^{\pi} T_n(x) \, dx$$

$$\leq \left(\frac{C_3 \varepsilon}{n} + \frac{C_4}{\delta^3 n^3} \right) \int_{-\pi}^{\pi} T_n(x) \, dx,$$

if n is large enough (depending on ε). Taking $n = b^k$, we get

$$a_n = \frac{1}{n} (C_3 \varepsilon + C_4 \delta^{-3} n^{-2}) = o\left(\frac{1}{n}\right),$$

for n large. This contradicts the fact that $|a_n| = b^{-\alpha n} = n^{-\alpha}$ for some $\alpha \leq 1$. Hence $f_{\alpha,b}$ could not have been differentiable at x_0. $\qquad \square$

The proof actually works for general lacunary Fourier series, not just the Weierstrass function. In fact, it even works more generally (see Exercise 5.26). It is also sharp in the sense that if f is a lacunary Fourier series whose coefficients are $a_n = o(1/n)$, then f must have finite derivatives on a dense set (although it may be a set of measure zero). See Exercises 5.24 and 5.25.

We proved that the Weierstrass function does not have a finite derivative at any point. It is possible, however, that the function has infinite derivatives at some points, i.e.,

$$\lim_{h \to 0} \frac{f(x+h) - f(x)}{h} = \pm\infty.$$

It is known that when $\alpha = 1$ this actually occurs for a set of dimension 1 (see the Notes). It is also known that $f_{\alpha,b}$ fails to have even infinite derivatives if α is small enough (depending on b); see Hardy (1916).

5.3 Lower Hölder estimates

We have already seen that if $0 < \alpha < 1$ the Weierstrass function $f_{\alpha,b}$ is Hölder of order α. In this section we will show that it satisfies a lower Hölder estimate of order α, which implies a lower bound on the Minkowski dimension of its graph.

Theorem 5.3.1 *If $0 < \alpha < 1$, then $f_{\alpha,b}$ satisfies a lower Hölder estimate of order α in $[-\pi,\pi]$.*

Proof Suppose $f = f_{\alpha,b}$ does not satisfy such an estimate. Then given $\varepsilon > 0$ we can find an interval J so that f varies by less than $\varepsilon^3 |J|^\alpha$ on J. By subtracting a constant from f we can assume f equals zero at the center of this interval. By translating f we can assume the interval J is centered at 0. Translating f only changes the arguments of its Fourier coefficients, not their absolute values, and this suffices for the proof below. Thus we may assume that J is centered at 0 and that $f(0) = 0$.

Define $r > 0$ by $r = \varepsilon|J|/2$. Our assumption on J then implies that

$$|f(x)| \le \varepsilon^3|J|^\alpha \le \varepsilon^{3-2\alpha}|x|^\alpha \le \varepsilon|x|^\alpha$$

for $|x| \in [2\varepsilon r, \frac{r}{\varepsilon}]$. We have already proven that f is Hölder of order α, so there is a $C < \infty$ so that $|f(x)| \le C|x|^\alpha$ for all x (here and below, C will denote various constants that may depend on α and b, but not on f). Choose k so that $b^{-k} \le r < b^{-k+1}$ and set $n = b^k - b^{k-1}$. Note that $r \asymp 1/n$.

We want to estimate the coefficients of f using (5.2.6), just as we did in the proof of Theorem 5.2.3. The numerator of (5.2.6) is bounded from above by

$$\left|\int_{-\pi}^{\pi} f(x)T_n(x)e^{in_kx}\,dx\right| \le \int_{|x|\le 2\varepsilon r} C|x|^\alpha T_n(x)\,dx + \int_{|x|<r} \varepsilon|x|^\alpha T_n(x)\,dx$$

$$+ \int_{|x|\ge r} \varepsilon|x|^\alpha T_n(x)\,dx + \int_{|x|>r/\varepsilon} C|x|^\alpha T_n(x)\,dx$$

$$= I + II + III + IV.$$

Equations (5.2.3) and (5.2.4) imply that

$$|T_n(x)| \le Cn^2 \text{ for } |x| \le r,$$

and

$$|T_n(x)| \le \frac{C}{n^2x^4} \text{ for } |x| \ge r.$$

We use these to estimate each term above. The first term is bounded by

$$I = C\int_{|x|\le 2\varepsilon r} |x|^\alpha n^2\,dx \le Cn^2(\varepsilon/n)^{1+\alpha} = Cn^{1-\alpha}\varepsilon^{1+\alpha}.$$

The second term is bounded by

$$II \le C\int_{|x|<r} \varepsilon n^2|x|^\alpha\,dx \le C\varepsilon n^{1-\alpha}.$$

The third term is bounded by

$$III \le \int_{r\le|x|} C\varepsilon|x|^{-4+\alpha}n^{-2}\,dx \le Cn^{-2}\varepsilon(1/n)^{-3+\alpha} \le C\varepsilon n^{1-\alpha}.$$

The final term is bounded by

$$IV \le \int_{r/\varepsilon \le |x|} C|x|^{-4+\alpha} n^{-2}\, dx \le Cn^{-2}(1/\varepsilon n)^{-3+\alpha} \le C\varepsilon^{3-\alpha} n^{1-\alpha}.$$

Since all four terms are bounded above by $C\varepsilon n^{1-\alpha}$ and since by (5.2.5) the denominator of (5.2.6) is bounded below by Cn, we get

$$b^{-k\alpha} = |a_{b_k}| \le C \frac{\varepsilon n^{1-\alpha}}{n} = C\varepsilon n^{-\alpha} \le C\varepsilon b^{-k\alpha}.$$

If ε is too small this is a contradiction, and so the theorem is proven. $\square$

By Lemmas 5.1.2 and 5.1.6 the Hölder lower bound gives:

Corollary 5.3.2 *For $0 < \alpha < 1$, the graph of $f_{\alpha,b}$ on $[-\pi, \pi]$ has Minkowski dimension $2 - \alpha$.*

Next we will prove a result of Przytycki and Urbański (1989) that implies that the Hausdorff dimension of the graph of $f_{\alpha,b}$ is strictly larger than 1. Recall from earlier that equality is known to hold in some cases, e.g., Barański et al. (2014).

Theorem 5.3.3 *Assume $0 < \alpha < 1$. Suppose that a function f is Hölder of order α on an interval in $\mathbb{R}$, and also satisfies a lower Hölder estimate of order α in the interval. Then the graph of f has Hausdorff dimension > 1.*

Proof For every interval I in the domain of f we assume

$$C_0|I|^\alpha \le \max_I f - \min_I f \le C_1|I|^\alpha.$$

We will show the graph of f has Hausdorff dimension $\ge 1 + \varepsilon$ where ε depends only on the ratio C_1/C_0 and α. In fact, by multiplying f by a constant we may assume $C_1 = 1$. Moreover, by making C_0 smaller if necessary, we may assume it has the form $C_0 = 4b^{-\alpha}$ for some integer $b > 1$.

The proof simply consists of projecting Lebesgue measure on $\mathbb{R}$ vertically onto the graph of f and showing that the resulting measure μ satisfies the Frostman estimate

$$\mu(B(x,r)) \le Cr^{1+\varepsilon}.$$

The Mass Distribution Principle then finishes the proof. To prove the desired estimate we fix a box intersecting the graph of f and use the lower Hölder estimate to show the graph must leave the box and use the upper estimate to show that it does not return to the box too quickly.

Consider a square Q that hits the graph G_f. Let I be the projection of Q onto the x-axis. Suppose b is as above and divide I into b equal subintervals

$\{I_1,\ldots,I_b\}$. By the lower Hölder estimate there is a point $x \in I$ such that the distance from $(x, f(x))$ to Q is at least $\frac{C_0}{2}|I|^\alpha - |I|$. Suppose $x \in I_k$. Then the upper estimate implies that the graph of f above I_k is disjoint from Q provided

$$\frac{C_0}{2}|I|^\alpha - |I| > \left(\frac{|I|}{b}\right)^\alpha. \tag{5.3.1}$$

For our choice of b and C_0,

$$\left(\frac{|I|}{b}\right)^\alpha = \frac{1}{4}C_0|I|^\alpha,$$

so (5.3.1) holds if

$$|I| < \left(\frac{|I|}{b}\right)^\alpha,$$

which is equivalent to

$$|I| < b^{-\alpha/(1-\alpha)}.$$

We now apply the same argument replacing I by each of the remaining subintervals $\{I_j\}$, $j \neq k$. Each such interval contains a subinterval of length $b^{-2}|I|$, above which G_f is disjoint from Q provided

$$\frac{C_0}{2}\left(\frac{|I|}{b}\right)^\alpha - |I| > \left(\frac{|I|}{b^2}\right)^\alpha.$$

As above, this holds if

$$|I| < \left(\frac{|I|}{b^2}\right)^\alpha,$$

which is equivalent to

$$|I| < b^{-2\alpha/(1-\alpha)}.$$

This procedure can be repeated for n steps as long as

$$\frac{C_0}{2}\left(\frac{|I|}{b^{n-1}}\right)^\alpha - |I| > \left(\frac{|I|}{b^n}\right)^\alpha.$$

In each step proportion $\frac{1}{b}$ of the measure is removed and we can continue as long as

$$|I| < b^{-n\alpha/(1-\alpha)}.$$

Choose n to be the maximal value for which this holds. By the maximality of n this fails for $n+1$, so we have

$$b^{-(n+1)\alpha/(1-\alpha)} \leq |I|$$

which implies

$$b^{-n\alpha/(1-\alpha)} \leq C_2|I|.$$

Choose $\varepsilon > 0$ so that

$$1 - \frac{1}{b} < b^{-\varepsilon\alpha/(1-\alpha)}.$$

Thus $\{x : (x, f(x)) \in Q\}$ has Lebesgue measure at most

$$|I|\left(1 - \frac{1}{b}\right)^n \leq |I|b^{-n\varepsilon\alpha/(1-\alpha)} \leq |I|C_2^\varepsilon|I|^\varepsilon \leq C_2^\varepsilon|I|^{1+\varepsilon},$$

which completes the proof. □

5.4 Notes

According to Weierstrass, Riemann had claimed that the function

$$f(x) = \sum_{n=1}^{\infty} n^{-2}\sin(n^2 x)$$

was nowhere differentiable. However, Gerver (1970) showed this was false (he proved $f'(p\pi/q) = -1/2$ whenever p and q are odd and relatively prime). See Figure 5.4.1. See Duistermaat (1991) for a survey of known results on Riemann's function. See also Exercise 5.27.

Figure 5.4.1 The graph of the Riemann function on $[0, 2\pi]$, $[\pi - .1, \pi + .1]$ and $[\pi - .01, \pi + .01]$.

The Weierstrass function $f_{1,b}$ has infinite derivative on a set of dimension 1. By this we mean that the graph of $f_{1,b}$, considered as curve Γ in the plane has a vertical tangent line on a set of positive $\mathcal{H}^1$ measure that projects to a length zero, but dimension 1, set on the real line. The proof of this is rather involved. First one verifies that the function is in the Zygmund class (see Exercise 5.21 for the definition). This implies that the graph Γ is a quasicircle, which in turn implies that there is a bi-Lipschitz map of the plane that pointwise fixes Γ, but swaps the two complementary components. This implies that almost every (for $\mathcal{H}^1$ measure) point of Γ that is a vertex of a cone in one of these components is also the vertex of a cone in the other. The McMillian twist point theorem implies there is a subset of Γ with positive $\mathscr{H}^1$ measure where this happens. A standard "sawtooth domain" argument then implies that almost every such point is a tangent point of the curve. Since we know that $f_{1,\alpha}$ has no tangent lines with bounded slope, they must all be vertical. The projection on the line has dimension 1 because Zygmund class functions have modulus of continuity $t \log 1/t$, and it has zero length because the set of tangent points on Γ has σ-finite $\mathscr{H}^1$ measure. The fact that Γ has vertical tangents is from the first author's 1987 Ph.D. thesis, but is unpublished except as Exercise VI.9 in Garnett and Marshall (2005); most of the facts needed in the proof are covered in this book.

Besicovitch and Ursell (1937) considered a variant of the Weierstrass function involving super-lacunary Fourier series. As a special case they showed that if

$$f(x) = \sum_{k=1}^{\infty} n_k^{-\alpha} \sin(n_k x),$$

where $n_{k+1}/n_k \to \infty$ and $\log n_{k+1}/\log n_k \to 1$, then f is Hölder of order α and G_f has Hausdorff dimension exactly $2 - \alpha$. The formula is also shown to be correct for the Weierstrass functions with random phases by Hunt (1998).

Kahane (1985) studies the properties of random Fourier series. A celebrated example is Brownian motion, which we will study in detail later. We shall show that it is Hölder of every order $< 1/2$ and that its graph has dimension $3/2$.

Przytycki and Urbański (1989) analyzed graphs of the **Rademacher series**

$$\sum_{n=1}^{\infty} 2^{-n\gamma} \varphi(2^n x),$$

where φ is the step function $\varphi(x) = (-1)^{\lfloor x \rfloor}$. These graphs are self-affine sets and their Minkowski dimension is easy to compute. They showed the Hausdorff and Minkowski dimensions differ when 2^{γ} is a Pisot number (a positive algebraic integer all of whose conjugates are in $(-1, 1)$), e.g., the golden mean $(1 + \sqrt{5})/2$. However, for almost every value of γ the Hausdorff dimension equals the Minkowski dimension. This fact depends on Solomyak's (1995) extension of a result of Erdős (1940) on infinite Bernoulli convolutions.

The Bernoulli convolution ν_λ, $\lambda \in (0, 1)$, is the probability measure on $\mathbb{R}$ that is the infinite convolution of the measures $\frac{1}{2}(\delta_{-\lambda^n} + \delta_{\lambda^n})$. It can also be viewed as the image of the standard product measure under the map $\pm^{\mathbb{N}} \to \sum \pm \lambda^n$. When $\lambda < 1/2$, ν_λ is just a standard singular measure on a Cantor set, and $\nu_{1/2}$ is half of Lebesgue measure on $[-1, 1]$, so the interesting problem is to describe ν_λ for $\lambda \in (\frac{1}{2}, 1)$. Erdős showed that it is absolutely continuous to Lebesgue measure for almost every λ close enough to 1 but singular when λ is the reciprocal of a Pisot number in $(1,2)$. Solomyak proved absolute continuity for almost every $\lambda \in (\frac{1}{2}, 1)$. For more on Bernoulli convolutions see Peres, Schlag and Solomyak (2000). A major advance was obtained by Hochman (2014), who showed that v_λ has dimension 1 for all λ except in a zero-dimensional set. Shmerkin (2014) improved this by showing v_λ is absolutely continuous outside a zero-dimensional set. It is still open whether the reciprocals of the Pisot numbers, found by Erdős, are the only singular examples. A proof of Solomyak's theorem, due to Peres and Solomyak (1996), is presented in Mattila (2015) along with further history and results.

5.5 Exercises

Exercise 5.1 Prove that if f is continuous, then its graph is a closed set. Is the converse true?

Exercise 5.2 Show that the graph of a function must have zero area.

Exercise 5.3 Construct a continuous, real-valued function f whose graph has dimension 2.

Exercise 5.4 Prove that for $E \subset [0,1]$,

$$\dim(E) = \sup\{\alpha : f(E) = [0,1] \text{ for some } f \text{ Hölder of order } \alpha\}.$$

Exercise 5.5 What is the dimension of the graph of the Cantor singular function? (See Example 5.1.4.)

• **Exercise 5.6** Refute a statement in a famous paper (Dvoretzky et al., 1961), page 105, lines 21-23, by showing that for any continuous $f : [0,1] \to \mathbb{R}$, the set of strict local maxima is countable.

Exercise 5.7 Suppose a function f is Hölder of order α and satisfies a lower Hölder estimate of order α, for $\alpha \in (0,1)$. Is it true that $\dim(f(K)) \geq \dim(K)$ for every compact set?

Exercise 5.8 Prove that if $\alpha > 1$ then the Weierstrass function $f_{\alpha,b}$ is differentiable (Example 5.1.7).

• **Exercise 5.9** (Riemann–Lebesgue Lemma) If $f \in L^1[0, 2\pi]$ then

$$a_n = \frac{1}{2\pi} \int_{-\pi}^{\pi} f(x) e^{inx} \, dx \to 0$$

as $|n| \to \infty$.

● **Exercise 5.10**   Suppose $f : [0, 1] \to \mathbb{R}^2$ satisfies

$$\frac{1}{C}|x - y|^\alpha \le |f(x) - f(y)| \le C|x - y|^\alpha$$

for some $\alpha \in [1/2, 1]$ and a positive, finite C. Show that $\dim(f(K)) = \frac{1}{\alpha}\dim(K)$ for every compact set K. Show that such maps exist for every $\alpha \in (1/2, 1]$.

● **Exercise 5.11**   Show that there is no map as in Exercise 5.10 if $\alpha = 1/2$.

Exercise 5.12   Suppose

$$f(x) = \sum_{n=-\infty}^{\infty} a_n e^{inx},$$

and that the series converges uniformly. Prove that if f is Hölder of order α, $0 < \alpha$, then $|a_n| = O(n^{-\alpha})$.

● **Exercise 5.13**   If we take $\alpha = 1$ in the previous exercise then we can improve this to $|a_n| = o(n^{-\alpha})$.

● **Exercise 5.14**   Prove that the Weierstrass function $f_{1,b}$ is not Lipschitz.

Exercise 5.15   Prove that if

$$f(x) = \sum_{n=-\infty}^{\infty} c_n e^{inx}, \quad g(x) = \sum_{n=-\infty}^{\infty} d_n e^{inx}$$

(assume the series converges absolutely), then

$$f * g(x) = \frac{1}{2\pi}\int_{-\pi}^{\pi} f(y)g(x - y)\,dy = \sum_{n=-\infty}^{\infty} c_n d_n e^{inx}.$$

● **Exercise 5.16**   Prove that

$$1 + 2\cos(x) + \cdots + 2\cos(nx) = \frac{\sin(n + \frac{1}{2})x}{\sin(x/2)}.$$

This function is the Dirichlet kernel and is denoted $D_n(x)$.

● **Exercise 5.17**   Define the Fejér kernel

$$F_n(x) = \frac{1}{n}(D_0(x) + \cdots + D_{n-1}(x)).$$

Prove that

$$F_n(x) = \frac{1}{n}\left(\frac{\sin(nx/2)}{\sin(x/2)}\right)^2.$$

Exercise 5.18 Use the previous exercise to deduce that

$$F_n(x) \le C \min\left(n, \frac{1}{nx^2}\right),$$

for all $x \in [-\pi, \pi]$ and

$$F_n(x) \ge Cn,$$

for $x \in [-\pi/2n, \pi/2n]$.

- **Exercise 5.19** Show that

$$\int_{-\pi}^{\pi} T_n(x)\, dx = \int_{-\pi}^{\pi} F_n^2(x)\, dx = 2\pi \left(\frac{1}{3}n + \frac{3}{2} + \frac{1}{6n}\right).$$

- **Exercise 5.20** (Weierstrass Approximation Theorem) If f is a continuous 2π-periodic function and $\varepsilon > 0$ show that there is a trigonometric polynomial p so that $\|f - p\|_\infty < \varepsilon$.

Exercise 5.21 A function f is in the Zygmund class, Λ_*, if there is a $C < \infty$ such that

$$\sup_{x,t} \left| \frac{f(x+t) + f(x-t) - 2f(x)}{2t} \right| \le C,$$

for all x and t and some C. Show that the Weierstrass function $f_{1,b}$ is in the Zygmund class.

Exercise 5.22 Show that if f is in the Zygmund class then it has modulus of continuity $x|\log x|$, i.e., for all x, y with $|x - y| \le 1$,

$$|f(x) - f(y)| \le C|x - y|\log|x - y|^{-1}.$$

Exercise 5.23 Show that if f is a real-valued function in the Zygmund class, then its graph has σ-finite one-dimensional measure. This is due to Mauldin and Williams (1986). Housworth (1994) showed this is false for complex-valued functions.

Exercise 5.24 Show that if $f(x) = \sum_{n=-\infty}^{\infty} a_n e^{inx}$ is lacunary and $a_n = o(1/n)$ then f is in the little Zygmund class, λ_*, i.e.,

$$\lim_{t \to 0} \left| \frac{f(x+t) + f(x-t) - 2f(x)}{2t} \right| = 0.$$

- **Exercise 5.25** Show that if f is in the little Zygmund class then f has a finite derivative on a dense set of points.

Exercise 5.26   Suppose

$$f(x) = \sum_{n=-\infty}^{\infty} a_n e^{inx},$$

where $a_n = 0$ unless $n \in \{n_k\}$ with $\lim_k (n_{k+1} - n_k) = \infty$. Prove that if f is differentiable at even one point, then $|a_n| = o(1/n)$ (the proof of Theorem 5.2.3 works here).

Exercise 5.27   Show that

$$f(x) = \sum_{n=1}^{\infty} n^{-2} \sin(n^3 x)$$

is nowhere differentiable.

Exercise 5.28   Here is an alternative choice of the kernel used in Section 5.2. For a positive integer m, set $T_{k,m}(x) = (D_k(x))^m$, where D_k is the Dirichlet kernel defined in Exercise 5.16. If $0 \le \alpha < m - 1$ show

$$\int_0^{\pi} x^{\alpha} T_{k,m}(x) \, dx = k^{-\alpha + m - 1} \left(\int_0^{\infty} u^{\alpha - m} \sin^m(u) \, du + o(1) \right) \text{ as } k \to \infty.$$

• **Exercise 5.29**   Suppose f is a continuous periodic function that has a derivative at 0. Prove that for any $\varepsilon > 0$ there is a trigonometric polynomial p so that $|f(x) - p(x)| < \varepsilon |x|$ for all x.

Exercise 5.30   Use the two previous exercises to give an alternate proof of Theorem 5.2.3. This is the approach of Izumi et al. (1965).

Exercise 5.31   If f has a Fourier series

$$f(x) = \sum_k a_k e^{in_k x},$$

so that

$$l_k = \min(n_{k+1} - n_k, n_k - n_{k-1}) \to \infty,$$

and there is a point x_0 where f admits a Taylor expansion of the form

$$f(x) = c_0 + c_1(x - x_0) + \cdots + c_p(x - x_0)^p + O(|x - x_0|^{\alpha}),$$

for some $p \le \alpha < p + 1$, then $a_k = O(l_k^{-\alpha})$. If we replace "O" by "o" in the hypothesis we also get "o" in the conclusion (Izumi et al., 1965).

Exercise 5.32   If a Fourier series f is lacunary and satisfies

$$|f(x) - f(x_0)| \le C|x - x_0|^{\alpha}, \text{ with } 0 < \alpha < 1$$

for all x and some x_0, then f is Hölder of order α. (The Weierstrass function $f_{1,b}$ shows this is not true if $\alpha = 1$.)

Exercise 5.33 Suppose that f is a lacunary Fourier series and that f vanishes on some interval. Prove f is the constant zero function.

• **Exercise 5.34** Suppose $\{n_k\}$ is not lacunary, i.e., $\liminf_k n_{k+1}/n_k = 1$ and suppose $\alpha \leq 1$. Then there is a function

$$f(x) = \sum_{n=-\infty}^{\infty} a_n e^{inx},$$

where $a_n = 0$ unless $n \in \{n_k\}$ such that $|f(x)| \leq C|x|^{\alpha}$, but such that f is not Hölder of order α. This is due to Izumi et al. (1965).

Exercise 5.35 We say that a (complex-valued) continuous function f is a Peano curve if its image covers an open set in the plane. Prove that

$$f(x) = \sum_{k=1}^{\infty} n^{-2} e^{in^n x},$$

is a Peano curve.

Exercise 5.36 The previous exercise is an easy special case of a result of Kahane et al. (1963): if $f(x) = \sum_{k=1}^{\infty} a_k e^{in_k x}$ is lacunary (meaning there is a $q > 1$ so that $n_{k+1} \geq q n_k$ for all k) and there is an A, depending only on q, so that

$$|a_n| \leq A \sum_{k=n+1}^{\infty} |a_k|,$$

for all n, then f is a Peano curve (they actually prove the image of a certain Cantor set covers an open set). This type of result was first proved by Salem and Zygmund (1945) when q is sufficiently large. Use the result of Kahane, Weiss and Weiss to show the following functions define Peano curves:

$$f(x) = \sum_{k=1}^{\infty} k^{-p} e^{in_k x}, \text{ any } q > 1, p > 1,$$

and

$$g(x) = \sum_{k=1}^{\infty} b^{-k\alpha} e^{ib^k x}, \text{ any } b,$$

if α is sufficiently small (depending on b). Prove the second function is not Peano if $\alpha > 1/2$.

Exercise 5.37 Show that for every complex number z with $|z| \leq 1$ there is a real x such that

$$z = \lim_{N \to \infty} \frac{1}{N} \sum_{n=1}^{N} e^{in^n x}.$$

Exercise 5.38 Using lacunary sequences, construct a real-valued function $f : [0,1] \to [0,1]$ so that for every $a \in (0,1)$,

$$\dim(\{x : f(x) = a\}) = 1.$$

Exercise 5.39 Interesting graphs can be constructed using self-affine sets. Fix integers $n > m$ and consider the patterns given by the following four matrices. Start with the pattern given by the matrix A_1. If there is a 0 we make no replacement. Otherwise we replace the rectangle with label j with the pattern A_j:

$$A_1 = \begin{pmatrix} 0 & 1 & 0 \\ 2 & 0 & 3 \end{pmatrix}, \qquad A_2 = \begin{pmatrix} 0 & 1 & 2 \\ 2 & 0 & 0 \end{pmatrix}, \qquad A_3 = \begin{pmatrix} 3 & 0 & 0 \\ 0 & 3 & 1 \end{pmatrix}.$$

The resulting set is drawn in Figure 5.5.1. Verify this is the graph of a continuous function.

Figure 5.5.1 A self-affine graph, Exercise 5.39.

Exercise 5.40 Another self-affine graph is based on the following patterns:

$$A_1 = \begin{pmatrix} 0 & 2 & 3 & 0 \\ 2 & 0 & 0 & 3 \end{pmatrix}, \qquad A_2 = \begin{pmatrix} 0 & 0 & 1 & 2 \\ 1 & 2 & 0 & 0 \end{pmatrix},$$

$$A_3 = \begin{pmatrix} 3 & 1 & 0 & 0 \\ 0 & 0 & 3 & 1 \end{pmatrix}.$$

The resulting set is drawn in Figure 5.5.2. Show that every horizontal cross section of this graph has dimension $\log_4 2 = 1/2$. Thus by Marstrand's Slicing Theorem (Theorem 1.6.1), the dimension of the graph is at least $3/2$. Prove the graph has exactly this dimension. Thus this graph gives an example where the Marstrand Slicing Theorem is sharp.

Figure 5.5.2 Another self-affine graph, Exercise 5.40.

Exercise 5.41 Prove that if $m \neq n$ then a self-affine graph must be nowhere differentiable.

Exercise 5.42 Next we consider the $m \times m^2$ matrix ($m = 3$ is illustrated),

$$A_1 = \begin{pmatrix} 0 & 0 & 0 & 0 & 0 & 0 & 0 & 0 & 1 \\ 0 & 0 & 0 & 0 & 0 & 0 & 0 & 1 & 0 \\ 1 & 2 & 1 & 2 & 1 & 2 & 1 & 0 & 0 \end{pmatrix},$$

$$A_2 = \begin{pmatrix} 2 & 0 & 0 & 0 & 0 & 0 & 0 & 0 & 0 \\ 0 & 2 & 0 & 0 & 0 & 0 & 0 & 0 & 0 \\ 0 & 0 & 2 & 1 & 2 & 1 & 2 & 1 & 2 \end{pmatrix}.$$

(The second is just the reflection of the first.) Show this defines a $\frac{1}{2}$-Hölder function whose graph has Hausdorff dimension $\log_m(m - 1 + \sqrt{m^2 - m + 1})$, and this tends to 1 as $m \to \infty$.

Exercise 5.43 Using self-affine graphs construct, for any $0 < k < n$, a continuous function $f : [0, 1] \to [0, 1]$ such that

$$\dim(\{x : f(x) = c\}) = \log_n k,$$

for all $0 \leq c \leq 1$.

Exercise 5.44 For any $0 < \alpha < 1$ construct $f : [0, 1] \to [0, 1]$ that is continuous and satisfies

$$\dim(\{x : f(x) = c\}) = \alpha,$$

for all $0 \leq c \leq 1$.

Exercise 5.45   For any continuous function $g : [0, 1] \to [0, 1]$, construct a function $f : [0, 1] \to [0, 1]$ such that

$$\dim(\{x : f(x) = c\}) = g(c),$$

for all $0 \leq c \leq 1$.

Exercise 5.46   If $\dim(G_f) > 1$, must G_f have a horizontal slice of dimension > 0?

Exercise 5.47   Consider the discontinuous function

$$f(x) = \limsup_{N \to \infty} \frac{1}{N} \sum_{n=1}^{N} x_n,$$

where $\{x_n\}$ is the binary expansion of x. What is the dimension of G_f?

• **Exercise 5.48**   Build a homeomorphism $h : [0, 1] \to [0, 1]$ and a set $E \subset [0, 1]$ of full Lebesgue measure so that $h(E)$ has measure zero. Such a mapping is called singular.

Exercise 5.49   For any $0 < \delta < 1$ construct a homeomorphism h from $[0, 1]$ to itself and $E \subset [0, 1]$ with $\dim(E) < \delta$ so that $\dim([0, 1] \setminus h(E)) < \delta$. See Bishop and Steger (1993), Rohde (1991), Tukia (1989) for examples that arise "naturally".

Exercise 5.50   Even more singular maps are possible. Construct a homeomorphism $h : [0, 1] \to [0, 1]$ and set $E \subset [0, 1]$ so that both E and $[0, 1] \setminus h(E)$ have dimension zero. Examples with this property arise in the theory of conformal welding, e.g., Bishop (2007).

Exercise 5.51   Construct a function $f : [0, 1] \to \mathbb{R}$ with lower Hölder estimate of order $\frac{1}{2}$, which is Hölder of every order less than $\frac{1}{2}$, but such that $\dim(G_f) = 1$. (Compare to Exercise 5.52.)

Exercise 5.52   Show that for each $0 < \alpha < 1$ there is a function $f : [0, 1] \to \mathbb{R}$ that satisfies a lower Hölder estimate of order α but such that its graph has Hausdorff dimension 1. Let $\varphi_n(x)$ be the 1-periodic piecewise linear function defined by

$$\varphi_n(x) = \max(0, 1 - 2^{-n^2} \operatorname{dist}(x, \mathbb{Z})).$$

This has a sharp spike of height 1 and width 2^{-n^2+1} at the integers and is zero elsewhere. For $0 < \alpha < 1$, let

$$f(x) = \sum_{n=1}^{\infty} 2^{-\alpha n} \varphi_n(2^n x).$$

Show that it satisfies a lower Hölder estimate of order α but the graph has dimension 1.

6

Brownian motion, Part I

This is the first of two chapters dealing with the most interesting and important continuous time random process: Brownian motion. We start with the basic definitions and construction of Brownian motion (following Paul Lévy) and then describe some of its geometrical properties: nowhere differentiability, the dimension of the graph of 1-dimensional paths, the dimension and measure of higher-dimensional paths, the size of the zero sets and the non-existence of points of increase.

6.1 Gaussian random variables

Brownian motion is at the meeting point of the most important categories of stochastic processes: it is a martingale, a strong Markov process, a process with independent and stationary increments and a Gaussian process. We will construct Brownian motion as a specific Gaussian process. We start with the definitions of Gaussian random variables. A **random variable** is a measurable real-valued function defined on a probability space $(\Omega, \mathscr{F}, \mathbb{P})$. A sequence of random variables is also called a **stochastic process**.

Definition 6.1.1 A real-valued random variable X on a probability space $(\Omega, \mathscr{F}, \mathbb{P})$ has a **standard Gaussian** (or **standard normal**) distribution if

$$\mathbb{P}(X > x) = \frac{1}{\sqrt{2\pi}} \int_x^{+\infty} e^{-u^2/2} \, du.$$

A vector-valued random variable X has an n-**dimensional standard Gaussian** distribution if its n coordinates are standard Gaussian and independent. A vector-valued random variable $Y : \Omega \to \mathbb{R}^p$ is **Gaussian** if there exists a vector-valued random variable X having an n-dimensional standard Gaussian distribution, a $p \times n$ matrix A and a p-dimensional vector b such that

$$Y = AX + b. \qquad (6.1.1)$$

We are now ready to define the Gaussian processes.

Definition 6.1.2 A stochastic process $\{X_t\}_{t \in I}$ is said to be a **Gaussian process** if for all k and $t_1, \ldots, t_k \in I$ the vector $(X_{t_1}, \ldots, X_{t_k})^t$ is Gaussian.

Recall that the **covariance matrix** of a random vector is defined as

$$\text{Cov}(Y) = \mathbb{E}\left[(Y - \mathbb{E}Y)(Y - \mathbb{E}Y)^t\right].$$

Then, by the linearity of expectation, the Gaussian vector Y in (6.1.1) has

$$\text{Cov}(Y) = AA^t.$$

Recall that an $n \times n$ matrix A is said to be **orthogonal** if $AA^t = I_n$. The following results show that the distribution of a Gaussian vector is determined by its mean and covariance.

Lemma 6.1.3 *If Θ is an orthogonal $n \times n$ matrix and X is an n-dimensional standard Gaussian vector, then ΘX is also an n-dimensional standard Gaussian vector.*

Proof As the coordinates of X are independent standard Gaussian, X has density given by

$$f(x) = (2\pi)^{-\frac{n}{2}} e^{-\|x\|^2/2},$$

where $\| \cdot \|$ denotes the Euclidean norm. Since Θ preserves this norm, the density of X is invariant under Θ. $\qquad \square$

Lemma 6.1.4 *If Y and Z are Gaussian vectors in $\mathbb{R}^n$ such that $\mathbb{E}Y = \mathbb{E}Z$ and $\text{Cov}(Y) = \text{Cov}(Z)$, then Y and Z have the same distribution.*

Proof It is sufficient to consider the case when $\mathbb{E}Y = \mathbb{E}Z = 0$. Then, using Definition 6.1.1, there exist standard Gaussian vectors X_1, X_2 and matrices A, C so that

$$Y = AX_1 \quad \text{and} \quad Z = CX_2.$$

By adding some columns of zeroes to A or C if necessary, we can assume that X_1, X_2 are both k-vectors for some k and A, C are both $n \times k$ matrices.

Let $\mathscr{A}, \mathscr{C}$ denote the vector spaces generated by the row vectors of A and C, respectively. To simplify notations, assume without loss of generality that the first ℓ row vectors of A form a basis for the space $\mathscr{A}$. For any matrix M let M_i denote the ith row vector of M, and define the linear map Θ from $\mathscr{A}$ to $\mathscr{C}$ by

$$A_i \Theta = C_i \quad \text{for } i = 1, \ldots, \ell.$$

We want to verify that Θ is an isomorphism. Assume there is a vector

$$v_1 A_1 + \cdots + v_\ell A_\ell$$

whose image is 0. Then the k-vector $v = (v_1, v_2, \ldots, v_\ell, 0, \ldots, 0)^t$ satisfies $v^t C = 0$, and so $\|v^t A\|^2 = v^t A A^t v = v^t C C^t v = 0$, giving $v^t A = 0$. This shows that Θ is one-to-one and, in particular, $\dim \mathscr{A} \le \dim \mathscr{C}$. By symmetry, $\mathscr{A}$ and $\mathscr{C}$ must have the same dimension, so Θ is an isomorphism.

As the coefficient (i, j) of the matrix AA^t is the scalar product of A_i and A_j, the identity $AA^t = CC^t$ implies that Θ is an orthogonal transformation from $\mathscr{A}$ to $\mathscr{C}$. We can extend it to map the orthocomplement of $\mathscr{A}$ to the orthocomplement of $\mathscr{C}$ orthogonally, getting an orthogonal map $\Theta : \mathbb{R}^k \to \mathbb{R}^k$. Then

$$Y = AX_1, \quad Z = CX_2 = A\Theta X_2,$$

and Lemma 6.1.4 follows from Lemma 6.1.3. $\qquad\square$

Thus, the first two moments of a Gaussian vector are sufficient to characterize its distribution, hence the introduction of the notation $\mathscr{N}(\mu, \Sigma)$ to designate the normal distribution with expectation μ and covariance matrix Σ. A useful corollary of this lemma is:

Corollary 6.1.5 *Let Z_1, Z_2 be independent $\mathscr{N}(0, \sigma^2)$ random variables. Then $Z_1 + Z_2$ and $Z_1 - Z_2$ are two independent random variables having the same distribution $\mathscr{N}(0, 2\sigma^2)$.*

Proof $\sigma^{-1}(Z_1, Z_2)$ is a standard Gaussian vector, and so, if

$$\Theta = \frac{1}{\sqrt{2}} \begin{bmatrix} 1 & 1 \\ 1 & -1 \end{bmatrix},$$

then Θ is an orthogonal matrix such that

$$(\sqrt{2}\sigma)^{-1}(Z_1 + Z_2, Z_1 - Z_2)^t = \Theta \sigma^{-1}(Z_1, Z_2)^t,$$

and our claim follows from Lemma 6.1.3. $\qquad\square$

As a conclusion of this section, we state the following tail estimate for the standard Gaussian distribution.

Lemma 6.1.6 *Let Z be distributed as $\mathscr{N}(0, 1)$. Then for all $x \ge 0$,*

$$\frac{x}{x^2 + 1} \frac{1}{\sqrt{2\pi}} e^{-x^2/2} \le \mathbb{P}(Z > x) \le \frac{1}{x} \frac{1}{\sqrt{2\pi}} e^{-x^2/2}.$$

Proof The right inequality is obtained by the estimate

$$\mathbb{P}(Z > x) \le \int_x^{+\infty} \frac{u}{x} \frac{1}{\sqrt{2\pi}} e^{-u^2/2} \, du$$

since, in the integral, $u \ge x$. The left inequality is proved as follows: let us define

$$f(x) := xe^{-x^2/2} - (x^2 + 1) \int_x^{+\infty} e^{-u^2/2} \, du.$$

We remark that $f(0) < 0$ and $\lim_{x \to +\infty} f(x) = 0$. Moreover,

$$f'(x) = (1 - x^2 + x^2 + 1)e^{-x^2/2} - 2x \int_x^{+\infty} e^{-u^2/2} \, du$$

$$= -2x \left(\int_x^{+\infty} e^{-u^2/2} \, du - \frac{1}{x} e^{-x^2/2} \right),$$

so the right inequality implies $f'(x) \ge 0$ for all $x \ge 0$. This implies $f(x) \le 0$, proving the lemma. $\qquad \square$

6.2 Lévy's construction of Brownian motion

Brownian motion is a precise way to define the idea of a random continuous function $[0, \infty) \to \mathbb{R}^d$. Strictly speaking, it is a function of two variables, $B(\omega, t)$ where ω lies in some probability space and $t \in [0, \infty)$ represents time. However, we shall write it as B_t or $B(t)$, suppressing variable ω. We will occasionally use W (for Wiener) instead of B. For example, $W(a, b)$ will mean the image of the interval (a, b) under a Brownian motion; $B(a, b)$ might be confused with a ball centered at a with radius b, and the correct notation $B((a, b))$ is awkward.

Standard Brownian motion on an interval $I = [0, a]$ or $I = [0, \infty)$ is defined as follows.

Definition 6.2.1 A real-valued stochastic process $\{B_t\}_{t \in I}$ is a **standard Brownian motion** if it is a Gaussian process such that:

(i) $B_0 = 0$;
(ii) for all k natural and for all $t_1 < \cdots < t_k$ in I: $B_{t_k} - B_{t_{k-1}}, \ldots, B_{t_2} - B_{t_1}$ are independent;
(iii) for all $t, s \in I$ with $t < s$, $B_s - B_t$ has $\mathcal{N}(0, s - t)$ distribution;
(iv) almost surely, $t \mapsto B_t$ is continuous on I.

As a corollary of this definition, one can already remark that for all $t, s \in I$:

$$\mathrm{Cov}(B_t, B_s) = s \wedge t$$

(where $s \wedge t = \min(s,t)$). Indeed, assume that $t \geq s$. Then

$$\mathrm{Cov}(B_t, B_s) = \mathrm{Cov}(B_t - B_s, B_s) + \mathrm{Cov}(B_s, B_s)$$

by bilinearity of the covariance. The first term vanishes by the independence of increments, and the second term equals s by properties (iii) and (i). Thus by Lemmas 6.1.3 and 6.1.4 we may replace properties (ii) and (iii) in the definition by:

- for all $t, s \in I$, $\mathrm{Cov}(B_t, B_s) = t \wedge s$;
- for all $t \in I$, B_t has $\mathcal{N}(0,t)$ distribution;

or by:

- for all $t, s \in I$ with $t < s$, $B_t - B_s$ and B_s are independent;
- for all $t \in I$, B_t has $\mathcal{N}(0,t)$ distribution.

Kolmogorov's Extension Theorem (see Durrett (1996)) implies the existence of any countable time set stochastic process $\{X_t\}$ if we know its finite-dimensional distributions and they are consistent. Thus, standard Brownian motion could be easily constructed on any countable time set. However knowing finite-dimensional distributions is not sufficient to get continuous paths in an interval, as the following example shows.

Example 6.2.2 Suppose that standard Brownian motion $\{B_t\}$ on $[0,1]$ has been constructed, and consider an independent random variable U uniformly distributed on $[0,1]$. Define

$$\tilde{B}_t = \begin{cases} B_t & \text{if } t \neq U \\ 0 & \text{otherwise.} \end{cases}$$

The finite-dimensional distributions of $\{\tilde{B}_t\}$ are the same as the ones of $\{B_t\}$. However, the process $\{\tilde{B}_t\}$ has almost surely discontinuous paths.

In measure theory, one often identifies functions with their equivalence class for almost-everywhere equality. As the above example shows, it is important not to make this identification in the study of continuous-time stochastic processes. Here we want to define a probability measure on the set of continuous functions.

The following construction, due to Paul Lévy, consists of choosing the "right" values for the Brownian motion at each dyadic point of $[0,1]$ and then interpolating linearly between these values. This construction is inductive, and at each step a process is constructed that has continuous paths. Brownian motion is then the uniform limit of these processes; hence its continuity. We will use the following basic lemma. The proof can be found, for instance, in Durrett (1996).

Lemma 6.2.3 (Borel–Cantelli) *Let $\{A_i\}_{i=0,\ldots,\infty}$ be a sequence of events, and let*

$$A_i \ i.o. = \limsup_{i \to \infty} A_i = \bigcap_{i=0}^{\infty} \bigcup_{j=i}^{\infty} A_j,$$

where "i.o." abbreviates "infinitely often".

(i) *If $\sum_{i=0}^{\infty} \mathbb{P}(A_i) < \infty$, then $\mathbb{P}(A_i \ i.o.) = 0$.*
(ii) *If $\{A_i\}$ are pairwise independent, and $\sum_{i=0}^{\infty} \mathbb{P}(A_i) = \infty$, then $\mathbb{P}(A_i \ i.o.) = 1$.*

We can now prove Brownian motion exists, a result of Wiener (1923), using the proof of Lévy (1948).

Theorem 6.2.4 *Standard Brownian motion on $[0, \infty)$ exists.*

Proof We first construct standard Brownian motion on $[0, 1]$. For $n \geq 0$, let $D_n = \{k/2^n : 0 \leq k \leq 2^n\}$, and let $D = \bigcup D_n$. Let $\{Z_d\}_{d \in D}$ be a collection of independent $\mathcal{N}(0, 1)$ random variables. We will first construct the values of B on D. Set $B_0 = 0$, and $B_1 = Z_1$. In an inductive construction, for each n we will construct B_d for all $d \in D_n$ so that:

(i) for all $r < s < t$ in D_n, the increment $B_t - B_s$ has $\mathcal{N}(0, t - s)$ distribution and is independent of $B_s - B_r$;
(ii) B_d for $d \in D_n$ are globally independent of the Z_d for $d \in D \setminus D_n$.

These assertions hold for $n = 0$. Suppose that they hold for $n - 1$. Define, for all $d \in D_n \setminus D_{n-1}$, a random variable B_d by

$$B_d = \frac{B_{d^-} + B_{d^+}}{2} + \frac{Z_d}{2^{(n+1)/2}}, \tag{6.2.1}$$

where $d^+ = d + 2^{-n}$, and $d^- = d - 2^{-n}$, and both are in D_{n-1}. Because $\frac{1}{2}(B_{d^+} - B_{d^-})$ is $\mathcal{N}(0, 1/2^{n+1})$ by induction, and $Z_d/2^{(n+1)/2}$ is an independent $\mathcal{N}(0, 1/2^{n+1})$, their sum and their difference, $B_d - B_{d^-}$ and $B_{d^+} - B_d$ are both $\mathcal{N}(0, 1/2^n)$ and independent by Corollary 6.1.5. Assertion (i) follows from this and the inductive hypothesis, and (ii) is clear.

Having thus chosen the values of the process on D, we now "interpolate" between them. Formally, let $F_0(x) = xZ_1$, and for $n \geq 1$, let us introduce the function

$$F_n(x) = \begin{cases} 2^{-(n+1)/2}Z_x & \text{for } x \in D_n \setminus D_{n-1}, \\ 0 & \text{for } x \in D_{n-1}, \\ \text{linear} & \text{between consecutive points in } D_n. \end{cases} \tag{6.2.2}$$

These functions are continuous on $[0,1]$, and for all n and $d \in D_n$

$$B_d = \sum_{i=0}^{n} F_i(d) = \sum_{i=0}^{\infty} F_i(d). \tag{6.2.3}$$

We prove this by induction. Suppose that it holds for $n-1$. Let $d \in D_n \setminus D_{n-1}$. Since for $0 \le i \le n-1$ the function F_i is linear on $[d^-, d^+]$, we get

$$\sum_{i=0}^{n-1} F_i(d) = \sum_{i=0}^{n-1} \frac{F_i(d^-) + F_i(d^+)}{2} = \frac{B_{d^-} + B_{d^+}}{2}. \tag{6.2.4}$$

Since $F_n(d) = 2^{-(n+1)/2}Z_d$, comparing (6.2.1) and (6.2.4) gives (6.2.3).

On the other hand, we have by definition of Z_d and by Lemma 6.1.6

$$\mathbb{P}\left(|Z_d| \ge c\sqrt{n}\right) \le \exp\left(-\frac{c^2 n}{2}\right)$$

for n large enough, so the series $\sum_{n=0}^{\infty} \sum_{d \in D_n} \mathbb{P}(|Z_d| \ge c\sqrt{n})$ converges as soon as $c > \sqrt{2\log 2}$. Fix such a c. By the Borel–Cantelli Lemma 6.2.3 we conclude that there exists a random but finite N so that for all $n > N$ and $d \in D_n$ we have $|Z_d| < c\sqrt{n}$, and so

$$|F_n|_\infty < c\sqrt{n}2^{-n/2}. \tag{6.2.5}$$

This upper bound implies that the series $\sum_{n=0}^{\infty} F_n(t)$ is uniformly convergent on $[0,1]$, and so it has a continuous limit, which we call $\{B_t\}$. All we have to check is that the increments of this process have the right finite-dimensional joint distributions. This is a direct consequence of the density of the set D in $[0,1]$ and the continuity of paths. Indeed, let $t_1 > t_2 > t_3$ be in $[0,1]$, then they are limits of sequences $t_{1,n}, t_{2,n}$ and $t_{3,n}$ in D, respectively. Now

$$B_{t_3} - B_{t_2} = \lim_{k \to \infty} (B_{t_{3,k}} - B_{t_{2,k}})$$

is a limit of Gaussian random variables, so itself is Gaussian with mean 0 and variance $\lim_{n \to \infty} (t_{3,n} - t_{2,n}) = t_3 - t_2$. The same holds for $B_{t_2} - B_{t_1}$; moreover, these two random variables are limits of independent random variables, since for n large enough, $t_{1,n} > t_{2,n} > t_{3,n}$. Applying this argument for any number of increments, we get that $\{B_t\}$ has independent increments such that for all $s < t$ in $[0,1]$, $B_t - B_s$ has $\mathcal{N}(0, t-s)$ distribution.

We have thus constructed Brownian motion on $[0,1]$. To conclude, if $\{B_t^n\}_n$ for $n \geq 0$ are independent Brownian motions on $[0,1]$, then

$$B_t = B_{t-\lfloor t \rfloor}^{\lfloor t \rfloor} + \sum_{0 \leq i < \lfloor t \rfloor} B_1^i$$

meets our definition of Brownian motion on $[0,\infty)$. $\square$

6.3 Basic properties of Brownian motion

We say that two random variables Y, Z have the same distribution, and write $Y \overset{d}{=} Z$, if $\mathbb{P}(Y \in A) = \mathbb{P}(Z \in A)$ for all Borel sets A. Let $\{B(t)\}_{t \geq 0}$ be a standard Brownian motion, and let $a \neq 0$. The following **scaling relation** is a simple consequence of the definitions.

$$\{\tfrac{1}{a}B(a^2 t)\}_{t \geq 0} \overset{d}{=} \{B(t)\}_{t \geq 0}.$$

Also, define the **time inversion** of $\{B_t\}$ as

$$W(t) = \begin{cases} 0 & t = 0; \\ tB(\tfrac{1}{t}) & t > 0. \end{cases}$$

We claim that W is a standard Brownian motion. Indeed,

$$\mathrm{Cov}(W(t), W(s)) = ts\,\mathrm{Cov}\left(B\left(\tfrac{1}{t}\right), B\left(\tfrac{1}{s}\right)\right) = ts\left(\tfrac{1}{t} \wedge \tfrac{1}{s}\right) = t \wedge s,$$

so W and B have the same finite-dimensional distributions, and they have the same distributions as processes on the rational numbers. Since the paths of $W(t)$ are continuous except maybe at 0, we have

$$\lim_{t \downarrow 0} W(t) = \lim_{t \downarrow 0, t \in \mathbb{Q}} W(t) = 0 \quad \text{a.s.}$$

(here $\mathbb{Q}$ is the set of rational numbers) so the paths of $W(t)$ are continuous on $[0,\infty)$ a.s. As a corollary, we get

Corollary 6.3.1 (Law of Large Numbers for Brownian motion)

$$\lim_{t \to \infty} \frac{B(t)}{t} = 0 \quad \text{a.s.}$$

Proof $\lim_{t \to \infty} \frac{B(t)}{t} = \lim_{t \to \infty} W(\tfrac{1}{t}) = 0$ a.s. $\square$

The symmetry inherent in the time inversion property becomes more apparent if one considers the Ornstein–Uhlenbeck diffusion, which is given by

$$X(t) = e^{-t}B(e^{2t}).$$

This is a stationary Markov chain where $X(t)$ has a standard normal distribution for all t. It is a diffusion with a drift toward the origin proportional to the distance from the origin. Unlike Brownian motion, the Ornstein–Uhlenbeck diffusion is time reversible. The time inversion formula gives

$$\{X(t)\}_{t\geq 0} \overset{d}{=} \{X(-t)\}_{t\geq 0}.$$

For t near $-\infty$, the process $X(t)$ relates to the Brownian motion near 0, and for t near ∞, the process $X(t)$ relates to the Brownian motion near ∞.

One of the advantages of Lévy's construction of Brownian motion is that it easily yields a modulus of continuity result. Following Lévy, we defined Brownian motion as an infinite sum $\sum_{n=0}^{\infty} F_n$, where each F_n is a piecewise linear function given in (6.2.2). The derivative of F_n exists except on a finite set, and by definition and (6.2.5)

$$\|F_n'\|_\infty \leq \frac{\|F_n\|_\infty}{2^{-n}} \leq C_1(\omega) + c\sqrt{n}\, 2^{n/2}. \tag{6.3.1}$$

The random constant $C_1(\omega)$ is introduced to deal with the finitely many exceptions to (6.2.5). Now for $t, t+h \in [0,1]$, we have

$$|B(t+h) - B(t)| \leq \sum_n |F_n(t+h) - F_n(t)| \leq \sum_{n \leq \ell} h\|F_n'\|_\infty + \sum_{n > \ell} 2\|F_n\|_\infty. \tag{6.3.2}$$

By (6.2.5) and (6.3.1) if $\ell > N$ for a random N, then the above is bounded by

$$h\left(C_1(\omega) + \sum_{n \leq \ell} c\sqrt{n}\, 2^{n/2}\right) + 2 \sum_{n > \ell} c\sqrt{n}\, 2^{-n/2}$$

$$\leq C_2(\omega)h\sqrt{\ell}\, 2^{\ell/2} + C_3(\omega)\sqrt{\ell}\, 2^{-\ell/2}.$$

The inequality holds because each series is bounded by a constant times its dominant term. Choosing $\ell = \lfloor \log_2(1/h) \rfloor$, and choosing $C(\omega)$ to take care of the cases when $\ell \leq N$, we get

$$|B(t+h) - B(t)| \leq C(\omega)\sqrt{h \log_2 \frac{1}{h}}. \tag{6.3.3}$$

The result is a (weak) form of Lévy's modulus of continuity. This is not enough to make $\{B_t\}$ a differentiable function since $\sqrt{h} \gg h$ for small h. But we still have

Corollary 6.3.2 *Brownian paths are α-Hölder a.s. for all $\alpha < \frac{1}{2}$.*

A Brownian motion is almost surely not $\frac{1}{2}$-Hölder; we leave this to the reader (Exercise 6.8). However, there does exist a $t = t(\omega)$ such that

$$|B(t + h) - B(t)| \leq C(\omega)h^{\frac{1}{2}}$$

for every h almost surely. The set of such t almost surely has measure 0. This is the slowest movement that is locally possible.

Having proven that Brownian paths are somewhat "regular", let us see why they are "bizarre". One reason is that the paths of Brownian motion have no intervals of monotonicity. Indeed, if $[a, b]$ is an interval of monotonicity, then dividing it up into n equal sub-intervals $[a_i, a_{i+1}]$ each increment $B(a_i) - B(a_{i+1})$ has to have the same sign. This has probability $2 \cdot 2^{-n}$, and taking $n \to \infty$ shows that the probability that $[a, b]$ is an interval of monotonicity must be 0. Taking a countable union gives that there is no interval of monotonicity with rational endpoints, but each monotone interval would have a monotone rational sub-interval.

We will now show that for any time t_0, Brownian motion is almost surely not differentiable at t_0. For this, we need a simple proposition.

Proposition 6.3.3 *Almost surely*

$$\limsup_{n \to \infty} \frac{B(n)}{\sqrt{n}} = +\infty, \quad \liminf_{n \to \infty} \frac{B(n)}{\sqrt{n}} = -\infty. \tag{6.3.4}$$

Comparing this with Corollary 6.3.1, it is natural to ask what sequence $B(n)$ should be divided by to get a lim sup that is greater than 0 but less than ∞. An answer is given by the Law of the Iterated Logarithm later in Section 7.2.

The proof of Proposition 6.3.3 relies on Hewitt–Savage 0–1 Law. Consider a probability measure on the space of real sequences, and let $X_1, X_2, \ldots$ be the sequence of random variables it defines. An event, i.e., a measurable set of sequences, A is **exchangeable** if $X_1, X_2, \ldots$ satisfy A implies that $X_{\sigma_1}, X_{\sigma_2}, \ldots$ satisfy A for all finite permutations σ. Finite permutation means that $\sigma_n = n$ for all sufficiently large n.

Proposition 6.3.4 (Hewitt–Savage 0–1 Law) *If A is an exchangeable event for an i.i.d. sequence then $\mathbb{P}(A)$ is 0 or 1.*

Sketch of Proof: Given i.i.d. variables $X_1, X_2, \ldots$, suppose that A is an exchangeable event for this sequence. Then for any $\varepsilon > 0$ there is an integer n and a Borel set $B_n \subset \mathbb{R}^n$ such that the event $A_n = \{\omega : (X_1, \ldots, X_n) \in B_n\}$ satisfies $\mathbb{P}(A_n \Delta A) < \varepsilon$. Now apply the permutation σ that transposes i with $i + n$ for $1 \leq i \leq n$. The event A is pointwise fixed by this transformation of the measure space (since A is exchangeable) and the probability of any event is invariant (the measure space is a product space with identical distributions in each coordinate and we are simply reordering the coordinates). Thus A_n is sent to a new event A_n^σ that has the same probability and $\mathbb{P}((A_n^\sigma \Delta A) = \mathbb{P}(A_n \Delta A) < \varepsilon$, hence $\mathbb{P}(A_n^\sigma \Delta A_n) < 2\varepsilon$ (since $\mathbb{P}(X \Delta Y)$ defines a metric on measurable sets). But A_n and A_n^σ are independent, so $\mathbb{P}(A_n \cap A_n^\sigma) = \mathbb{P}(A_n)\mathbb{P}(A_n^\sigma) = \mathbb{P}(A_n)^2$. Thus

$$\mathbb{P}(A) = \mathbb{P}(A_n \cap A_n^\sigma) + O(\varepsilon) = \mathbb{P}(A_n)^2 + O(\varepsilon) = \mathbb{P}(A)^2 + O(\varepsilon).$$

Taking $\varepsilon \to 0$ shows $\mathbb{P}(A) \in \{0, 1\}$. The result is from Hewitt and Savage (1955). Also see Durrett (1996). $\qquad\qquad\qquad\qquad\qquad\qquad\qquad\qquad\qquad\qquad\square$

Proof of Proposition 6.3.3. In general, the probability that infinitely many events $\{A_n\}$ occur satisfies

$$\mathbb{P}(A_n \text{ i.o.}) = \mathbb{P}\left(\bigcap_{n=1}^\infty \bigcup_{k=n}^\infty A_k\right) = \lim_{n\to\infty} \mathbb{P}\left(\bigcup_{k=n}^\infty A_k\right) \geq \limsup_{n\to\infty} \mathbb{P}(A_n),$$

where "i.o." means "infinitely often". So, in particular,

$$\mathbb{P}(B(n) > c\sqrt{n} \text{ i.o.}) \geq \limsup_{n\to\infty} \mathbb{P}(B(n) > c\sqrt{n}).$$

By the scaling property, the expression in the lim sup equals $\mathbb{P}(B(1) > c)$, which is positive. Let $X_n = B(n) - B(n-1)$. These are i.i.d. random variables, $\{\sum_{k=1}^n X_k > c\sqrt{n} \text{ i.o.}\} = \{B(n) > c\sqrt{n} \text{ i.o.}\}$ is exchangeable and has positive probability, so the Hewitt–Savage 0–1 law says it has probability 1. Taking the intersection over all natural numbers c gives the first part of Proposition 6.3.3, and the second is proved similarly. $\qquad\qquad\qquad\qquad\qquad\qquad\qquad\square$

The two claims of Proposition 6.3.3 together mean that $B(t)$ crosses 0 for arbitrarily large values of t. If we use time inversion $W(t) = tB(\frac{1}{t})$, we get that Brownian motion crosses 0 for arbitrarily small values of t. Letting $Z_B = \{t : B(t) = 0\}$, this means that 0 is an accumulation point from the right for Z_B. But we get even more. For a function f, define the upper and lower right derivatives

$$D^* f(t) = \limsup_{h\downarrow 0} \frac{f(t+h) - f(t)}{h},$$

$$D_* f(t) = \liminf_{h\downarrow 0} \frac{f(t+h) - f(t)}{h}.$$

Then

$$D^* W(0) \geq \limsup_{n\to\infty} \frac{W(\frac{1}{n}) - W(0)}{\frac{1}{n}} \geq \limsup_{n\to\infty} \sqrt{n}\, W(\tfrac{1}{n}) = \limsup_{n\to\infty} \frac{B(n)}{\sqrt{n}},$$

which is infinite by Proposition 6.3.3. Similarly, $D_* W(0) = -\infty$, showing that W is not differentiable at 0.

Corollary 6.3.5 *Fix* $t_0 \geq 0$. *The Brownian motion W almost surely satisfies* $D^*W(t_0) = +\infty, D_*W(t_0) = -\infty$, *and* t_0 *is an accumulation point from the right for the level set* $\{s : W(s) = W(t_0)\}$.

Proof $t \mapsto W(t_0 + t) - W(t_0)$ is a standard Brownian motion. $\qquad\square$

Does this imply that a.s. each t_0 is an accumulation point from the right for the level set $\{s : W(s) = W(t_0)\}$? Certainly not; consider, for example, the last 0 of $\{B_t\}$ before time 1. However, Z_B almost surely has no isolated points, as we will see later. Also, the set of exceptional t_0 must have Lebesgue measure 0. This is true in general. Suppose A is a measurable event (set of paths) such that

$$\forall t_0, \ \mathbb{P}(t \to W(t_0 + t) - W(t_0) \text{ satisfies } A) = 1.$$

Let Θ_t be the operator that shifts paths by t. Then $\mathbb{P}(\bigcap_{t_0 \in \mathbb{Q}} \Theta_{t_0}(A)) = 1$. In fact, the Lebesgue measure of points t_0 so that W does not satisfy $\Theta_{t_0}(A)$ is 0 a.s. To see this, apply Fubini's Theorem to the double integral

$$\int \int_0^\infty \mathbf{1}_{W \notin \Theta_{t_0}(A)} \, dt_0 \, d\mathbb{P}(W).$$

We leave it to the reader to apply this idea to show that for all t,

$$\mathbb{P}(t \text{ is a local maximum}) = 0,$$

but almost surely, the set of local maxima of a 1-dimensional Brownian motion is a countable dense set in $(0, \infty)$; see Exercise 6.9.

Nowhere differentiability of Brownian motion therefore requires a more careful argument than almost sure non-differentiability at a fixed point.

Theorem 6.3.6 (Paley et al., 1933) *Almost surely, Brownian motion is nowhere differentiable. Furthermore, almost surely for all t either* $D^*B(t) = +\infty$ *or* $D_*B(t) = -\infty$.

For local maxima we have $D^*B(t) \leq 0$, and for local minima, $D_*B(t) \geq 0$, so it is important to have the either-or in the statement.

Proof (Dvoretzky et al., 1961) Suppose that there is a $t_0 \in [0, 1]$ such that $-\infty < D_*B(t_0) \leq D^*B(t_0) < \infty$. Then for some finite constant M we would have

$$\sup_{h \in [0,1]} \frac{|B(t_0 + h) - B(t_0)|}{h} \leq M. \tag{6.3.5}$$

If t_0 is contained in the binary interval $[(k-1)/2^n, k/2^n]$ for $n > 2$, then for all $1 \leq j \leq n$ the triangle inequality gives

$$|B((k+j)/2^n) - B((k+j-1)/2^n)| \leq M(2j+1)/2^n. \tag{6.3.6}$$

Let $\Omega_{n,k}$ be the event that (6.3.6) holds for $j = 1, 2$, and 3. Then by the scaling property

$$\mathbb{P}(\Omega_{n,k}) \leq \mathbb{P}\left(|B(1)| \leq 7M/\sqrt{2^n}\right)^3,$$

which is at most $(7M2^{-n/2})^3$, since the normal density is less than $1/2$. Hence

$$\mathbb{P}\left(\bigcup_{k=1}^{2^n} \Omega_{n,k}\right) \leq 2^n(7M2^{-n/2})^3 = (7M)^3 2^{-n/2}.$$

Therefore by the Borel–Cantelli Lemma,

$$\mathbb{P}\left((6.3.5) \text{ holds}\right) \leq \mathbb{P}\left(\bigcup_{k=1}^{2^n} \Omega_{n,k} \text{ holds for infinitely many } n\right) = 0. \qquad \square$$

Exercise 6.10 asks the reader to show that if $\alpha > 1/2$, then a.s. for all $t > 0$, there exists $h > 0$ such that $|B(t + h) - B(t)| > h^{\alpha}$.

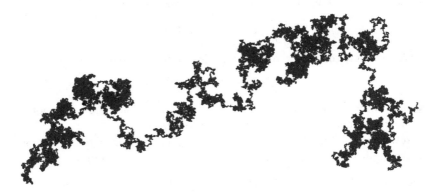

Figure 6.3.1 A 2-dimensional Brownian path. This was drawn by taking 100,000 unit steps in uniformly random directions.

6.4 Hausdorff dimension of the Brownian path and graph

We have shown in Corollary 6.3.2 that Brownian motion is β-Hölder for any $\beta < 1/2$ a.s. This will allow us to infer an upper bound on the Hausdorff dimension of its image and graph. Recall that the **graph** G_f of a function f is the set of points $(t, f(t))$ as t ranges over the domain of f. As corollaries of Lemmas 5.1.2 and 5.1.3 and Corollary 6.3.2 we obtain upper bounds on the dimension of the graph and the image of Brownian motion.

Corollary 6.4.1

$$\dim(G_B) \leq \underline{\dim}_{\mathscr{M}}(G_B) \leq \overline{\dim}_{\mathscr{M}}(G_B) \leq 3/2 \quad a.s.$$

Corollary 6.4.2   *For $A \subset [0,\infty)$, we have $\dim(B(A)) \leq (2\dim(A)) \wedge 1$ a.s.*

The nowhere differentiability of Brownian motion established in the previous section suggests that its graph has dimension higher than one. Taylor (1953) showed that the graph of Brownian motion has Hausdorff dimension $3/2$.

Define the d-dimensional standard Brownian motion whose coordinates are independent 1-dimensional standard Brownian motions. Its distribution is invariant under orthogonal transformations of $\mathbb{R}^d$, since Gaussian random variables are invariant to such transformations by Lemmas 6.1.3 and 6.1.4. For $d \geq 2$ it is interesting to look at the image set of Brownian motion. We will see that planar Brownian motion is neighborhood recurrent; that is, it visits every neighborhood in the plane infinitely often. In this sense, the image of planar Brownian motion is comparable to the plane itself; another sense in which this happens is that of Hausdorff dimension: the image of planar and higher-dimensional Brownian motion has Hausdorff dimension two. Summing up, we will prove

Theorem 6.4.3 (Taylor, 1953)   *Let B be d-dimensional Brownian motion defined on the time set $[0,1]$. If $d = 1$ then*

$$\dim G_B = 3/2 \quad a.s.$$

Moreover, if $d \geq 2$, then

$$\dim B[0,1] = 2 \quad a.s.$$

Higher-dimensional Brownian motion therefore doubles the dimension of the time line. Naturally, the question arises whether this holds for subsets of the time line as well. In a certain sense, this even holds for $d = 1$: note the "$\wedge d$" in the following theorem.

Theorem 6.4.4 (McKean, 1955)   *For every subset A of $[0,\infty)$, the image of A under d-dimensional Brownian motion has Hausdorff dimension $(2\dim A) \wedge d$ almost surely.*

Theorem 6.4.5 (Uniform Dimension Doubling Kaufman (1969))   *Let B be Brownian motion in dimension at least 2. Almost surely, for every $A \subset [0,\infty)$, we have $\dim B(A) = 2\dim(A)$.*

Notice the difference between the last two results. In Theorem 6.4.4, the null probability set depends on A, while Kaufman's Theorem has a much stronger claim: it states dimension doubling uniformly for all sets. For this theorem, we must assume $d \geq 2$: we will see later that the zero set of 1-dimensional Brownian motion has dimension half, while its image is the single point 0. We will prove Kaufman's Theorem in the next chapter (Theorem 7.1.5).

For Theorem 6.4.3 we need the following result due to Frostman (it is essentially the same as Theorem 3.4.2; that result was stated for $\mathbb{R}^d$, but the proof of its part (ii) works for any metric space).

Theorem 6.4.6 (Frostman's Energy Method (Frostman, 1935)) *Given a metric space* (X, ρ), *if* μ *is a finite Borel measure supported on* $A \subset X$ *and*

$$\mathscr{E}_\alpha(\mu) \overset{\text{def}}{=} \iint \frac{d\mu(x)d\mu(y)}{\rho(x,y)^\alpha} < \infty,$$

then $\mathscr{H}^\alpha_\infty(A) = \infty$, *and hence* $\dim(A) \geq \alpha$.

Proof of Theorem 6.4.3, Part 2. From Corollary 6.3.2 we have that B_d is β Hölder for every $\beta < 1/2$ a.s. Therefore Lemma 5.1.3 implies that

$$\dim B_d[0,1] \leq 2 \quad a.s.$$

For the other inequality, we will use Frostman's Energy Method. A natural measure on $B_d[0,1]$ is the occupation measure $\mu_B \overset{\text{def}}{=} \mathscr{L}B^{-1}$, which means that $\mu_B(A) = \mathscr{L}B^{-1}(A)$, for all measurable subsets A of $\mathbb{R}^d$, or, equivalently,

$$\int_{\mathbb{R}^d} f(x)\, d\mu_B(x) = \int_0^1 f(B_t)\, dt$$

for all measurable functions f. We want to show that for any $0 < \alpha < 2$,

$$\mathbb{E} \int_{\mathbb{R}^d} \int_{\mathbb{R}^d} \frac{d\mu_B(x)d\mu_B(y)}{|x-y|^\alpha} = \mathbb{E} \int_0^1 \int_0^1 \frac{ds\, dt}{|B(t)-B(s)|^\alpha} < \infty. \tag{6.4.1}$$

Let us evaluate the expectation:

$$\mathbb{E}|B(t)-B(s)|^{-\alpha} = \mathbb{E}\left((|t-s|^{1/2}|Z|)^{-\alpha}\right)$$

$$= |t-s|^{-\alpha/2} \int_{\mathbb{R}^d} \frac{c_d}{|z|^\alpha} e^{-|z|^2/2}\, dz.$$

Here Z denotes the d-dimensional standard Gaussian random variable. The integral can be evaluated using polar coordinates, but all we need is that, for $d \geq 2$, it is a finite constant c depending on d and α only. Substituting this expression into (6.4.1) and using Fubini's Theorem we get

$$\mathbb{E}\mathscr{E}_\alpha(\mu_B) = c \int_0^1 \int_0^1 \frac{ds\, dt}{|t-s|^{\alpha/2}} \leq 2c \int_0^1 \frac{du}{u^{\alpha/2}} < \infty. \tag{6.4.2}$$

Therefore $\mathscr{E}_\alpha(\mu_B) < \infty$ almost surely and we are done by Frostman's method.

□

Remark 6.4.7 Lévy showed earlier (Lévy, 1940) that when $d = 2$ we have $\mathscr{H}^2(B[0,1]) = 0$ a.s. (Theorem 6.8.2). The statement is actually also true for all $d \geq 2$.

Now let us turn to the graph G_B of Brownian motion. We will show a proof of the first half of Taylor's Theorem for one-dimensional Brownian motion.

Proof of Theorem 6.4.3, Part 1. We have shown in Corollary 6.4.1 that

$$\dim G_B \leq 3/2.$$

For the other inequality, let $\alpha < 3/2$ and let A be a subset of the graph. Define a measure on the graph using projection to the time axis:

$$\mu(A) \stackrel{\text{def}}{=} \mathscr{L}(\{0 \leq t \leq 1 : (t, B(t)) \in A\}).$$

Changing variables, the α energy of μ can be written as

$$\iint \frac{d\mu(x) d\mu(y)}{|x-y|^\alpha} = \int_0^1 \int_0^1 \frac{ds\,dt}{(|t-s|^2 + |B(t)-B(s)|^2)^{\alpha/2}}.$$

Bounding the integrand, taking expectations and applying Fubini's Theorem we get that

$$\mathbb{E}\mathscr{E}_\alpha(\mu) \leq 2 \int_0^1 \mathbb{E}\left((t^2 + B(t)^2)^{-\alpha/2}\right) dt. \qquad (6.4.3)$$

Let $n(z)$ denote the standard normal density. By scaling, the expected value above can be written as

$$2 \int_0^{+\infty} (t^2 + tz^2)^{-\alpha/2} n(z)\,dz. \qquad (6.4.4)$$

Comparing the size of the summands in the integration suggests separating $z \leq \sqrt{t}$ from $z > \sqrt{t}$. Then we can bound (6.4.4) above by twice

$$\int_0^{\sqrt{t}} (t^2)^{-\alpha/2}\,dz + \int_{\sqrt{t}}^\infty (tz^2)^{-\alpha/2} n(z)\,dz = t^{\frac{1}{2}-\alpha} + t^{-\alpha/2} \int_{\sqrt{t}}^\infty z^{-\alpha} n(z)\,dz.$$

Furthermore, we separate the last integral at 1. We get

$$\int_{\sqrt{t}}^\infty z^{-\alpha} n(z)\,dz \leq c_\alpha + \int_{\sqrt{t}}^1 z^{-\alpha}\,dz.$$

The latter integral is of order $t^{(1-\alpha)/2}$. Substituting these results into (6.4.3), we see that the expected energy is finite when $\alpha < 3/2$. Therefore $\mathscr{E}_\alpha(\mu_B) < \infty$ almost surely. The claim now follows from Frostman's Energy Method. □

6.5 Nowhere differentiability is prevalent

Lévy (1953) asks whether it is true that

$$\mathbb{P}[\forall t,\, D^*B(t) \in \{\pm\infty\}] = 1?$$

The following proposition gives a negative answer to this question.

Proposition 6.5.1 *Almost surely there is an uncountable set of times t at which the upper right derivative $D^*B(t)$ is zero.*

We sketch a proof below. Stronger and more general results can be found in Barlow and Perkins (1984).

Sketch of Proof.   Put

$$I = \left[B(1),\, \sup_{0 \le s \le 1} B(s)\right],$$

and define a function $g : I \to [0, 1]$ by setting

$$g(x) = \sup\{s \in [0, 1] : B(s) = x\}.$$

It is easy to check that a.s. the interval I is non-degenerate, g is strictly decreasing, left continuous and satisfies $B(g(x)) = x$. Furthermore, a.s. the set of discontinuities of g is dense in I since a.s. B has no interval of monotonicity. We restrict our attention to the event of probability 1 on which these assertions hold. Let

$$V_n = \{x \in I : g(x - h) - g(x) > nh \text{ for some } h \in (0, n^{-1})\}.$$

Since g is left continuous and strictly decreasing, one readily verifies that V_n is open; it is also dense in I as every point of discontinuity of g is a limit from the right of points of V_n. By the Baire Category Theorem, $V := \bigcap_n V_n$ is uncountable and dense in I. Now if $x \in V$ then there is a sequence $x_n \uparrow x$ such that $g(x_n) - g(x) > n(x - x_n)$. Setting $t = g(x)$ and $t_n = g(x_n)$, we have $t_n \downarrow t$ and $t_n - t > n(B(t) - B(t_n))$, from which it follows that $D^*B(t) \ge 0$. On the other hand $D^*B(t) \le 0$ since $B(s) \le B(t)$ for all $s \in (t, 1)$ by definition of $t = g(x)$. $\square$

Is the "typical" function in $C([0, 1])$ nowhere differentiable? It is an easy application of the Baire Category Theorem to show that nowhere differentiability is a generic property for $C([0, 1])$. This result leaves something to be desired, perhaps, as topological and measure theoretic notions of a "large" set need not coincide. For example, the set of points in $[0, 1]$ whose binary expansion has zeros with asymptotic frequency $1/2$ is a meager set, yet it has Lebesgue measure 1. We consider a related idea proposed by Christensen (1972) and by Hunt et al. (1993).

Let X be a separable Banach space. If X is infinite-dimensional, then there is no locally finite, translation invariant analog of Lebesgue measure, but there is a translation invariant analog of measure zero sets. We say that a Borel set $A \subset X$ is **prevalent** if there exists a Borel probability measure μ on X such that $\mu(x + A) = 1$ for every $x \in X$. A general set A is called prevalent if it contains a Borel prevalent set. A set is called **negligible** if its complement is prevalent. In other words, a set is negligible if there is a measure μ and a Borel subset A that has zero measure for every translate of μ. Obviously translates of a negligible set are negligible.

Proposition 6.5.2 *If $A_1, A_2, \ldots$ are negligible subsets of X then $\bigcup_{i \geq 1} A_i$ is also negligible.*

Proof For each $i \geq 1$ let μ_{A_i} be a Borel probability measure on X satisfying $\mu_{A_i}(x + A_i) = 0$ for all $x \in X$. Using separability we can find for each i a ball D_i of radius 2^{-i} centered at $x_i \in X$ with $\mu_{A_i}(D_i) > 0$. Define probability measures μ_i, $i \geq 1$, by setting $\mu_i(E) = \mu_{A_i}((E + x_i) \cap D_i)$ for each Borel set E, so that $\mu_i(x + A_i) = 0$ for all x and for all i. Let $(Y_i; i \geq 0)$ be a sequence of independent random variables with distributions equal to μ_i. For all i we have $\mu_i[|Y_i| \leq 2^{-i}] = 1$. Therefore, $S = \sum_i Y_i$ converges almost surely. Writing μ for the distribution of S and putting $v_j = \text{dist}(S - Y_j)$, we have $\mu = \mu_j * v_j$, and hence $\mu(x + A_j) = \mu_j * v_j(x + A_j) = 0$ for all x and for all j. Thus $\mu(x + \bigcup_{i \geq 1} A_i) = 0$ for all x. $\square$

Proposition 6.5.3 *A subset A of $\mathbb{R}^d$ is negligible if and only if $\mathscr{L}_d(A) = 0$.*

Proof ($\Rightarrow$) Assume A is negligible. Let μ_A be a (Borel) probability measure such that $\mu_A(x + A) = 0$ for all $x \in \mathbb{R}^d$. Since $\mathscr{L}_d * \mu_A = \mathscr{L}_d$ (indeed the equality $\mathscr{L}_d * \mu = \mathscr{L}_d$ holds for any Borel probability measure μ on $\mathbb{R}^d$) we must have $0 = \mathscr{L}_d * \mu_A(x + A) = \mathscr{L}_d(x + A)$ for all $x \in \mathbb{R}^d$.

($\Leftarrow$) If $\mathscr{L}_d(A) = 0$ then the restriction of $\mathscr{L}_d$ to the unit cube is a probability measure that vanishes on every translate of A. $\square$

Theorem 6.5.4 *If $f \in C([0, 1])$, then $B(t) + f(t)$ is nowhere differentiable almost surely.*

The proof of this is left to the reader (Exercise 6.12); using Proposition 6.5.3, it is just a matter of checking that the proof of Theorem 6.3.6 goes through without changes. Using Wiener measure on the Banach space of continuous functions on $[0, 1]$ (with the supremum norm) then gives:

Corollary 6.5.5 *The set of nowhere differentiable functions is prevalent in* $C([0,1])$.

6.6 Strong Markov property and the reflection principle

For each $t \geq 0$ let $\mathscr{F}_0(t) = \sigma\{B(s) : s \leq t\}$ be the smallest σ-field making every $B(s)$, $s \leq t$, measurable, and set $\mathscr{F}_+(t) = \bigcap_{u>t} \mathscr{F}_0(u)$ (the right-continuous filtration). It is known (see, for example, Durrett (1996), Theorem 7.2.4) that $\mathscr{F}_0(t)$ and $\mathscr{F}_+(t)$ have the same completion. A filtration $\{\mathscr{F}(t)\}_{t\geq 0}$ is a **Brownian filtration** if for all $t \geq 0$ the process $\{B(t+s) - B(t)\}_{s\geq 0}$ is independent of $\mathscr{F}(t)$ and $\mathscr{F}(t) \supset \mathscr{F}_0(t)$. A random variable τ is a **stopping time** for a Brownian filtration $\{\mathscr{F}(t)\}_{t\geq 0}$ if $\{\tau \leq t\} \in \mathscr{F}(t)$ for all t. For any random time τ we define the pre-τ σ-field

$$\mathscr{F}(\tau) := \{A : \forall t, \, A \cap \{\tau \leq t\} \in \mathscr{F}(t)\}.$$

Proposition 6.6.1 (Markov property) *For every $t \geq 0$ the process*

$$\{B(t+s) - B(t)\}_{s\geq 0}$$

is standard Brownian motion independent of $\mathscr{F}_0(t)$ and $\mathscr{F}_+(t)$.

It is evident from independence of increments that $\{B(t+s) - B(t)\}_{s\geq 0}$ is standard Brownian motion independent of $\mathscr{F}_0(t)$. That this process is independent of $\mathscr{F}_+(t)$ follows from continuity; see, e.g., Durrett (1996), 7.2.1 for details.

The main result of this section is the **strong Markov property** for Brownian motion, established independently by Hunt (1956) and Dynkin and Yushkevich (1956):

Theorem 6.6.2 *Suppose that τ is a stopping time for the Brownian filtration* $\{\mathscr{F}(t)\}_{t\geq 0}$. *Then* $\{B(\tau + s) - B(\tau)\}_{s\geq 0}$ *is Brownian motion independent of* $\mathscr{F}(\tau)$.

Sketch of Proof. Suppose first that τ is an integer-valued stopping time with respect to a Brownian filtration $\{\mathscr{F}(t)\}_{t\geq 0}$. For each integer j the event $\{\tau = j\}$ is in $\mathscr{F}(j)$ and the process $\{B(t+j) - B(j)\}_{t\geq 0}$ is independent of $\mathscr{F}(j)$, so the result follows from the Markov property in this special case. It also holds if the values of τ are integer multiples of some $\varepsilon > 0$, and approximating τ by such discrete stopping times gives the conclusion in the general case. See, e.g., Durrett (1996), 7.3.7 for more details. $\qquad\square$

Given real random variables Y, Z, we write $Y \overset{\mathrm{d}}{=} Z$ if $\mathbb{P}[Y \in B] = \mathbb{P}[Z \in B]$ for all Borel sets B in $\mathbb{R}$. This definition extends to random variables taking values in any metric space. The following is an elementary fact we need below:

Lemma 6.6.3   *Let X, Y, Z be random variables with X, Y independent and X, Z independent. If $Y \overset{\mathrm{d}}{=} Z$ then $(X, Y) \overset{\mathrm{d}}{=} (X, Z)$.*

Proof   We must show that every Borel set A in the plane satisfies

$$\mathbb{P}[(X, Y) \in A] = \mathbb{P}[(X, Z) \in A].$$

If A is a Borel rectangle (a product of two Borel sets) then this clearly follows from the hypothesis. By the Carathéodory Extension Theorem, a Borel measure on the plane is determined by its values on the algebra of disjoint finite unions of Borel rectangles; thus the assertion holds for all planar Borel sets A.   □

One important consequence of the strong Markov property is the following:

Theorem 6.6.4 (Reflection Principle)   *If τ is a stopping time then*

$$B^*(t) := B(t)\mathbf{1}_{(t \leq \tau)} + (2B(\tau) - B(t))\mathbf{1}_{(t > \tau)}$$

(Brownian motion reflected at time τ) is also standard Brownian motion.

Proof   The strong Markov property states that $\{B(\tau + t) - B(\tau)\}_{t \geq 0}$ is Brownian motion independent of $\mathscr{F}(\tau)$, and by symmetry this also holds for

$$\{-(B(\tau + t) - B(\tau))\}_{t \geq 0}.$$

We see from Lemma 6.6.3 that

$$(\{B(t)\}_{0 \leq t \leq \tau}, \{B(t + \tau) - B(\tau)\}_{t \geq 0})$$

$$\overset{\mathrm{d}}{=} (\{B(t)\}_{0 \leq t \leq \tau}, \{(B(\tau) - B(t + \tau))\}_{t \geq 0}),$$

and the reflection principle follows immediately.   □

To see that τ needs to be a stopping time, consider

$$\tau = \inf\{t : B(t) = \max_{0 \leq s \leq 1} B(s)\}.$$

Almost surely $\{B(\tau + t) - B(\tau)\}_{t \geq 0}$ is non-positive on some right neighborhood of $t = 0$, and hence is *not* Brownian motion. The strong Markov property does not apply here because τ is not a stopping time for any Brownian filtration. We will later see that Brownian motion almost surely has no point of increase. Since τ is a point of increase of the reflected process $\{B^*(t)\}$, it follows that the distributions of Brownian motion and of $\{B^*(t)\}$ are singular.

A simple example of a stopping time is the time when a Brownian motion first enters a closed set (Exercise 6.13). More generally, if A is a Borel set then the hitting time τ_A is a stopping time (see Bass, 1995).

Set $M(t) = \max\limits_{0 \le s \le t} B(s)$. Our next result says $M(t) \overset{d}{=} |B(t)|$.

Theorem 6.6.5   *If $a > 0$, then $\mathbb{P}[M(t) > a] = 2\mathbb{P}[B(t) > a]$.*

Proof   Set $\tau_a = \min\{t \ge 0 : B(t) = a\}$ and let $\{B^*(t)\}$ be Brownian motion reflected at τ_a. Then $\{M(t) > a\}$ is the disjoint union of the events $\{B(t) > a\}$ and $\{M(t) > a, B(t) \le a\}$, and since $\{M(t) > a, B(t) \le a\} = \{B^*(t) \ge a\}$ the desired conclusion follows immediately.   □

6.7 Local extrema of Brownian motion

Proposition 6.7.1   *Almost surely, every local maximum of Brownian motion is a strict local maximum.*

For the proof we shall need

Lemma 6.7.2   *Given two disjoint closed time intervals, the maxima of Brownian motion on them are different almost surely.*

Proof   Suppose the intervals are $[a_1, b_1]$ and $[a_2, b_2]$ and suppose $b_1 < a_2$. For $i = 1, 2$, let m_i denote the maximum of Brownian motion on $[a_i, b_i]$. We claim that $B(a_2) - B(b_1)$ is independent of the pair $m_1 - B(b_1)$ and $m_2 - B(a_2)$. Note that

$$m_i - B(b_i) = \sup\{B(q) - B(b_i) : q \in \mathbb{Q} \cap [a_i, b_i]\},$$

so the claim follows from independent increments of Brownian motion. Thus if $m_1 = m_2$, we have

$$B(a_2) - B(b_1) = m_1 - B(b_1) - (m_2 - B(a_2)).$$

i.e., we have equality of two independent random variables, and the left side is Gaussian, so the probability they are equal is zero.   □

Proof of Proposition 6.7.1   The statement of the lemma holds jointly for all disjoint pairs of intervals with rational endpoints. The Proposition follows, since if Brownian motion had a non-strict local maximum, then there were two disjoint rational intervals where Brownian motion has the same maximum.   □

Corollary 6.7.3 *The set M of times where Brownian motion assumes a local maximum is countable and dense almost surely.*

Proof Consider the function from the set of non-degenerate closed intervals with rational endpoints to $\mathbb{R}$ given by

$$[a,b] \mapsto \inf \left\{ t \geq a : B(t) = \max_{a \leq s \leq b} B(s) \right\}.$$

The image of this map contains the set M almost surely by Lemma 6.7.2. This shows that M is countable almost surely. We already know that B has no interval of increase or decrease almost surely. It follows that B almost surely has a local maximum in every interval with rational endpoints, implying the corollary. □

6.8 Area of planar Brownian motion

We have seen that the image of Brownian motion is always 2-dimensional, so one might ask what its 2-dimensional Hausdorff measure is. It turns out to be 0 in all dimensions; we will prove it for the planar case. We will need the following lemma.

Lemma 6.8.1 *If $A_1, A_2 \subset \mathbb{R}^2$ are Borel sets with positive area, then*

$$\mathscr{L}_2(\{x \in \mathbb{R}^2 : \mathscr{L}_2(A_1 \cap (A_2 + x)) > 0\}) > 0.$$

Proof One proof of this fact relies on (outer) regularity of Lebesgue measure. The proof below is more streamlined.

We may assume A_1 and A_2 are bounded. By Fubini's Theorem,

$$
\begin{aligned}
\int_{\mathbb{R}^2} \mathbf{1}_{A_1} * \mathbf{1}_{-A_2}(x)\, dx &= \int_{\mathbb{R}^2} \int_{\mathbb{R}^2} \mathbf{1}_{A_1}(w) \mathbf{1}_{A_2}(w - x)\, dw\, dx \\
&= \int_{\mathbb{R}^2} \mathbf{1}_{A_1}(w) \left(\int_{\mathbb{R}^2} \mathbf{1}_{A_2}(w - x)\, dx \right) dw \\
&= \mathscr{L}_2(A_1) \mathscr{L}_2(A_2) \\
&> 0.
\end{aligned}
$$

Thus $\mathbf{1}_{A_1} * \mathbf{1}_{-A_2}(x) > 0$ on a set of positive area. But $\mathbf{1}_{A_1} * \mathbf{1}_{-A_2}(x)$ is exactly the area of $A_1 \cap (A_2 + x)$, so this proves the Lemma. □

Throughout this section B denotes planar Brownian motion. We are now ready to prove Lévy's Theorem on the area of its image.

Theorem 6.8.2 (Lévy, 1940) *Almost surely $\mathscr{L}_2(B[0,1]) = 0$.*

Proof Let X denote the area of $B[0,1]$, and M be its expected value. First we check that $M < \infty$. If $a \geq 1$ then

$$\mathbb{P}[X > a] \leq 2\mathbb{P}[|W(t)| > \sqrt{a}/2 \text{ for some } t \in [0,1]] \leq 8e^{-a/8},$$

where W is standard one-dimensional Brownian motion. Thus

$$M = \int_0^\infty \mathbb{P}[X > a]\,da \leq 8\int_0^\infty e^{-a/8}\,da + 1 < \infty.$$

Note that $B(3t)$ and $\sqrt{3}B(t)$ have the same distribution, and hence

$$\mathbb{E}\mathscr{L}_2(B[0,3]) = 3\mathbb{E}\mathscr{L}_2(B[0,1]) = 3M.$$

Note that we have $\mathscr{L}_2(B[0,3]) \leq \sum_{j=0}^{2} \mathscr{L}_2(B[j,j+1])$ with equality if and only if for $0 \leq i < j \leq 2$ we have $\mathscr{L}_2(B[i,i+1] \cap B[j,j+1]) = 0$. On the other hand, for $j = 0,1,2$, we have $\mathbb{E}\mathscr{L}_2(B[j,j+1]) = M$ and

$$3M = \mathbb{E}\mathscr{L}_2(B[0,3]) \leq \sum_{j=0}^{2} \mathbb{E}\mathscr{L}_2(B[j,j+1]) = 3M,$$

whence the intersection of any two of the $B[j,j+1]$ has measure zero almost surely. In particular, $\mathscr{L}_2(B[0,1] \cap B[2,3]) = 0$ almost surely.

For $x \in \mathbb{R}^2$, let $R(x)$ denote the area of $B[0,1] \cap (x + B[2,3] - B(2) + B(1))$. If we condition on the values of $B[0,1], B[2,3] - B(2)$, then in order to evaluate the expected value of $\mathscr{L}_2(B[0,1] \cap B[2,3])$ we should integrate $R(x)$ where x has the distribution of $B(2) - B(1)$. Thus

$$0 = \mathbb{E}[\mathscr{L}_2(B[0,1] \cap B[2,3])] = (2\pi)^{-1} \int_{\mathbb{R}^2} e^{-|x|^2/2} \mathbb{E}[R(x)]\,dx,$$

where we average with respect to the Gaussian distribution of $B(2) - B(1)$. Thus $R(x) = 0$ a.s. for $\mathscr{L}_2$-almost all x, or, by Fubini's Theorem, the area of the set where $R(x)$ is positive is a.s. zero. From the lemma we get that a.s.

$$\mathscr{L}_2(B[0,1]) = 0 \quad \text{or} \quad \mathscr{L}_2(B[2,3]) = 0.$$

The observation that $\mathscr{L}_2(B[0,1])$ and $\mathscr{L}_2(B[2,3])$ are identically distributed and independent completes the proof that $\mathscr{L}_2(B[0,1]) = 0$ almost surely. $\square$

This also follows from the fact that Brownian motion has probability 0 of hitting a given point (other than its starting point), a fact we will prove using potential theory in the next chapter.

6.9 General Markov processes

In this section we define general Markov processes. Then we prove that Brownian motion, reflected Brownian motion and a process that involves the maximum of Brownian motion are Markov processes.

Definition 6.9.1 A function $p(t,x,A)$, $p : \mathbb{R} \times \mathbb{R}^d \times \mathcal{B} \to \mathbb{R}$, where $\mathcal{B}$ is the Borel σ-algebra in $\mathbb{R}^d$, is a **Markov transition kernel** provided:

(1) $p(\cdot,\cdot,A)$ is measurable as a function of (t,x), for each $A \in \mathcal{B}$,
(2) $p(t,x,\cdot)$ is a Borel probability measure for all $t \in \mathbb{R}$ and $x \in \mathbb{R}^d$,
(3) $\forall A \in \mathcal{B}$, $x \in \mathbb{R}^d$ and $t,s > 0$,

$$p(t+s,x,A) = \int_{y \in \mathbb{R}^d} p(t,y,A)\,dp(s,x,\cdot).$$

Definition 6.9.2 A process $\{X(t)\}$ is a **Markov process** with transition kernel $p(t,x,A)$ if for all $t > s$ and Borel set $A \in \mathcal{B}$ we have

$$\mathbb{P}(X(t) \in A | \mathcal{F}_s) = p(t-s, X(s), A),$$

where $\mathcal{F}_s = \sigma(X(u),\, u \leq s)$.

The next two examples are trivial consequences of the Markov property for Brownian motion.

Example 6.9.3 A d-dimensional Brownian motion is a Markov process and its transition kernel $p(t,x,\cdot)$ has $N(x,t)$ distribution in each component.

Suppose Z has $N(x,t)$ distribution. Define $|N(x,t)|$ to be the distribution of $|Z|$.

Example 6.9.4 The reflected one-dimensional Brownian motion $|B(t)|$ is a Markov process. Moreover, its kernel $p(t,x,\cdot)$ has $|N(x,t)|$ distribution.

Theorem 6.9.5 (Lévy, 1948) *Let $M(t)$ be the maximum process of a one-dimensional Brownian motion $B(t)$, i.e. $M(t) = \max_{0 \leq s \leq t} B(s)$. Then, the process $Y(t) = M(t) - B(t)$ is Markov and its transition kernel $p(t,x,\cdot)$ has $|N(x,t)|$ distribution.*

Proof For $t > 0$, consider the two random processes $\hat{B}(t) = B(s+t) - B(s)$ and $\hat{M}(t) = \max_{0 \leq u \leq t} \hat{B}(u)$. Define $\mathcal{F}_B(s) = \sigma(B(t), 0 \leq t \leq s)$. To prove the theorem it suffices to check that conditional on $\mathcal{F}_B(s)$ and $Y(s) = y$, we have $Y(s+t) \stackrel{\mathrm{d}}{=} |y + \hat{B}(t)|$.

To prove the claim note that $M(s+t) = M(s) \vee (B(s) + \hat{M}(t))$ (recall that $s \vee t = \max(s,t)$) and so we have

$$Y(s+t) = M(s) \vee (B(s) + \hat{M}(t)) - (B(s) + \hat{B}(t)).$$

Using the fact that $a \vee b - c = (a-c) \vee (b-c)$, we have

$$Y(s+t) = Y(s) \vee \hat{M}(t) - \hat{B}(t).$$

To finish, it suffices to check for every $y \geq 0$ that $y \vee \hat{M}(t) - \hat{B}(t) \overset{d}{=} |y + \hat{B}(t)|$. For any $a \geq 0$ write

$$\mathbb{P}(y \vee \hat{M}(t) - \hat{B}(t) > a) = I + II,$$

where

$$I = \mathbb{P}(y - \hat{B}(t) > a)$$

and

$$II = \mathbb{P}(y - \hat{B}(t) \leq a \text{ and } \hat{M}(t) - \hat{B}(t) > a).$$

Since $\hat{B} \overset{d}{=} -\hat{B}$ we have

$$I = \mathbb{P}(y + \hat{B}(t) > a).$$

To study the second term it is useful to define the "time reversed" Brownian motion

$$W(u) = \hat{B}(t-u) - \hat{B}(t),$$

for $0 \leq u \leq t$. Note that W is also a Brownian motion for $0 \leq u \leq t$ since it is continuous and its finite-dimensional distributions are Gaussian with the right covariances.

Let $M_W(t) = \max_{0 \leq u \leq t} W(u)$. This implies $M_W(t) = \hat{M}(t) - \hat{B}(t)$. Because $W(t) = -\hat{B}(t)$, we have

$$II = \mathbb{P}(y + W(t) \leq a \text{ and } M_W(t) > a).$$

If we use the reflection principle by reflecting $W(u)$ at the first time it hits a we get another Brownian motion $W^*(u)$. In terms of this Brownian motion we have $II = \mathbb{P}(W^*(t) \geq a + y)$. Since $W^* \overset{d}{=} \hat{B}$, it follows $II = \mathbb{P}(y + \hat{B}(t) \leq -a)$. The Brownian motion $\hat{B}(t)$ has continuous distribution, and so, by adding I and II, we get

$$\mathbb{P}(y \vee \hat{M}(t) - \hat{B}(t) > a) = \mathbb{P}(|y + \hat{B}(t)| > a).$$

This proves the claim and, consequently, the theorem. $\qquad\qquad\square$

Proposition 6.9.6 *Two Markov processes in $\mathbb{R}^d$ with continuous paths, with the same initial distribution and transition kernel are identical in law.*

Outline of Proof. The finite-dimensional distributions are the same. From this we deduce that the restriction of both processes to rational times agree in distribution. Finally we can use continuity of paths to prove that they agree, as processes, in distribution (see Freedman (1971) for more details). □

Since the process $Y(t)$ is continuous and has the same distribution as $|B(t)|$ (they have the same Markov transition kernel and same initial distribution), this proposition implies $\{Y(t)\} \overset{\mathrm{d}}{=} \{|B(t)|\}$.

6.10 Zeros of Brownian motion

In this section, we start the study of the properties of the zero set Z_B of one-dimensional Brownian motion. We will prove that this set is an uncountable closed set with no isolated points. This is, perhaps, surprising since almost surely, a Brownian motion has isolated zeros from the left (for instance, the first zero after $1/2$) or from the right (the last zero before $1/2$). However, according to the next theorem, with probability 1, it does not have any isolated zero.

Theorem 6.10.1 *Let B be a one-dimensional Brownian motion and Z_B be its zero set, i.e.,*

$$Z_B = \{t \in [0, +\infty) : B(t) = 0\}.$$

Then, almost surely, Z_B is an uncountable closed set with no isolated points.

Proof Clearly, with probability 1, Z_B is closed because B is continuous almost surely. To prove that no point of Z_B is isolated we consider the following construction: for each rational $q \in [0, \infty)$ consider the first zero after q, i.e., $\tau_q = \inf\{t > q : B(t) = 0\}$. Note that $\tau_q < \infty$ a.s. and, since Z_B is closed, the inf is a.s. a minimum. By the strong Markov property we have that for each q, a.s. τ_q is not an isolated zero from the right. But, since there are only countably many rationals, we conclude that a.s., for all q rational, τ_q is not an isolated zero from the right. Our next task is to prove that the remaining points of Z_B are not isolated from the left. So we claim that any $0 < t \in Z_B$ that is different from τ_q for all rational q is not an isolated point from the left. To see this take a sequence $q_n \uparrow t, q_n \in \mathbb{Q}$. Define $t_n = \tau_{q_n}$. Clearly $q_n \le t_n < t$ ($t \ne t_n$ since t is not of the form T_q for any $q \in \mathbb{Q}$) and so $t_n \uparrow t$. Thus t is not isolated from the left.

Finally, recall (see, for instance, Hewitt and Stromberg (1975)) that a closed set with no isolated points is uncountable; this finishes the proof. $\square$

Next we will prove that, with probability 1, the Hausdorff dimension of Z_B is $1/2$. It turns out that it is relatively easy to bound from below the dimension of the zero set of $Y(t)$ (also known as set of record values of B). Then, by the results in the last section, this dimension must be the same as of Z_B since these two (random) sets have the same distribution.

Definition 6.10.2   A time t is a **record time** for B if $Y(t) = M(t) - B(t) = 0$, i.e., if t is a global maximum from the left.

The next lemma gives a lower bound on the Hausdorff dimension of the set of record times.

Lemma 6.10.3   *With probability 1,* $\dim\{t \in [0,1] : Y(t) = 0\} \geq 1/2$.

Proof   Since $M(t)$ is an increasing function, we can regard it as a distribution function of a measure μ, with $\mu(a,b] = M(b) - M(a)$. This measure is supported on the set of record times. We know that, with probability 1, the Brownian motion is Hölder continuous with any exponent $\alpha < 1/2$. Thus

$$M(b) - M(a) \leq \max_{0 \leq h \leq b-a} B(a+h) - B(a) \leq C_\alpha (b-a)^\alpha,$$

where $\alpha < 1/2$ and C_α is some random constant that doesn't depend on a or b. The Mass Distribution Principle implies that $\dim\{t \in [0,1] : Y(t) = 0\} \geq \alpha$ almost surely. By choosing a sequence $\alpha_n \uparrow 1/2$ we finish the proof. $\square$

Recall that the upper Minkowski dimension of a set is an upper bound for the Hausdorff dimension. To estimate the Minkowski dimension of Z_B we will need to know

$$\mathbb{P}(\exists t \in (a, a+\varepsilon) : B(t) = 0). \tag{6.10.1}$$

Finding the exact value is Exercise 6.14. However, for our purposes, the following estimate will suffice.

Lemma 6.10.4   *For any* $a, \varepsilon > 0$ *we have*

$$\mathbb{P}(\exists t \in (a, a+\varepsilon) : B(t) = 0) \leq C\sqrt{\frac{\varepsilon}{a+\varepsilon}},$$

for some appropriate positive constant C.

Proof Consider the event A given by $|B(a+\varepsilon)| \leq \sqrt{\varepsilon}$. By the scaling property of the Brownian motion, we can give the upper bound

$$\mathbb{P}(A) = \mathbb{P}\left(|B(1)| \leq \sqrt{\frac{\varepsilon}{a+\varepsilon}}\right) \leq 2\sqrt{\frac{\varepsilon}{a+\varepsilon}}. \tag{6.10.2}$$

However, knowing that Brownian motion has a zero in $(a, a+\varepsilon)$ makes the event A very likely. Indeed, we certainly have

$$\mathbb{P}(A) \geq \mathbb{P}(A \text{ and } 0 \in B[a, a+\varepsilon]),$$

and the strong Markov property implies that

$$\mathbb{P}(A) \geq \tilde{c}\mathbb{P}(0 \in B[a, a+\varepsilon]), \tag{6.10.3}$$

where

$$\tilde{c} = \min_{a \leq t \leq a+\varepsilon} \mathbb{P}(A|B(t) = 0).$$

Because the minimum is achieved when $t = a$, we have

$$\tilde{c} = \mathbb{P}(|B(1)| \leq 1) > 0,$$

by the scaling property of the Brownian motion. From inequalities (6.10.2) and (6.10.3), we conclude

$$\mathbb{P}(0 \in B[a, a+\varepsilon]) \leq \frac{2}{\tilde{c}}\sqrt{\frac{\varepsilon}{a+\varepsilon}}. \qquad \square$$

For any, possibly random, closed set $A \subset [0, 1]$, define a function

$$N_m(A) = \sum_{k=1}^{2^m} \mathbf{1}_{\left\{A \cap \left[\frac{k-1}{2^m}, \frac{k}{2^m}\right] \neq \emptyset\right\}}.$$

This function counts the number of intervals of the form $\left[\frac{k-1}{2^m}, \frac{k}{2^m}\right]$ intersected by the set A and so is a natural object if we want to compute the Minkowski dimension of A. In the special case where $A = Z_B$ we have

$$N_m(Z_B) = \sum_{k=1}^{2^m} \mathbf{1}_{\left\{0 \in B\left[\frac{k-1}{2^m}, \frac{k}{2^m}\right]\right\}}.$$

The next lemma shows that estimates on the expected value of $N_m(A)$ will give us bounds on the Minkowski dimension (and hence on the Hausdorff dimension).

Lemma 6.10.5 *Suppose A is a closed random subset of $[0, 1]$ such that*

$$\mathbb{E}N_m(A) \leq c2^{m\alpha},$$

for some $c, \alpha > 0$. Then $\overline{\dim}_{\mathcal{M}}(A) \leq \alpha$.

Proof Consider

$$\mathbb{E} \sum_{m=1}^{\infty} \frac{N_m(A)}{2^{m(\alpha+\varepsilon)}},$$

for $\varepsilon > 0$. Then, by the Monotone Convergence Theorem,

$$\mathbb{E} \sum_{m=1}^{\infty} \frac{N_m(A)}{2^{m(\alpha+\varepsilon)}} = \sum_{m=1}^{\infty} \frac{\mathbb{E} N_m(A)}{2^{m(\alpha+\varepsilon)}} < \infty.$$

This estimate implies that

$$\sum_{m=1}^{\infty} \frac{N_m(A)}{2^{m(\alpha+\varepsilon)}} < \infty \quad \text{a.s.,}$$

and so, with probability 1,

$$\limsup_{m\to\infty} \frac{N_m(A)}{2^{m(\alpha+\varepsilon)}} = 0.$$

From the last equation follows

$$\overline{\dim}_{\mathcal{M}}(A) \leq \alpha + \varepsilon, \quad \text{a.s.}$$

Let $\varepsilon \to 0$ through some countable sequence to get

$$\overline{\dim}_{\mathcal{M}}(A) \leq \alpha, \quad \text{a.s.}$$

And this completes the proof of the lemma.   □

To get an upper bound on the Hausdorff dimension of Z_B note that

$$\mathbb{E} N_m(Z_B) \leq C \sum_{k=1}^{2^m} \frac{1}{\sqrt{k}} \leq \tilde{C} 2^{m/2},$$

since $\mathbb{P}\left(\exists t \in \left[\frac{k-1}{2^m}, \frac{k}{2^m}\right] : B(t) = 0\right) \leq \frac{C}{\sqrt{k}}$. Therefore, by the previous lemma, $\overline{\dim}_{\mathcal{M}}(Z_B) \leq 1/2$ a.s. This implies immediately $\dim(Z_B) \leq 1/2$ a.s. Combining this estimate with Lemma 6.10.3 we have

Theorem 6.10.6 *With probability 1 we have*

$$\dim(Z_B) = \frac{1}{2}.$$

6.11 Harris' inequality and its consequences

Lemma 6.11.1 (Harris (1960)) *Suppose that $\mu_1, \ldots, \mu_d$ are Borel probability measures on $\mathbb{R}$ and $\mu = \mu_1 \times \mu_2 \times \cdots \times \mu_d$. Let $f, g : \mathbb{R}^d \to \mathbb{R}$ be measurable functions that are non-decreasing in each coordinate. Then,*

$$\int_{\mathbb{R}^d} f(x)g(x)\,d\mu \geq \left(\int_{\mathbb{R}^d} f(x)\,d\mu \right) \left(\int_{\mathbb{R}^d} g(x)\,d\mu \right), \tag{6.11.1}$$

provided the above integrals are well defined.

Proof One can argue, using the Monotone Convergence Theorem, that it suffices to prove the result when f and g are bounded. We assume f and g are bounded and proceed by induction. Suppose $d = 1$. Note that

$$(f(x) - f(y))(g(x) - g(y)) \geq 0$$

for all $x, y \in \mathbb{R}$. Therefore,

$$0 \leq \int_{\mathbb{R}} \int_{\mathbb{R}} (f(x) - f(y))(g(x) - g(y))\,d\mu(x)\,d\mu(y)$$
$$= 2 \int_{\mathbb{R}} f(x)g(x)\,d\mu(x) - 2 \left(\int_{\mathbb{R}} f(x)\,d\mu(x) \right) \left(\int_{\mathbb{R}} g(y)\,d\mu(y) \right),$$

and (6.11.1) follows easily. Now, suppose (6.11.1) holds for $d - 1$. Define

$$f_1(x_1) = \int_{\mathbb{R}^{d-1}} f(x_1, \ldots, x_d)\,d\mu_2(x_2) \ldots d\mu_d(x_d),$$

and define g_1 similarly. Note that $f_1(x_1)$ and $g_1(x_1)$ are non-decreasing functions of x_1. Since f and g are bounded, we may apply Fubini's Theorem to write the left-hand side of (6.11.1) as

$$\int_{\mathbb{R}} \left(\int_{\mathbb{R}^{d-1}} f(x_1, \ldots, x_d)g(x_1, \ldots, x_d)\,d\mu_2(x_2) \ldots d\mu_d(x_d) \right) d\mu_1(x_1). \tag{6.11.2}$$

The integral in the parenthesis is at least $f_1(x_1)g_1(x_1)$ by the induction hypothesis. Thus, by using the result for the $d = 1$ case we can bound (6.11.2) below by

$$\left(\int_{\mathbb{R}} f_1(x_1)\,d\mu_1(x_1) \right) \left(\int_{\mathbb{R}} g_1(x_1)\,d\mu_1(x_1) \right),$$

which equals the right-hand side of (6.11.1), completing the proof. $\square$

Example 6.11.2 We say an event $A \subset \mathbb{R}^d$ is an **increasing event** if

$$(x_1, \ldots, x_{i-1}, x_i, x_{i+1}, \ldots x_d) \in A,$$

and $\tilde{x}_i \geq x_i$ imply that

$$(x_1, \ldots, x_{i-1}, \tilde{x}_i, x_{i+1}, \ldots x_d) \in A.$$

If A and B are increasing events, then it is easy to see by applying Harris' Inequality to the indicator functions $\mathbf{1}_A$ and $\mathbf{1}_B$ that $\mathbb{P}(A \cap B) \geq \mathbb{P}(A)\mathbb{P}(B)$.

Example 6.11.3 Let $X_1, \ldots, X_n$ be an i.i.d. sample, where each X_i has distribution μ. Given any $(x_1, \ldots, x_n) \in \mathbb{R}^n$, relabeling the points in decreasing order $x_{(1)} \geq x_{(2)} \geq \cdots \geq x_{(n)}$. Fix i and j, and define $f(x_1, \ldots, x_n) = x_{(i)}$ and $g(x_1, \ldots, x_n) = x_{(j)}$. Then f and g are measurable and non-decreasing in each component. Let $X_{(i)}$ and $X_{(j)}$ denote the ith and jth order statistics of $X_1, \ldots, X_n$. It follows from Harris' Inequality that $\mathbb{E}[X_{(i)}X_{(j)}] \geq \mathbb{E}[X_{(i)}]\mathbb{E}[X_{(j)}]$, provided these expectations are well defined. See Lehmann (1966) and Bickel (1967) for further discussion.

For the rest of this section, let $X_1, X_2, \ldots$ be i.i.d. random variables, and let $S_k = \sum_{i=1}^k X_i$ be their partial sums. Denote

$$p_n = \mathbb{P}(S_i \geq 0 \text{ for all } 1 \leq i \leq n). \tag{6.11.3}$$

Observe that the event that $\{S_n \text{ is the largest among } S_0, S_1, \ldots, S_n\}$ is precisely the event that the reversed random walk $X_n + \cdots + X_{n-k+1}$ is non-negative for all $k = 1, \ldots, n$; thus this event also has probability p_n. The following theorem gives the order of magnitude of p_n.

Theorem 6.11.4 *If the increments X_i have a symmetric distribution (that is, $X_i \overset{d}{=} -X_i$) or have mean zero and finite variance, then there are positive constants C_1 and C_2 such that $C_1 n^{-1/2} \leq p_n \leq C_2 n^{-1/2}$ for all $n \geq 1$.*

Proof For the general argument, see Feller (1966), Section XII.8. We prove the result here for the simple random walk; that is, when each X_i takes values ± 1 with probability half each.

Define the stopping time $\tau_{-1} = \min\{k : S_k = -1\}$. Then

$$p_n = \mathbb{P}(S_n \geq 0) - \mathbb{P}(S_n \geq 0, \tau_{-1} < n).$$

Let $\{S_j^*\}$ denote the random walk reflected at time τ_{-1}, that is

$$\begin{aligned} S_j^* &= S_j && \text{for } j \leq \tau_{-1}, \\ S_j^* &= (-1) - (S_j + 1) && \text{for } j > \tau_{-1}. \end{aligned}$$

Note that if $\tau_{-1} < n$ then $S_n \geq 0$ if and only if $S_n^* \leq -2$, so

$$p_n = \mathbb{P}(S_n \geq 0) - \mathbb{P}(S_n^* \leq -2).$$

Using symmetry and the reflection principle, we have

$$p_n = \mathbb{P}(S_n \geq 0) - \mathbb{P}(S_n \geq 2) = \mathbb{P}(S_n \in \{0,1\}),$$

which means that

$$
\begin{aligned}
p_n &= \mathbb{P}(S_n = 0) &= \binom{n}{n/2}2^{-n} & \quad \text{for } n \text{ even,} \\
p_n &= \mathbb{P}(S_n = 1) &= \binom{n}{(n-1)/2}2^{-n} & \quad \text{for } n \text{ odd.}
\end{aligned}
$$

Recall that Stirling's Formula gives $m! \sim \sqrt{2\pi} m^{m+1/2} e^{-m}$, where the symbol $\sim$ means that the ratio of the two sides approaches 1 as $m \to \infty$. One can deduce from Stirling's Formula that

$$p_n \sim \sqrt{\frac{2}{\pi n}},$$

which proves the theorem. $\qquad\square$

The following theorem expresses, in terms of the p_i, the probability that S_j stays between 0 and S_n for j between 0 and n. It will be used in the next section.

Theorem 6.11.5 *We have* $p_n^2 \leq \mathbb{P}(0 \leq S_j \leq S_n \text{ for all } 1 \leq j \leq n) \leq p_{\lfloor n/2 \rfloor}^2$.

Proof The two events

$$
\begin{aligned}
A &= \{0 \leq S_j \text{ for all } j \leq n/2\} \text{ and} \\
B &= \{S_j \leq S_n \text{ for all } j \geq n/2\}
\end{aligned}
$$

are independent, since A depends only on $X_1,\ldots,X_{\lfloor n/2 \rfloor}$ and B depends only on the remaining $X_{\lfloor n/2 \rfloor+1},\ldots,X_n$. Therefore,

$$\mathbb{P}(0 \leq S_j \leq S_n \text{ for all } 0 < j < n) \leq \mathbb{P}(A \cap B) = \mathbb{P}(A)\mathbb{P}(B) \leq p_{\lfloor n/2 \rfloor}^2,$$

which proves the upper bound.

To prove the lower bound, we let $f(x_1,\ldots,x_n) = 1$ if $x_1 + \cdots + x_k \geq 0$ for every $k = 1,\ldots,n$, and $f(x_1,\ldots,x_n) = 0$ otherwise. Reversing the order of the variables, we define $g(x_1,\ldots,x_n) = f(x_n,\ldots,x_1)$. Then f and g are non-decreasing in each component. Let μ_j be the distribution of X_j, and let $\mu = \mu_1 \times \cdots \times \mu_n$. By Harris' Inequality,

$$\int_{\mathbb{R}^n} fg \, d\mu \geq \left(\int_{\mathbb{R}^n} f \, d\mu\right)\left(\int_{\mathbb{R}^n} g \, d\mu\right) = p_n^2.$$

Also, let $X \subset \mathbb{R}^n$ be the set such that for all j,

$$x_1 + \cdots + x_j \geq 0 \text{ and } x_{j+1} + \cdots + x_n \geq 0.$$

Then

$$\int_{\mathbb{R}^n} fg\, d\mu = \int_{\mathbb{R}^n} \mathbf{1}_X \, d\mu = \mathbb{P}(0 \le S_j \le S_n \text{ for all } 1 \le j \le n),$$

which proves the lower bound.   □

6.12 Points of increase

The material in this section has been taken, with minor modifications, from Peres (1996a).

A real-valued function f has a **global point of increase** in the interval $(\mathbf{a}, \mathbf{b})$ if there is a point $t_0 \in (a, b)$ such that $f(t) \le f(t_0)$ for all $t \in (a, t_0)$ and $f(t_0) \le f(t)$ for all $t \in (t_0, b)$. We say t_0 is a **local point of increase** if it is a global point of increase in some interval. Dvoretzky et al. (1961) proved that Brownian motion almost surely has no global points of increase in any time interval, or, equivalently, that Brownian motion has no local points of increase. Knight (1981) and Berman (1983) noted that this follows from properties of the local time of Brownian motion; direct proofs were given by Adelman (1985) and Burdzy (1990). Here we show that the non-increase phenomenon holds for arbitrary symmetric random walks, and can thus be viewed as a combinatorial consequence of fluctuations in random sums.

Definition   Say that a sequence of real numbers $s_0, s_1, \ldots, s_n$ has a (global) **point of increase** at index k if $s_i \le s_k$ for $i = 0, 1, \ldots, k-1$ and $s_k \le s_j$ for $j = k+1, \ldots, n$.

Theorem 6.12.1   *Let $S_0, S_1, \ldots, S_n$ be a random walk where the i.i.d. increments $X_i = S_i - S_{i-1}$ have a symmetric distribution, or have mean 0 and finite variance. Then*

$$\mathbb{P}(S_0, \ldots, S_n \text{ has a point of increase}) \le \frac{C}{\log n},$$

for $n > 1$, where C does not depend on n.

We will now see how this result implies the following

Corollary 6.12.2   *Brownian motion almost surely has no points of increase.*

Proof   To deduce this, it suffices to apply Theorem 6.12.1 to a simple random walk on the integers. Indeed it clearly suffices to show that the Brownian motion $\{B(t)\}_{t \ge 0}$ almost surely has no global points of increase in a fixed rational time interval (a, b). Sampling the Brownian motion when it visits a lattice yields a simple random walk; by refining the lattice, we may make this walk as long as we wish, which will complete the proof. More precisely, for any vertical spacing $h > 0$ define τ_0 to be the first $t \geq a$ such that $B(t)$ is an integral multiple of h, and for $i \geq 0$ let τ_{i+1} be the minimal $t \geq \tau_i$ such that $|B(t) - B(\tau_i)| = h$. Define $N_b = \max\{k \in \mathbb{Z} : \tau_k \leq b\}$. For integers i satisfying $0 \leq i \leq N_b$, define

$$S_i = \frac{B(\tau_i) - B(\tau_0)}{h}.$$

Then, $\{S_i\}_{i=1}^{N_b}$ is a finite portion of a simple random walk. If the Brownian motion has a (global) point of increase in (a, b) at t_0, and if k is an integer such that $\tau_{k-1} \leq t_0 \leq \tau_k$, then this random walk has points of increase at $k - 1$ and k. Similarly, if $t_0 < \tau_0$ or $t_0 > \tau_N$, then $k = 0$, resp. $k = N_b$, is a point of increase for the random walk. Therefore, for all n,

$\mathbb{P}($ Brownian motion has a global point of increase in $(a, b))$

$$\leq \mathbb{P}(N_b \leq n) + \sum_{m=n+1}^{\infty} \mathbb{P}(S_0, \ldots, S_m \text{ has a point of increase, and } N_b = m).$$

$$(6.12.1)$$

Note that $N_b \leq n$ implies $|B(b) - B(a)| \leq (n+1)h$, so

$$\mathbb{P}(N_b \leq n) \leq \mathbb{P}(|B(b) - B(a)| \leq (n+1)h) = \mathbb{P}\left(|Z| \leq \frac{(n+1)h}{\sqrt{b-a}}\right),$$

where Z has a standard normal distribution. Since $S_0, \ldots, S_m$, conditioned on $N_b = m$ is a finite portion of a simple random walk, it follows from Theorem 6.12.1 that for some constant C, we have

$$\sum_{m=n+1}^{\infty} \mathbb{P}(S_0, \ldots, S_m \text{ has a point of increase, and } N_b = m)$$

$$\leq \sum_{m=n+1}^{\infty} \mathbb{P}(N_b = m) \frac{C}{\log m} \leq \frac{C}{\log(n+1)}.$$

Thus, the probability in (6.12.1) can be made arbitrarily small by first taking n large and then picking $h > 0$ sufficiently small. $\qquad\square$

To prove Theorem 6.12.1, we prove first

Theorem 6.12.3 *For any random walk $\{S_j\}$ on the line,*

$$\mathbb{P}(S_0, \ldots, S_n \text{ has a point of increase}) \leq 2 \frac{\sum_{k=0}^{n} p_k p_{n-k}}{\sum_{k=0}^{\lfloor n/2 \rfloor} p_k^2},$$

$$(6.12.2)$$

where p_n are defined as in (6.11.3).

Proof   The idea is simple. The expected number of points of increase is the numerator in (6.12.2), and given that there is at least one such point, the expected number is bounded below by the denominator; the ratio of these expectations bounds the required probability.

To carry this out, denote by $I_n(k)$ the event that k is a point of increase for $S_0, S_1, \ldots, S_n$ and by $F_n(k) = I_n(k) \setminus \bigcup_{i=0}^{k-1} I_n(i)$ the event that k is the first such point. The events that $\{S_k$ is largest among $S_0, S_1, \ldots, S_k\}$ and that $\{S_k$ is smallest among $S_k, S_{k+1}, \ldots, S_n\}$ are independent, and therefore $\mathbb{P}(I_n(k)) = p_k p_{n-k}$.

Observe that if S_j is minimal among $S_j, \ldots, S_n$, then any point of increase for $S_0, \ldots, S_j$ is automatically a point of increase for $S_0, \ldots, S_n$. Therefore for $j \leq k$, $F_n(j) \cap I_n(k)$ is equal to

$$F_j(j) \cap \{S_j \leq S_i \leq S_k \ \forall \, i \in [j,k]\} \cap \{S_k = \min(S_k, \ldots, S_n)\}. \quad (6.12.3)$$

The three events on the right-hand side are independent, as they involve disjoint sets of summands; the second of these events is of the type considered in Theorem 6.11.5. Thus,

$$\mathbb{P}(F_n(j) \cap I_n(k)) \geq \mathbb{P}(F_j(j)) \, p_{k-j}^2 \, p_{n-k}$$

$$\geq p_{k-j}^2 \, \mathbb{P}(F_j(j)) \, \mathbb{P}(S_j \text{ is minimal among } S_j, \ldots, S_n) \,,$$

since $p_{n-k} \geq p_{n-j}$. Here the two events on the right are independent, and their intersection is precisely $F_n(j)$. Consequently, $\mathbb{P}(F_n(j) \cap I_n(k)) \geq p_{k-j}^2 \mathbb{P}(F_n(j))$.

Decomposing the event $I_n(k)$ according to the first point of increase gives

$$\sum_{k=0}^{n} p_k p_{n-k} = \sum_{k=0}^{n} \mathbb{P}(I_n(k)) = \sum_{k=0}^{n} \sum_{j=0}^{k} \mathbb{P}(F_n(j) \cap I_n(k)) \quad (6.12.4)$$

$$\geq \sum_{j=0}^{\lfloor n/2 \rfloor} \sum_{k=j}^{j+\lfloor n/2 \rfloor} p_{k-j}^2 \mathbb{P}(F_n(j)) = \sum_{j=0}^{\lfloor n/2 \rfloor} \mathbb{P}(F_n(j)) \sum_{i=0}^{\lfloor n/2 \rfloor} p_i^2 \,.$$

This yields an upper bound on the probability that $\{S_j\}_{j=0}^{n}$ has a point of increase by time $n/2$; but this random walk has a point of increase at time k if and only if the "reversed" walk $\{S_n - S_{n-i}\}_{i=0}^{n}$ has a point of increase at time $n - k$. Thus, doubling the upper bound given by (6.12.4) proves the theorem. $\qquad \square$

Proof of Theorem 6.12.1.   To bound the numerator in (6.12.2), we can use symmetry to deduce from Theorem 6.11.4 that

$$\sum_{k=0}^{n} p_k p_{n-k} \leq 2 + 2 \sum_{k=1}^{\lfloor n/2 \rfloor} p_k p_{n-k}$$

$$\leq 2 + 2C_2^2 \sum_{k=1}^{\lfloor n/2 \rfloor} k^{-1/2}(n-k)^{-1/2}$$

$$\leq 2 + 4C_2^2 n^{-1/2} \sum_{k=1}^{\lfloor n/2 \rfloor} k^{-1/2},$$

which is bounded above because the last sum is $O(n^{1/2})$. Since Theorem 6.11.4 implies that the denominator in (6.12.2) is at least $C_1^2 \log \lfloor n/2 \rfloor$, this completes the proof. $\qquad \square$

The following theorem shows that we can obtain a lower bound on the probability that a random walk has a point of increase that differs from the upper bound only by a constant factor.

Theorem 6.12.4   *For any random walk on the line,*

$$\mathbb{P}(S_0, \dots, S_n \text{ has a point of increase}) \geq \frac{\sum_{k=0}^{n} p_k p_{2n-k}}{2 \sum_{k=0}^{\lfloor n/2 \rfloor} p_k^2}. \tag{6.12.5}$$

In particular if the increments have a symmetric distribution, or have mean 0 and finite variance, then $\mathbb{P}(S_0, \dots, S_n \text{ has a point of increase}) \asymp 1/\log n$ *for* $n > 1$, *where the symbol* $\asymp$ *means that the ratio of the two sides is bounded above and below by positive constants that do not depend on* n.

Proof   Using (6.12.4), we get

$$\sum_{k=0}^{n} p_k p_{2n-k} = \sum_{k=0}^{n} \mathbb{P}(I_{2n}(k)) = \sum_{k=0}^{n} \sum_{j=0}^{k} \mathbb{P}(F_{2n}(j) \cap I_{2n}(k)).$$

Using Theorem 6.11.5, we see that for $j \leq k \leq n$, we have

$$\mathbb{P}(F_{2n}(j) \cap I_{2n}(k)) \leq \mathbb{P}(F_n(j) \cap \{S_j \leq S_i \leq S_k \text{ for } j \leq i \leq k\})$$
$$\leq \mathbb{P}(F_n(j)) p_{\lfloor (k-j)/2 \rfloor}^2.$$

Thus,

$$\sum_{k=0}^{n} p_k p_{2n-k} \leq \sum_{k=0}^{n} \sum_{j=0}^{k} \mathbb{P}(F_n(j)) p_{\lfloor (k-j)/2 \rfloor}^2 \leq \sum_{j=0}^{n} \mathbb{P}(F_n(j)) \sum_{i=0}^{n} p_{\lfloor i/2 \rfloor}^2.$$

This implies (6.12.5). The assertion concerning symmetric or mean 0, finite variance walks follows from Theorem 6.11.4 and the proof of Theorem 6.12.1. $\qquad \square$

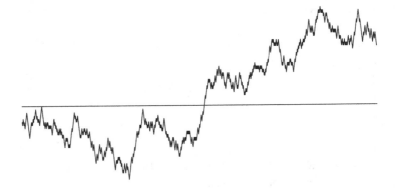

Figure 6.12.1 A simple random walk (1000 steps) with a point of increase (shown by the horizontal line). After generating 10,000 1000-step random walks, 2596 had at least one point of increase. Equations (6.12.5) and (6.12.2) put the probability of a 1000-step walk having a point of increase between .2358 and .9431, so the lower bound seems more accurate.

6.13 Notes

The physical phenomenon of Brownian motion was observed by the botanist Robert Brown (1828), who reported on the random movement of particles suspended in water. The survey by Duplantier (2006) gives a detailed account of subsequent work in physics by Sutherland, Einstein, Perrin and others. In particular, Einstein (1905) was crucial in establishing the atomic view of matter. The first mathematical study of Brownian motion is due to Bachelier (1900) in the context of modelling stock market fluctuations. The first rigorous construction of mathematical Brownian motion is due to Wiener (1923); indeed, the distribution of standard Brownian motion on the space of continuous functions equipped with the Borel σ-algebra (arising from the topology of uniform convergence) is called **Wiener measure**.

Mathematically, Brownian motion describes the macroscopic picture emerging from a random walk if its increments are sufficiently tame not to cause jumps which are visible in the macroscopic description.

The construction of Brownian motion via interpolation given in the text is due to Paul Lévy (see (Lévy, 1948)). An alternative is to first show that a uniformly continuous Markov process with the correct transition probabilities at rational times can be constructed, using, e.g., Kolmogorov's criterion and then extending to a continuous process defined at real times. See, for example,

Revuz and Yor (1994), Karatzas and Shreve (1991) and Kahane (1985) for further alternative constructions.

Theorem 6.4.3 states that $B[0,1]$ has Hausdorff dimension 2 for Brownian motion in $\mathbb{R}^d$, $d \geq 2$. A more precise result is that $B[0,1]$ has finite, positive $\mathscr{H}^\varphi$ measure for $\varphi(t) = t^2 \log \frac{1}{t} \log\log\log \frac{1}{t}$ when $d = 2$ and for $\varphi(t) = t^2 \log\log \frac{1}{t}$ when $d \geq 3$. Moreover, $\mathscr{H}^\varphi$ measure on the path is proportional to the occupation time (the image of Lebesgue measure on $[0, 1]$). The higher-dimensional case was established in Ciesielski and Taylor (1962), and the planar case in Ray (1963) (for the lower bound) and Taylor (1964) (for the upper bound). Le Gall (1987) gives the corresponding gauge functions for k-multiple points of Brownian motion (points hit by Brownian motion k times). In $d = 2$ the formula is

$$\varphi(t) = t^2 \left[\log \frac{1}{t} \log\log\log \frac{1}{t} \right]^k,$$

and for $d = 3$ and $k = 2$

$$\varphi_3(t) = t^2 \left[\log\log \frac{1}{t} \right]^2.$$

For $d = 3$, triple points do not exist (Dvoretzky et al., 1950) and for $d \geq 4$, double points almost surely do not exist.

From the proof of Theorem 6.10.6 we can also infer that $\mathscr{H}^{1/2}(Z_B) < \infty$ almost surely. If we set

$$\varphi(r) = \sqrt{r} \log\log \frac{1}{r},$$

then Taylor and Wendel (1966) prove that $0 < \mathscr{H}^\varphi(Z) < \infty$ almost surely. Also see Mörters and Peres (2010), Theorem 6.43.

Theorem 6.5.5 is due to Hunt (1994).

The notion of negligible sets was invented by Christensen (1972) in the context of Polish groups (separable topological groups with a compatible complete metric) using the name "sets of Haar measure zero" and was rediscovered twenty years later by Hunt et al. (1992) in the more specialized setting of infinite-dimensional vector spaces. Negligible sets are also known as shy sets.

There are two possible definitions of prevalent in a Banach space X:

(1) There is a Borel probability measure μ and a Borel subset B of A such that $\mu(B + x) = 1$ for all $x \in X$. (This is the one in the text.)

(2) There is a finite Borel measure μ such that for every $x \in X$ there is a Borel subset $B_x \subset A + x$ with $\mu(B_x) = 1$.

Clearly (1) implies (2), and Solecki (1996) proves these are same for analytic sets A (for the definition of analytic, see Appendix B), but Elekes and

Steprāns (2014) give an example of a set $A \subset \mathbb{R}$ that satisfies (2) but not (1). Co-analytic examples in certain non-locally compact groups are given in Elekes and Vidnyánszky (2015).

The second author heard the following story from Shizuo Kakutani about the paper Dvoretzky et al. (1961). Erdős was staying at Dvoretzky's apartment in New York City and Kakutani drove down from Yale and the three of them constructed a proof that Brownian motion has points of increase. Satisfied with the proof, Kakutani left to go home, but, despite repeated attempts, could not get his car to start. It was too late to find a mechanic, so he spent the night at Dvoretzky's apartment with the others, filling in the remaining details of the existence proof until it turned into a non-existence proof around 2am. In the morning the car started perfectly on the first attempt (but if it hadn't, we might not have Exercise 5.6).

A natural refinement of the results in Section 6.12 is the question whether, for Brownian motion in the plane, there exists a line such that the Brownian motion path, projected onto that line, has a global point of increase. An equivalent question is whether the Brownian motion path admits cut-lines. (We say a line ℓ is a **cut-line** for the Brownian motion if, for some t_0, $B(t)$ lies on one side of ℓ for all $t < t_0$ and on the other side of ℓ for all $t > t_0$.) It was proved by Bass and Burdzy (1999) that Brownian motion almost surely does not have cut-lines. The same paper also shows that for Brownian motion in three dimensions, there almost surely exist cut-planes, where we say P is a **cut-plane** if, for some t_0, $B(t)$ lies on one side of the plane for $t < t_0$ and on the other side for $t > t_0$ (this was first proved by R. Pemantle (unpublished)).

Burdzy (1989) showed that Brownian motion in the plane almost surely does have **cut-points**; these are points $B(t_0)$ such that the Brownian motion path with the point $B(t_0)$ removed is disconnected. Lawler et al. (2001c) proved that the Hausdorff dimension of the set of cut-points is $3/4$.

Pemantle (1997) has shown that a Brownian motion path almost surely does not cover any straight line segment. Which curves can be covered by a Brownian motion path is, in general, an open question. Also unknown is the minimal Hausdorff dimension of curves in a typical Brownian motion path. Burdzy and Lawler (1990) showed this minimal dimension to be at most $3/2 - 1/4\pi^2 \approx 1.47$ and this was improved to $4/3$ by Lawler et al. (2001a), who computed the dimension of the Brownian frontier (the boundary of the unbounded component in the complement of the planar Brownian path $B([0, 1])$). This was improved further to $5/4$ by Zhan (2011).

The lower bound argument in Theorem 6.11.5 is from Peres (1996a).

6.14 Exercises

Exercise 6.1   Let $A \subset [0,1]$ be closed and let B_1, B_2 be independent standard Brownian motions in $\mathbb{R}$. Show that $\dim B_1(B_2(A)) = 4\dim(A) \wedge 1$ a.s. Does the same hold for $B_1(B_1(A))$?

Exercise 6.2   Prove that double points are dense in $B[0,1]$ for 2-dimensional Brownian motion. (A double point is a point visited at least twice by the Brownian path.)

• **Exercise 6.3**   Use the previous exercise to show that double points are dense on boundary of each complementary component of planar $B[0,1]$.

Exercise 6.4   We say that two density functions f, g have a monotone likelihood ratio (MLR) if $\frac{f}{g}$ is non-decreasing ($\frac{0}{0} := 0$). We say that a function $h : \mathbb{R} \to \mathbb{R}$ is log-concave if $\log h$ is concave. Let f, g be two densities that have a MLR. Use the Harris inequality to show that f stochastically dominates g, i.e., for any $c \in \mathbb{R}$, $\int_c^\infty g(x)\,dx \leq \int_c^\infty f(x)\,dx$.

Exercise 6.5   Let f, g be continuous densities that have a MLR. Let $h \in C^1$ be a log-concave density function with $\int_{-\infty}^\infty |h'(y)|\,dy < \infty$. Show that the convolutions $f * h$ and $g * h$ also have a MLR. (Here $f * h(y) := \int_{-\infty}^\infty f(x)h(y-x)\,dx$.)

Exercise 6.6   Let $h_1, h_2 : [0,1] \to \mathbb{R}$ be continuous functions such that $h_1 \leq h_2$. Show that a Brownian motion conditioned to remain above h_2 will stochastically dominate at time 1 a Brownian motion conditioned to remain above h_1. In other words, if $A_i := \{B(x) \geq h_i(x), 0 \leq x \leq 1\}, i = 1, 2$, then for any $c \in \mathbb{R}$, $\mathbb{P}[B(1) \geq c|A_1] \leq \mathbb{P}[B(1) \geq c|A_2])$.

Exercise 6.7   Prove the Law of Large Numbers for Brownian motion (Corollary 6.3.1) directly. Use the usual Law of Large Numbers to show that

$$\lim_{n \to \infty} \frac{B(n)}{n} = 0.$$

Then show that $B(t)$ does not oscillate too much between n and $n + 1$.

• **Exercise 6.8**   Show that a Brownian motion is a.s. not $\frac{1}{2}$-Hölder.

Exercise 6.9   Show that for all t, $\mathbb{P}(t$ is a local maximum$) = 0$, but almost surely local maxima are a countable dense set in $(0, \infty)$.

• **Exercise 6.10**   Let $\alpha > 1/2$. Show that a.s. for all $t > 0$, there exists $h > 0$ such that $|B(t+h) - B(t)| > h^\alpha$.

Exercise 6.11   Show that almost surely, if $B(t_0) = \max_{0 \leq t \leq 1} B(t)$ then $D^* B(t_0) = -\infty$.

● **Exercise 6.12**   Prove Corollary 6.5.5: the set of nowhere differentiable functions is prevalent in $C([0,1])$.

● **Exercise 6.13**   Prove that if A is a closed set then $\tau_A = \inf\{t : B(t) \in A\}$ is a stopping time.

● **Exercise 6.14**   Compute the exact value of (6.10.1), i.e., show

$$\mathbb{P}(\exists t \in (a, a + \varepsilon) : B(t) = 0) = \frac{2}{\pi}\arctan\sqrt{\frac{\varepsilon}{a}}.$$

● **Exercise 6.15**   Numerically estimate the ratio that appears in both equations (6.12.5) and (6.12.2) for $n = 10^3, \ldots, 10^6$.

7

Brownian motion, Part II

We continue the discussion of Brownian motion, focusing more on higher dimensions and potential theory in this chapter. We start with a remarkable result of Robert Kaufman stating that with probability 1, Brownian motion in dimensions ≥ 2 simultaneously doubles the dimension of every time set. We then prove the Law of the Iterated Logarithm for Brownian motion and use Skorokhod's representation to deduce the LIL for discrete random walks. Next comes Donsker's Invariance Principle that implies Brownian motion is the limit of a wide variety of i.i.d. random walks. We end the chapter discussing the close connection between Brownian motion and potential theory. In particular, we will solve the Dirichlet problem, discuss the recurrence/transience properties of Brownian motion in $\mathbb{R}^d$ and prove it is conformally invariant in two dimensions.

7.1 Dimension doubling

We introduced the idea of dimension doubling in the previous chapter when computing the dimension of the Brownian paths and graphs. Now we return to the topic to give the proofs of results stated earlier. In particular, we apply the following converse to the Frostman's Energy Method (Theorem 6.4.6).

Theorem 7.1.1 (Frostman, 1935)   *If $K \subset \mathbb{R}^d$ is closed and $\dim(K) > \alpha$, then there exists a Borel probability measure μ on K such that*

$$\mathscr{E}_\alpha(\mu) = \int_{\mathbb{R}^d} \int_{\mathbb{R}^d} \frac{d\mu(x)d\mu(y)}{|x-y|^\alpha} < \infty.$$

The above theorem can be deduced by applying Theorem 3.4.2(i) to intersections of K and closed balls of diameter tending to infinity.

Theorem 7.1.2 (McKean, 1955)   *Let B_d denote Brownian motion in $\mathbb{R}^d$. Let $A \subset [0, \infty)$ be a closed set such that $\dim(A) \leq d/2$. Then $\dim B(A) = 2\dim(A)$ almost surely.*

Proof   Let $\alpha < \dim(A)$. By Theorem 7.1.1, there exists a Borel probability measure μ on A such that $\mathscr{E}_\alpha(\mu) < \infty$. Denote by μ_B the random measure on $\mathbb{R}^d$ defined by

$$\mu_B(D) = \mu(B_d^{-1}(D)) = \mu(\{t : B_d(t) \in D\})$$

for all Borel sets D. Then

$$\mathbb{E}[\mathscr{E}_{2\alpha}(\mu_B)] = \mathbb{E}\left[\int_{\mathbb{R}^d}\int_{\mathbb{R}^d} \frac{d\mu_B(x)d\mu_B(y)}{|x-y|^{2\alpha}}\right] = \mathbb{E}\left[\int_{\mathbb{R}}\int_{\mathbb{R}} \frac{d\mu(t)d\mu(s)}{|B_d(t) - B_d(s)|^{2\alpha}}\right],$$

where the second equality can be verified by a change of variables. Note that the denominator on the right-hand side has the same distribution as $|t-s|^\alpha |Z|^{2\alpha}$, where Z is a d-dimensional standard normal random variable. Since $2\alpha < d$ we have:

$$\mathbb{E}[|Z|^{-2\alpha}] = \frac{1}{(2\pi)^{d/2}} \int_{\mathbb{R}^d} |y|^{-2\alpha} e^{-|y|^2/2}\, dy < \infty.$$

Hence, using Fubini's Theorem,

$$\mathbb{E}[\mathscr{E}_{2\alpha}(\mu_B)] = \int_{\mathbb{R}}\int_{\mathbb{R}} \mathbb{E}[|Z|^{-2\alpha}] \frac{d\mu(t)d\mu(s)}{|t - s|^\alpha}$$

$$\leq \mathbb{E}[|Z|^{-2\alpha}]\mathscr{E}_\alpha(\mu) < \infty.$$

Thus, $\mathbb{E}[\mathscr{E}_{2\alpha}(\mu_B)] < \infty$, whence $\mathscr{E}_{2\alpha}(\mu_B) < \infty$ a.s. Moreover, μ_B is supported on $B_d(A)$ since μ is supported on A. It follows from the Energy Theorem 6.4.6 that $\dim B_d(A) \geq 2\alpha$ a.s. By letting $\alpha \to \dim(A)$, we see that, almost surely, $\dim(B_d(A)) \geq 2\dim(A)$.

Using the fact that B_d is almost surely γ-Hölder for all $\gamma < 1/2$, it follows from Lemma 5.1.3 that $\dim(B_d(A)) \leq 2\dim(A)$ a.s. This finishes the proof of Theorem 7.1.2. $\square$

Suppose $2\alpha < d$. Our proof of Theorem 7.1.2 shows that if $\mathrm{Cap}_\alpha(A) > 0$, then $\mathrm{Cap}_{2\alpha}(B_d(A)) > 0$ almost surely. The converse of this statement is also true, but much harder to prove.

We have just seen the dimension doubling property of Brownian motion for $d \geq 1$: for a single set A, we have $\dim(W(A)) = 2\dim(A)$ almost surely. However, for $d \geq 2$ this is true for *all sets* almost surely. To highlight the distinction, consider the zero set of 1-dimensional Brownian motion; these are sets of dimension $1/2$ whose images under Brownian motion have dimension 0. This cannot happen in higher dimensions; if $d \geq 2$ then almost surely every set of dimension $1/2$ must have an image of dimension 1. We will start by proving the result for $d \geq 3$, where the transience of Brownian motion can be used; after that we deal with $d = 2$.

Lemma 7.1.3 *Consider a cube $Q \subset \mathbb{R}^d$ centered at a point x and having diameter $2r$. Let W be Brownian motion in $\mathbb{R}^d$, with $d \geq 3$. Define recursively*

$$\tau_1^Q = \inf\{t \geq 0 : W(t) \in Q\}$$
$$\tau_{k+1}^Q = \inf\{t \geq \tau_k^Q + r^2 : W(t) \in Q\}$$

with the usual convention that $\inf \emptyset = \infty$. There exists a positive $\theta = \theta_d < 1$ that does not depend on r and z, such that $\mathbb{P}_z(\tau_{n+1}^Q < \infty) \leq \theta^n$.

Proof It is sufficient to show that for some θ as above,

$$\mathbb{P}_z(\tau_{k+1}^Q = \infty \mid \tau_k^Q < \infty) > 1 - \theta.$$

But the quantity on the left can be bounded below by

$$\mathbb{P}_z(\tau_{k+1}^Q = \infty \mid |W(\tau_k^Q + r^2) - x| > 4r, \tau_k^Q < \infty)\mathbb{P}_z(|W(\tau_k^Q + r^2) - x| > 4r \mid \tau_k^Q < \infty).$$

The second factor is clearly positive, and the first is also positive since W is transient (this means that $B(t) \to \infty$ a.s. when $d \geq 3$; this fact will be proven later in the chapter as Theorem 7.5.7). Both probabilities are invariant under changing the scaling factor r and do not depend on z. □

Corollary 7.1.4 *Let D_m denote the set of binary cubes of side length 2^{-m} inside $[-\frac{1}{2}, \frac{1}{2}]^d$. A.s. there exists a random variable $C(\omega)$ so that for all m and for all cubes $Q \in D_m$ we have $\tau_{K+1}^Q = \infty$ with $K = C(\omega)m$.*

Proof

$$\sum_m \sum_{Q \in D_m} \mathbb{P}(\tau_{\lceil cm+1 \rceil}^Q < \infty) \leq \sum_m 2^{dm} \theta^{cm}.$$

Choose c so that $2^d \theta^c < 1$. Then by Borel–Cantelli, for all but finitely many m we have $\tau_{\lceil cm+1 \rceil+1}^Q = \infty$ for all $Q \in D_m$. Finally, we can choose a random $C(\omega) > c$ to handle the exceptional cubes. □

Theorem 7.1.5 (Kaufman's Uniform Dimension Doubling for $d \geq 3$)

$$\mathbb{P}(\dim W(A) = 2 \dim A \text{ for all } A \subset [0, \infty]) = 1. \tag{7.1.1}$$

Proof The $\leq$ direction holds in all dimensions by the Hölder property of Brownian motion (see Corollary 6.3.2 and Lemma 5.1.3). For the other direction, fix L. We will show that with probability 1, for all subsets S of $[-L, L]^d$ we have

$$\dim W^{-1}(S) \leq \frac{1}{2} \dim S. \tag{7.1.2}$$

Applying this to $S = W(A) \cap [-L, L]^d$ for a countable unbounded set of L we get the desired conclusion. By scaling, it is sufficient to prove (7.1.2) for $L = \frac{1}{2}$. We will verify this for the paths satisfying Corollary 7.1.4; these have full measure. The rest of the argument is deterministic, we fix an ω to be such a path. For $\beta > \dim S$ and for all ε there exist covers of S by binary cubes $\{Q_j\}$ in $\bigcup_m D_m$ so that $\sum |Q_j|^\beta < \varepsilon$ (recall $|Q| = \text{diam}(Q)$). If N_m denotes the number of cubes from D_m in such a cover, then

$$\sum_m N_m 2^{-m\beta} < \varepsilon.$$

Consider the W-inverse image of these cubes. Since we chose ω so that Corollary 7.1.4 is satisfied, this yields a cover of $W^{-1}(S)$, that for each $m \geq 1$ uses at most $C(\omega) m N_m$ intervals of length $r^2 = d2^{-2m-2}$.

For $\beta_1 > \beta$ we can bound the $\beta_1/2$-dimensional Hausdorff content of $W^{-1}(S)$ above by

$$\sum_{m=1}^{\infty} C(\omega) m N_m (d2^{-2m})^{\beta_1/2} = C(\omega) d^{\beta_1/2} \sum_{m=1}^{\infty} m N_m 2^{-m\beta_1}.$$

This is small if ε is small enough. Thus $W^{-1}(S)$ has Hausdorff dimension at most $\beta/2$ for all $\beta > \dim S$, and so $\dim W^{-1}(S) \leq \frac{1}{2} \dim S$. $\square$

In two dimensions we cannot rely on transience of Brownian motion. To get around this problem, we can look at the Brownian path up to a stopping time. A convenient one is

$$\tau_R^* = \min\{t \,:\, |W(t)| = R\}.$$

For the two-dimensional version of Kaufman's Theorem it is sufficient to show that

$$\mathbb{P}(\dim W(A) = 2\dim(A \cap [0, \tau_R^*]) \text{ for all } A \subset [0, \infty]) = 1.$$

Lemma 7.1.3 has to be changed accordingly. Define τ_k as in the lemma, and assume that the cube Q is inside the ball of radius R about 0. Then we have

Lemma 7.1.6

$$\mathbb{P}_z(\tau_k < \tau_R^*) \leq \left(1 - \frac{c}{m}\right)^k \leq e^{-ck/m}. \tag{7.1.3}$$

Figure 7.1.1 The image of the middle thirds Cantor set under a 2-dimensional Brownian path. By dimension doubling, this set has Hausdorff dimension $\log 4/\log 3$.

Here $c = c(R) > 0$, $2^{-m-1} < r < 2^{-m}$, and z is any point in $\mathbb{R}^d$.

Proof We start by bounding $\mathbb{P}_z(\tau_{k+1} \geq \tau_R^* \mid \tau_k < \tau_R^*)$ from below by

$$\mathbb{P}_z(\tau_{k+1} \geq \tau_R^* \mid |W(\tau_k + r^2) - x| > 2r, \tau_k < \tau_R^*)$$
$$\times \mathbb{P}_z(|W(\tau_k + r^2) - x| > 2r \mid \tau_k < \tau_R^*). \quad (7.1.4)$$

The second factor does not depend on r and R, and it can be bounded below by a constant. The first factor is bounded below by the probability that planar Brownian motion started at distance $2r$ from the origin hits the sphere of radius $2R$ before the sphere of radius r (both centered at the origin). Using (7.8.1), this is given by

$$\frac{\log_2 \frac{2r}{r}}{\log_2 \frac{2R}{r}} \geq \frac{1}{\log_2(2R) + m}.$$

This is at least c_1/m for some c_1 that depends on R only. □

The bound (7.1.3) on $\mathbb{P}_z(\tau_k < \infty)$ in two dimensions is worse by a linear factor than the bound in higher dimensions. This, however, does not make a significant difference in the proof of the two-dimensional version of Theorem 7.1.5. Completing the proof is left to the reader (Exercise 7.1).

7.2 The Law of the Iterated Logarithm

We know that at time t, Brownian motion is roughly of size $t^{1/2}$. The following result quantifies how close this average size is to an almost sure upper bound.

Theorem 7.2.1 (The Law of the Iterated Logarithm) *For $\psi(t) = \sqrt{2t \log\log t}$*

$$\limsup_{t \to \infty} \frac{B(t)}{\psi(t)} = 1 \quad a.s.$$

By symmetry it follows that

$$\liminf_{t \to \infty} \frac{B(t)}{\psi(t)} = -1 \quad \text{a.s.}$$

Khinchin (1924) proved the Law of the Iterated Logarithm for simple random walks, Kolmogorov (1929) for other walks, and Lévy for Brownian motion. The proof for general random walks is much simpler through Brownian motion than directly.

Proof The main idea is to first consider a geometric sequence of times. We start by proving the upper bound. Fix $\varepsilon > 0$ and $q > 1$. Let

$$A_n = \left\{ \max_{0 \le t \le q^n} B(t) \ge (1+\varepsilon)\psi(q^n) \right\}.$$

By Theorem 6.6.5 the maximum of Brownian motion up to a fixed time t has the same distribution as $|B(t)|$. Therefore

$$\mathbb{P}(A_n) = \mathbb{P}\left[\frac{|B(q^n)|}{\sqrt{q^n}} \ge \frac{(1+\varepsilon)\psi(q^n)}{\sqrt{q^n}} \right].$$

We can use the tail estimate for a standard normal random variable (Lemma 6.1.6) $\mathbb{P}(Z > x) \le e^{-x^2/2}$ for $x > 1$ to conclude that for large n:

$$\mathbb{P}(A_n) \le 2\exp\left(-(1+\varepsilon)^2 \log\log q^n\right) = \frac{2}{(n\log q)^{(1+\varepsilon)^2}},$$

which is summable in n. Since $\sum_n \mathbb{P}(A_n) < \infty$, by the Borel–Cantelli Lemma we get that only finitely many of these events occur. For large t write $q^{n-1} \le t < q^n$. We have

$$\frac{B(t)}{\psi(t)} = \frac{B(t)}{\psi(q^n)} \frac{\psi(q^n)}{q^n} \frac{t}{\psi(t)} \frac{q^n}{t} \le (1+\varepsilon)q,$$

since $\psi(t)/t$ is decreasing in t. Thus

$$\limsup_{t \to \infty} \frac{B(t)}{\psi(t)} \le (1+\varepsilon)q \quad \text{a.s.}$$

Since this holds for any $\varepsilon > 0$ and $q > 1$ we have proved that

$$\limsup_{t \to \infty} \frac{B(t)}{\psi(t)} \leq 1.$$

For the lower bound, fix $q > 1$. In order to use the Borel–Cantelli lemma in the other direction, we need to create a sequence of *independent* events. Let

$$D_n = \left\{ B(q^n) - B(q^{n-1}) \geq \psi(q^n - q^{n-1}) \right\}.$$

We will now use Lemma 6.1.6 for large x:

$$\mathbb{P}(Z > x) \geq \frac{ce^{-x^2/2}}{x}.$$

Using this estimate we get

$$\mathbb{P}(D_n) = \mathbb{P}\left[Z \geq \frac{\psi(q^n - q^{n-1})}{\sqrt{q^n - q^{n-1}}} \right] \geq c\frac{\exp(-\log\log(q^n - q^{n-1}))}{\sqrt{2\log\log(q^n - q^{n-1})}}$$

$$\geq \frac{c\exp(-\log(n\log q))}{\sqrt{2\log(n\log q)}} > \frac{c'}{n\log n}$$

and therefore $\sum_n \mathbb{P}(D_n) = \infty$. Thus for infinitely many n

$$B(q^n) \geq B(q^{n-1}) + \psi(q^n - q^{n-1}) \geq -2\psi(q^{n-1}) + \psi(q^n - q^{n-1}),$$

where the second inequality follows from applying the previously proven upper bound to $-B(q^{n-1})$. From the above we get that for infinitely many n

$$\frac{B(q^n)}{\psi(q^n)} \geq \frac{-2\psi(q^{n-1}) + \psi(q^n - q^{n-1})}{\psi(q^n)} \geq \frac{-2}{\sqrt{q}} + \frac{q^n - q^{n-1}}{q^n}. \qquad (7.2.1)$$

To obtain the second inequality first note that

$$\frac{\psi(q^{n-1})}{\psi(q^n)} = \frac{\psi(q^{n-1})}{\sqrt{q^{n-1}}} \frac{\sqrt{q^n}}{\psi(q^n)} \frac{1}{\sqrt{q}} \leq \frac{1}{\sqrt{q}}$$

since $\psi(t)/\sqrt{t}$ is increasing in t for large t. For the second term we just use the fact that $\psi(t)/t$ is decreasing in t.

Now (7.2.1) implies that

$$\limsup_{t \to \infty} \frac{B(t)}{\psi(t)} \geq -\frac{2}{\sqrt{q}} + 1 - \frac{1}{q} \quad \text{a.s.}$$

and letting $q \uparrow \infty$ concludes the proof of the upper bound. $\qquad\qquad\square$

Corollary 7.2.2 *If $\{\lambda_n\}$ is a sequence of random times (not necessarily stopping times) satisfying $\lambda_n \to \infty$ and $\lambda_{n+1}/\lambda_n \to 1$ almost surely, then*

$$\limsup_{n\to\infty} \frac{B(\lambda_n)}{\psi(\lambda_n)} = 1 \quad a.s.$$

Furthermore, if $\lambda_n/n \to a$ almost surely, then

$$\limsup_{n\to\infty} \frac{B(\lambda_n)}{\psi(an)} = 1 \quad a.s.$$

Proof The upper bound follows from the upper bound for continuous time. To prove the lower bound, we might run into the problem that λ_n and q^n may not be close for large n; we have to exclude the possibility that λ_n is a sequence of times where the value of Brownian motion is too small. To get around this problem define

$$D_k^* = D_k \cap \left\{ \min_{q^k \le t \le q^{k+1}} B(t) - B(q^k) \ge -\sqrt{q^k} \right\} \overset{\text{def}}{=} D_k \cap \Omega_k.$$

Note that D_k and Ω_k are independent events. Moreover, by scaling, $\mathbb{P}(\Omega_k)$ is a constant $c_q > 0$ that does not depend on k. Thus $\mathbb{P}(D_k^*) = c_q \mathbb{P}(D_k)$, so the sum of these probabilities is infinite. The events $\{D_{2k}^*\}$ are independent, so by the Borel–Cantelli lemma, for infinitely many (even) k,

$$\min_{q^k \le t \le q^{k+1}} B(t) \ge \psi(q^k)\left(1 - \frac{1}{q} - \frac{2}{\sqrt{q}}\right) - \sqrt{q^k}.$$

Now define $n(k) = \min\{n : \lambda_n > q^k\}$. Since the ratios λ_{n+1}/λ_n tend to 1, it follows that $q^k \le \lambda_{n(k)} < q^{k+1}$ for all large k. Thus for infinitely many k

$$\frac{B(\lambda_{n(k)})}{\psi(\lambda_{n(k)})} \ge \frac{\psi(q^k)}{\psi(\lambda_{n(k)})}\left[1 - \frac{1}{q} - \frac{2}{\sqrt{q}}\right] - \frac{\sqrt{q^k}}{\psi(\lambda_{n(k)})}.$$

Since $q^k/\lambda_{n(k)}) \to 1$ (so $\psi(q^k)/\psi(\lambda_{n(k)})) \to 1$ as well) and $\sqrt{q^k}/\psi(q^k) \to 0$ as $k \to \infty$ we conclude that

$$\limsup_{n\to\infty} \frac{B(\lambda_n)}{\psi(\lambda_n)} \ge 1 - \frac{1}{q} - \frac{2}{\sqrt{q}}$$

and, as the left-hand side does not depend on q, we arrive at the desired conclusion.

For the last part, note that if $\lambda_n/n \to a$ then $\psi(\lambda_n)/\psi(an) \to 1$. $\square$

Corollary 7.2.3 *If $\{S_n\}$ is a simple random walk on $\mathbb{Z}$, then almost surely*

$$\limsup_{n\to\infty} \frac{S_n}{\psi(n)} = 1.$$

This immediately follows from the previous corollary by setting:

$$\lambda_0 = 0, \quad \lambda_n = \min\{t > \lambda_{n-1} : |B(t) - B(\lambda_{n-1})| = 1\}.$$

The waiting times $\{\lambda_n - \lambda_{n-1}\}$ are i.i.d. random variables with mean 1; see (7.3.4) below. By the Markov property, $\mathbb{P}[\lambda_1 > k] \leq \mathbb{P}[|B(1)| < 2]^k$ for every positive integer k, so λ_1 has finite variance. By the Law of Large Numbers λ_n/n will converge to 1, and the corollary follows.

7.3 Skorokhod's Representation

Brownian motion stopped at a fixed time t has a normal distribution with mean zero. On the other hand, if we stop Brownian motion by the stopping time τ given by the first hitting of $[1, \infty)$ then the expected value is clearly not zero. The crucial difference between these examples is the value of $\mathbb{E}(\tau)$.

Lemma 7.3.1 (Wald's Lemma for Brownian Motion) *Let τ be a stopping time for Brownian motion in $\mathbb{R}^d$.*

(i) $\mathbb{E}[|B_\tau|^2] = d\,\mathbb{E}[\tau]$ *(possibly both ∞)*,
(ii) *if* $\mathbb{E}[\tau] < \infty$*, then* $\mathbb{E}[B_\tau] = 0$.

Proof (i) We break the proof of (i) into three steps: (a) τ is integer valued and bounded, (b) τ is integer valued and unbounded and (c) general τ.

(a) Let τ be bounded with integer values. Write

$$|B_\tau|^2 = \sum_{k=1}^{\infty} (|B_k|^2 - |B_{k-1}|^2)\mathbf{1}_{\tau \geq k}. \tag{7.3.1}$$

If $Z_k = B_k - B_{k-1}$ and $\mathscr{F}_t$ is the σ-field determined by Brownian motion in $[0, t]$, then

$$\mathbb{E}[|B_k|^2 - |B_{k-1}|^2 | \mathscr{F}_{k-1}] = \mathbb{E}[|Z_k|^2 + 2B_{k-1} \cdot Z_k | \mathscr{F}_{k-1}] = d.$$

The expectation of $|Z_k|^2$ is d, since Z_k is Brownian motion run for time 1. Given the conditioning, B_{k-1} is constant and Z_k has mean zero, so their product has zero expectation. Since the event $\{k \leq \tau\}$ is $\mathscr{F}_{k-1}$ measurable, we deduce that $\mathbb{E}[(|B_k|^2 - |B_{k-1}|^2)\mathbf{1}_{k \leq \tau}] = d\,\mathbb{P}(k \leq \tau)$. Therefore by (7.3.1),

$$\mathbb{E}(|B_\tau|^2) = d\sum_{k=1}^{\infty} \mathbb{P}(k \leq \tau) = d\,\mathbb{E}(\tau).$$

(b) Next, suppose that τ is integer valued but unbounded. Applying (7.3.1) to $\tau \wedge n$ and letting $n \to \infty$ yields $\mathbb{E}|B_\tau|^2 \leq d\,\mathbb{E}\tau$ by Fatou's Lemma. On the other hand, the strong Markov property yields independence of $B_{\tau \wedge n}$ and $B_\tau - B_{\tau \wedge n}$, which implies that $\mathbb{E}(|B_\tau|^2) \geq d\mathbb{E}(\tau \wedge n)$. Letting $n \to \infty$ proves (7.3.1) in this case. By scaling, (7.3.1) also holds if τ takes values that are multiples of a fixed ε.

(c) Now suppose just that $\mathbb{E}\tau < \infty$ and write $\tau_\ell := 2^{-\ell}\lceil \tau 2^\ell \rceil$. Then (7.3.1) holds for the stopping times τ_ℓ, which decrease to τ as $\ell \to \infty$. Since we have $0 \leq \tau_\ell - \tau \leq 2^{-\ell}$ and $\mathbb{E}[|B(\tau_\ell) - B(\tau)|^2] \leq 2^{-\ell}$ by the strong Markov property, (7.3.1) follows.

(ii) We also break the proof of (ii) into the integer-valued and general cases.

(a) Suppose τ is integer valued and has finite expectation. Then

$$B_\tau = \sum_{k=1}^{\infty}(B_k - B_{k-1})\mathbf{1}_{\tau \geq k}.$$

The terms in the sum are orthogonal in L^2 (with respect to the Wiener measure), and the second moment of the kth term is $d\mathbb{P}(\tau \geq k)$. The second moment of the sum is the sum of the second moments of the individual terms,

$$\mathbb{E}[B_\tau^2] = d\sum_{k=1}^{\infty}\mathbb{P}(\tau \geq k) = d\sum_{j=1}^{\infty}j\mathbb{P}(\tau = j) = d\mathbb{E}[\tau] < \infty.$$

Thus

$$\mathbb{E}[B_\tau] = \sum_{k=1}^{\infty}\mathbb{E}[(B_k - B_{k-1})\mathbf{1}_{\tau \geq k}],$$

since taking the expectation of an L^2 function is just taking the inner product with the constant 1 function, and this is continuous with respect to convergence in L^2. Finally, every term on the right-hand side is 0 by independence of increments.

(b) We will apply the Lebesgue dominated convergence theorem to deduce the general case. Suppose τ is any stopping time with finite expectation and define $\tau_\ell := 2^{-\ell}\lceil 2^\ell \tau \rceil$. Note that $B(\tau_\ell) \to B(\tau)$ almost surely; thus we just want to show this sequence is dominated in L^1. Define

$$Y = \max\{|B(\tau + s) - B(\tau_1)| : 0 \leq s \leq 1\}$$

and note that

$$Y \leq 2\max\{|B(\tau + s) - B(\tau)| : 0 \leq s \leq 1\},$$

by the triangle inequality. The right-hand side is in L^2 (hence L^1) by Theorem 6.9.5. Moreover $|B(\tau_\ell)| \leq |B(\tau_1)| + Y$, for every ℓ, so the sequence $\{B(\tau_\ell)\}_{\ell \geq 1}$ is dominated by a single L^1 function. Thus $\mathbb{E}[B(\tau)] = \lim_\ell \mathbb{E}[B(\tau_\ell)]$ and the limit is zero by part (a). $\square$

Note that if we run Brownian motion until it hits a fixed point $a \neq 0$, then the $\mathbb{E}[B_\tau] = a \neq 0$, so we must have $\mathbb{E}[\tau] = \infty$.

Also note that Wald's lemma implies that the expected time for Brownian motion in $\mathbb{R}^d$ to hit the sphere $S(0, r)$ is r^2/d.

Next we consider the converse to Wald's Lemma: given a distribution with mean zero and finite variance, can we find a stopping time τ of finite mean so that $B(\tau)$ has this distribution?

If the distribution is on two points $a < 0 < b$, then this is easy. Suppose X gives mass p to a and mass $q = 1 - p$ to b. In order for X to have zero mean, we must have $ap + b(1 - p) = 0$ or $p = b/(b - a)$. Define the stopping time

$$\tau = \tau_{a,b} := \min\{t : B_t \in \{a, b\}\}. \tag{7.3.2}$$

Part (i) of Wald's Lemma implies that $\mathbb{E}\tau = \mathbb{E}B_\tau^2 \leq \max\{|a|^2, |b|^2\}$. Thus, by part (ii) of the lemma,

$$0 = \mathbb{E}B_\tau = a\mathbb{P}(B_\tau = a) + b\mathbb{P}(B_\tau = b).$$

Therefore,

$$\mathbb{P}(B_\tau = a) = \frac{b}{b - a} = \frac{b}{b + |a|}, \qquad \mathbb{P}(B_\tau = b) = \frac{-a}{b - a} = \frac{|a|}{b + |a|}. \tag{7.3.3}$$

Appealing to Part (i) of the lemma again, we conclude that

$$\mathbb{E}\tau = \mathbb{E}B_\tau^2 = \frac{a^2 b}{|a| + b} + \frac{b^2 |a|}{|a| + b} = |a|b, \tag{7.3.4}$$

so we have $B_\tau = X$ in distribution. In fact, we can generalize this considerably.

Theorem 7.3.2 (Skorokhod's Representation, Skorokhod (1965)) *Let B be the standard Brownian motion on $\mathbb{R}$.*

(i) *If X is a real random variable, then there exists a stopping time τ, that is finite a.s., such that B_τ has the same distribution as X.*

(ii) *If $\mathbb{E}[X] = 0$ and $\mathbb{E}[X^2] < \infty$, then τ can be chosen so $\mathbb{E}[\tau] < \infty$.*

Only part (ii) of the theorem is useful.

Proof (i) Pick X according to its distribution. Define $\tau = \min\{t : B(t) = X\}$. Since almost surely the range of Brownian motion consists of all the real numbers, it is clear τ is finite almost surely.

(ii) Let X have distribution ν on $\mathbb{R}$. We first reduce to the case when ν has no mass on $\{0\}$. If $\nu(\{0\}) > 0$ then we write $\nu = \nu(\{0\})\delta_0 + (1 - \nu(\{0\}))\tilde{\nu}$, where the distribution $\tilde{v}$ has no mass on $\{0\}$. Let stopping time $\tilde{\tau}$ be the solution of the problem for the distribution $\tilde{v}$. The solution for the distribution v is

$$\tau = \begin{cases} \tilde{\tau} & \text{with probability } 1 - v(\{0\}) \\ 0 & \text{with probability } v(\{0\}). \end{cases}$$

Then $\mathbb{E}\tau = (1 - v(\{0\}))\mathbb{E}\tilde{\tau} < \infty$ and $B(\tau)$ has distribution v. Thus it suffices to assume $v(\{0\}) = 0$.

From $\mathbb{E}X = 0$ it follows that:

$$M \overset{\text{def}}{=} \int_0^\infty x\,dv = -\int_{-\infty}^0 x\,dv.$$

Let $\phi : \mathbb{R} \longrightarrow \mathbb{R}$ be a non-negative measurable function. Then

$$M\int_{-\infty}^\infty \phi(x)\,dv = M\int_0^\infty \phi(y)\,dv(y) + M\int_{-\infty}^0 \phi(z)\,dv(z)$$

$$= \int_{-\infty}^0 (-z)\,dv(z)\int_0^\infty \phi(y)\,dv(y)$$

$$+ \int_0^\infty y\,dv(y)\int_{-\infty}^0 \phi(z)\,dv(z)$$

$$= \int_{-\infty}^0\int_0^\infty (y\phi(z) - z\phi(y))\,dv(y)dv(z).$$

In the last step we applied Fubini to the second integral. By the definition of the distribution $\mu_{z,y}$ in (7.3.3), we can write

$$y\phi(z) - z\phi(y) = (|z| + y)\int_{\{z,y\}} \phi(x)\,d\mu_{z,y}(x).$$

Then,

$$\int_{-\infty}^\infty \phi(x)\,dv = \frac{1}{M}\int_{-\infty}^0\int_0^\infty \left(\int_{\{z,y\}} \phi(x)\,d\mu_{z,y}(x)\right)(|z| + y)\,dv(y)dv(z).$$

$$\tag{7.3.5}$$

Consider the random variable (Z,Y) on the space $(-\infty,0) \times (0,\infty)$ with the distribution defined by

$$\mathbb{P}((Z,Y) \in A) \overset{\text{def}}{=} \frac{1}{M}\int_A (|z| + y)\,dv(y)dv(z) \tag{7.3.6}$$

for all Borel sets A on $(-\infty,0) \times (0,\infty)$. It is easy to verify that (7.3.6) defines a probability measure. In particular, let $\phi(x) = 1$, and by (7.3.5),

$$\frac{1}{M}\int_{-\infty}^0\int_0^\infty (|z| + y)\,dv(y)dv(z) = 1.$$

Once (Z,Y) is defined, (7.3.5) can be rewritten as

$$\mathbb{E}\phi(X) = \int_{-\infty}^{\infty} \phi(x)\,dv = \mathbb{E}\left[\int_{\{Z,Y\}} \phi\,d\mu_{Z,Y}\right]. \qquad (7.3.7)$$

In the last term above, the expectation is taken with respect to the distribution of (Z,Y). The randomness comes from (Z,Y). When ϕ is any bounded measurable function, apply (7.3.7) to the positive and negative part of ϕ separately. We conclude that (7.3.7) holds for any bounded measurable function.

The stopping time τ is defined as follows. Let the random variable (Z,Y) be independent of the Brownian motion B. Now let $\tau = \tau_{Z,Y}$ be as in (7.3.2). In words, the stopping rule is to first pick the values for Z,Y independent of the Brownian motion, according to the distribution defined by (7.3.6). Stop when the Brownian motion reaches either Z or Y for the first time. Notice that τ is a stopping time with respect to the Brownian filtration $\mathscr{F}_t = \sigma\{\{B(s)\}_{s\leq t}, Z, Y\}$.

Next, we show $B(\tau) \stackrel{d}{=} X$. Indeed, for any bounded measurable function ϕ:

$$\mathbb{E}\phi(B(\tau)) = \mathbb{E}[\mathbb{E}[\phi(B(\tau_{Z,Y}))|Z,Y]]$$
$$= \mathbb{E}[\int_{\{Z,Y\}} \phi\,d\mu_{Z,Y}] = \mathbb{E}\phi(X).$$

Here the second equality is due to the definition of $\tau_{Z,Y}$, and the third one is due to (7.3.7).

The expectation of τ can be computed similarly:

$$\mathbb{E}\tau = \mathbb{E}[\mathbb{E}[\tau_{Z,Y}|Z,Y]] = \mathbb{E}[\int_{\{Z,Y\}} x^2\,d\mu_{Z,Y}] = \int x^2\,dv(x).$$

The second equality follows from Wald's Lemma (Lemma 7.3.1), and the third one, from (7.3.7), by letting $\phi(x) = x^2$. $\qquad\square$

7.3.1 Root's Method *

Root (1969) proved that for a random variable X with $\mathbb{E}X = 0$ and $\mathbb{E}X^2 < \infty$, there exists a closed set $A \subset \mathbb{R}^2$, such that $B(\tau) \stackrel{d}{=} X$ and with $\mathbb{E}\tau = \mathbb{E}X^2$, where $\tau = \min\{t : (t, B(t)) \in A\}$. In words, τ is the first time the Brownian graph hits the set A (see Figure 7.3.1). This beautiful result is not useful in practice since the proof is based on a topological existence theorem, and does not provide a construction of the set A.

To illustrate the difference between Skorokhod's Method and Root's Method, let the random variable X take values in $\{-2, -1, 1, 2\}$, each with probability $1/4$. Since this is a very simple case, it is not necessary to go through the procedure shown in the proof of the theorem. A variant of Skorokhod's stopping rule simply says: with probability $1/2$ stop at the first time $|B(t)| = 1$

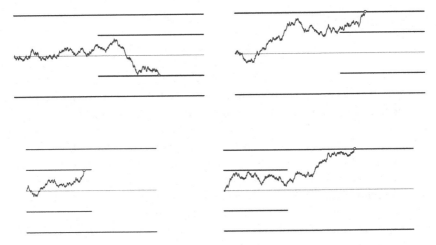

Figure 7.3.1 Root's Approach. On the top are two sample paths for Root's Method when the distribution to approximated lives on $\{\pm 1, \pm 2\}$ and the set A consists of two infinite rays at height ± 2 and two infinite rays at height ± 1. The bottom pictures show an alternative choice where the set A consists of two horizontal rays and two bounded horizontal segments.

and with probability $1/2$ stop at the first time $|B(t)| = 2$. In Root's stopping rule, the two-dimensional set A consists of four horizontal lines represented by $\{(x,y) : x \geq M, |y| = 1\} \cup \{(x,y) : x \geq 0, |y| = 2\}$, for some $M > 0$. This is intuitively clear by the following argument. Let M approach 0. The Brownian motion takes the value of 1 or -1, each with probability $1/2$, at the first time the Brownian graph hits the set A. Let M approach ∞. The Brownian motion takes value of 2 or -2, each with probability $1/2$, at the first time the Brownian graph hits the set A. Since the probability assignment is a continuous function of M, by the intermediate value theorem, there exists an $M > 0$ such that

$$\mathbb{P}(B(\tau) = 2) = \mathbb{P}(B(\tau) = 1) = \mathbb{P}(B(\tau) = -1) = \mathbb{P}(B(\tau) = -2) = 1/4.$$

However, it is difficult to compute M explicitly.

7.3.2 Dubins' Stopping Rule *

Skorokhod's stopping rule depends on random variables (i.e., Z, Y in the proof of the theorem) independent of the Brownian motion. Since the Brownian motion contains a lot of randomness, it seems possible not to introduce the extra randomness. Dubins (1968) developed a method for finding the stopping time following this idea. We use the same X above as an example.

First, run the Brownian motion until $|B(t)| = 3/2$. Then, stop when it hits one of the original four lines. Figure 7.3.2 gives the graphical demonstration of this procedure. To generalize it to the discrete case, let X have discrete distribution v. Suppose $v(\{0\}) = 0$. First, find the centers of mass for the positive and negative part of the distribution separately. For example, for the positive part, the center of mass is

$$\frac{\int_0^\infty x \, dv}{v([0, \infty])}.$$

Run the Brownian motion until it reaches one of the centers of mass, either positive or negative. Then shift the distribution so that the center of mass is at 0. Normalize the distribution (the positive or negative part corresponding to the center of mass). Then repeat the procedure until exactly one line lies above the center of mass and another one lies below it, or until the center of mass overlaps with the last line left. Stop the Brownian motion when it hits one of these two lines in the former case, or when it hits the last center of mass.

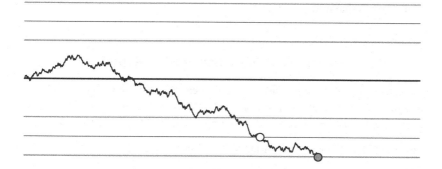

Figure 7.3.2 Dubins' Approach – The white dot shows the first time the Brownian path hits $\pm\frac{3}{2}$, and the gray dot shows the first hit on $\{\pm 1, \pm 2\}$ after this.

In the case where X has a continuous distribution, it needs to be approximated by discrete distributions. See Dudley (2002) for details.

7.3.3 Skorokhod's representation for a sequence

Let $\{X_i\}_{i \geq 1}$ be independent random variables with mean 0 and finite variances. Let τ_1 be a stopping time with $\mathbb{E}\tau_1 = \mathbb{E}X_1^2$ and $B(\tau_1) \stackrel{d}{=} X_1$. By the strong Markov property, $\{B(\tau_1 + t) - B(\tau_1)\}_{t \geq 0}$ is again a Brownian motion. We can find a stopping time τ_2 with $\mathbb{E}\tau_2 = \mathbb{E}X_2^2$, and $B(\tau_1 + \tau_2) - B(\tau_1) \stackrel{d}{=} X_2$ and is independent of $\mathscr{F}_{\tau_1}$. Repeat the procedure for $\tau_3, \tau_4, \ldots$, etc. Define $T_1 = \tau_1$, and

$T_n = \tau_1 + \tau_2 + \cdots + \tau_n$. Then, $B(T_k + \tau_{k+1}) - B(T_k) \overset{d}{=} X_{k+1}$ and is independent of $\mathscr{F}_{T_k}$. We get,

$$B(T_n) \overset{d}{=} X_1 + X_2 + \cdots + X_n,$$

$$\mathbb{E}T_n = \sum_{i=1}^{n} \mathbb{E}\tau_i = \sum_{i=1}^{n} \mathbb{E}X_i^2.$$

This is a very useful formulation. For example, if $\{X_i\}_{i \geq 1}$ is an i.i.d. sequence of random variables with zero expectation and bounded variance, then $\{\tau_i\}_{i \geq 1}$ is also i.i.d. By the Strong Law of Large Numbers, $\frac{1}{n}T_n \longrightarrow \mathbb{E}\tau_1 = \mathbb{E}X_1^2$ almost surely, as $n \longrightarrow \infty$. Let $S_n = \sum_{i=1}^{n} X_i = B(T_n)$. By the Corollary 7.2.2 of the Law of the Iterated Logarithm (LIL) for the Brownian motion, we have,

$$\limsup_{n \to \infty} \frac{S_n}{\sqrt{2n \log \log n}\sqrt{\mathbb{E}X_1^2}} = 1.$$

This result was first proved by Hartman and Wintner (1941), and the proof given above is due to Strassen (1964).

7.4 Donsker's Invariance Principle

Let $\{X_i\}_{i \geq 1}$ be i.i.d. random variables with mean 0 and finite variances. By normalization, we can assume the variance $\text{Var}(X_i) = 1$, for all i. Let $S_n = \sum_{i=1}^{n} X_i$, and interpolate it linearly to get the continuous paths $\{S_t\}_{t \geq 0}$ (Figure 7.4.1).

Theorem 7.4.1 (Donsker's Invariance Principle)   *As* $n \longrightarrow \infty$,

$$\left\{ \frac{S_{tn}}{\sqrt{n}} \right\}_{0 \leq t \leq 1} \overset{\text{in law}}{\Longrightarrow} \{B_t\}_{0 \leq t \leq 1},$$

i.e., if $\psi : \tilde{C}[0,1] \longrightarrow \mathbb{R}$, *where* $\tilde{C}[0,1] = \{f \in C[0,1] : f(0) = 0\}$, *is a bounded continuous function with respect to the sup norm, then, as* $n \longrightarrow \infty$,

$$\mathbb{E}\psi\left(\left\{ \frac{S_{tn}}{\sqrt{n}} \right\}_{0 \leq t \leq 1}\right) \longrightarrow \mathbb{E}\psi(\{B_t\}_{0 \leq t \leq 1}).$$

The proof of this shows we may replace the assumption of continuity of ψ by the weaker assumption that ψ is continuous at almost all Brownian paths.

Consider Figure 7.4.1. Each picture on the left shows 100 steps of an i.i.d. random walk with a different distribution and each picture on the right is 10,000 steps with the same rule. The rules from top to bottom are:

(1) the standard random walk with step sizes ± 1,

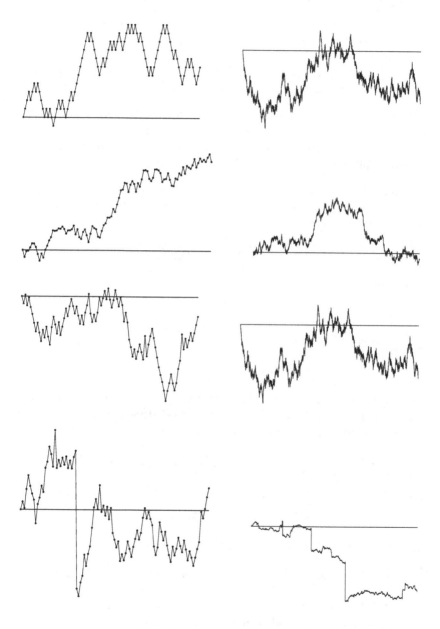

Figure 7.4.1 An illustration of Donsker's Invariance Principle.

(2) steps chosen uniformly in $[-1, 1]$,

(3) absolute step sizes chosen uniformly in $[0, 1]$ and then raised to the $-.3$ power and the sign chosen uniformly in ± 1 (since $.3 < .5$ this has finite variance),

(4) the same as (3), but with $-.3$ replaced by $-.6$ (this distribution does not have finite variance).

By Donsker's Principle the first three all converge to (suitable scaled) Brownian motion and the right-hand pictures seem to confirm this. However, the bottom right picture does not "look like" Brownian motion; Donsker's Principle fails in this case.

Example 7.4.2 For our first application we estimate the chance that a random walk travels distance $\sqrt{n}$ in time n using the analogous estimate for Brownian motion. More precisely, we claim that as $n \longrightarrow \infty$,

$$\frac{\max_{1 \leq k \leq n} S_k}{\sqrt{n}} \overset{\text{in law}}{\Longrightarrow} \max_{0 \leq t \leq 1} B(t)$$

i.e., for any constant a,

$$\mathbb{P}(\max_{1 \leq k \leq n} S_k \geq a\sqrt{n}) \longrightarrow \frac{2}{\sqrt{2\pi}} \int_a^\infty e^{-u^2/2} \, du,$$

because by Theorem 6.6.5 we have

$$\mathbb{P}(\max_{0 \leq t \leq 1} B(t) \geq a) = 2\mathbb{P}(B(1) \geq a).$$

To prove this, let $\phi : \mathbb{R} \longrightarrow \mathbb{R}$ be a bounded continuous function. Take the function $\psi(f) = \phi(\max_{[0,1]} f)$. Then ψ is a bounded and continuous function on $\tilde{C}[0, 1]$. By the construction of $\{S_t\}_{t \geq 0}$, we have

$$\mathbb{E}\psi\left(\left\{\frac{S_{tn}}{\sqrt{n}}\right\}_{0 \leq t \leq 1}\right) = \mathbb{E}\phi\left(\max_{0 \leq t \leq 1}\left\{\frac{S_{tn}}{\sqrt{n}}\right\}\right) = \mathbb{E}\phi\left(\frac{\max_{1 \leq k \leq n} S_k}{\sqrt{n}}\right).$$

Also,

$$\mathbb{E}\psi(\{B(t)\}_{0 \leq t \leq 1}) = \mathbb{E}\phi(\max_{0 \leq t \leq 1} B(t)).$$

Then, by Donsker's Theorem,

$$\mathbb{E}\phi\left(\frac{\max_{1 \leq k \leq n} S_k}{\sqrt{n}}\right) \longrightarrow \mathbb{E}\phi(\max_{0 \leq t \leq 1} B(t)).$$

Example 7.4.3 Next we compare the last time a random walk crosses zero to the last time a Brownian motion crosses zero, i.e., as $n \longrightarrow \infty$,

$$\frac{1}{n} \max\{k \in [1, n] : S_k S_{k-1} \leq 0\} \overset{\text{in law}}{\Longrightarrow} \max\{t \in [0, 1] : B(t) = 0\}. \qquad (7.4.1)$$

The distribution for the Brownian motion problem can be explicitly calculated.

To prove (7.4.1), define the function ψ by

$$\psi(f) = \max\{t \le 1 : f(t) = 0\}.$$

The function ψ is not a continuous function on $\tilde{C}[0,1]$, but it is continuous at every $f \in \tilde{C}[0,1]$ with the property that

$$f(\psi(f) - \delta, \psi(f) + \delta)$$

contains a neighborhood of 0 for every $\delta > 0$. To elaborate this, suppose $f(t) > 0$ for $\psi(f) < t \le 1$. For any given $\delta > 0$, let $\varepsilon_0 = \min_{[\psi(f)+\delta,1]} f(t)$. Choose $\varepsilon_1 > 0$ so that $(-\varepsilon_1, \varepsilon_1) \subseteq f(\psi(f) - \delta, \psi(f) + \delta)$. Choose a positive $\varepsilon < \min\{\varepsilon_0, \varepsilon_1\}$. Then, $\psi(f - \varepsilon) - \psi(f) < \delta$, and $\psi(f) - \psi(f + \varepsilon) < \delta$ (Figure 7.4.2). Let $\tilde{f} \in \tilde{C}[0,1]$ such that $\|\tilde{f} - f\|_\infty < \varepsilon$. Then, for every t, $f(t) - \varepsilon \le \tilde{f}(t) \le f(t) + \varepsilon$. Hence, $|\psi(\tilde{f}) - \psi(f)| < \delta$. That is, ψ is continuous at f. Since the last zero of a Brownian path on $[0,1]$ almost surely is strictly less than 1, and is an accumulation point of zeroes from the left, the Brownian path almost surely has the property that f has. Hence, ψ is continuous at almost all Brownian paths.

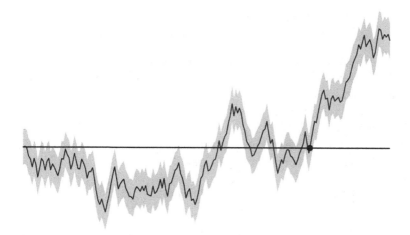

Figure 7.4.2 A Brownian graph and a ε-neighborhood of the graph. The function ψ is continuous at f if the rightmost zero t of f in $[0,1]$ is not an extreme value. In that case, any nearby function g also has its rightmost zero s close to t (otherwise s could be $\ll t$).

Proof of Theorem 7.4.1.   Let

$$F_n(t) = \frac{S_{tn}}{\sqrt{n}}, \ 0 \le t \le 1.$$

By Skorokhod embedding, there exist stopping times $T_k, k = 1, 2, \ldots,$ for some standard Brownian motion B such that the sequence $\{B(T_k)\}_{k \ge 1}$ has the same joint distribution as $\{S_k\}_{k \ge 1}$. Thus we may assume that $S_k = B(T_k)$ for all k. Define $W_n(t) = B(nt)/\sqrt{n}$. Note that W_n is also a standard Brownian motion. We will show that for any $\varepsilon > 0$, as $n \to \infty$,

$$\mathbb{P}(\sup_{0 \le t \le 1} |F_n - W_n| > \varepsilon) \to 0. \tag{7.4.2}$$

The theorem will follow since by (7.4.2) if $\psi : \tilde{C}[0,1] \to \mathbb{R}$ is bounded by M and is continuous on almost every Brownian motion path, then for any $\delta > 0$,

$$\mathbb{P}(\exists f : ||W - f|| < \varepsilon, |\psi(W) - \psi(f)| > \delta) \tag{7.4.3}$$

converges to 0 as $\varepsilon \to 0$. Now

$$|\mathbb{E}\psi(F_n) - \mathbb{E}\psi(W_n)| \le \mathbb{E}|\psi(F_n) - \psi(W_n)|$$

and the right-hand side is bounded above by

$$2M\mathbb{P}(||W_n - F_n|| \ge \varepsilon) +$$
$$2M\mathbb{P}(||W_n - F_n|| < \varepsilon, |\psi(W_n) - \psi(F_n)| > \delta) + \delta.$$

The second term is bounded by (7.4.3), so by setting δ small, then setting ε small and then setting n large, the three terms of the last expression may be made arbitrarily small.

To prove (7.4.2), let A_n be the event that $|F_n(t) - W_n(t)| > \varepsilon$ for some t. We will show that $\mathbb{P}(A_n) \to 0$.

Let $k = k(t)$ designate the integer such that $\frac{k-1}{n} \le t < \frac{k}{n}$. Then, since $F_n(t)$ is linearly interpolated between $F_n(\frac{k-1}{n})$ and $F_n(\frac{k}{n})$,

$$A_n \subset \left\{ \exists t : \left| \frac{S(k)}{\sqrt{n}} - W_n(t) \right| > \varepsilon \right\} \bigcup \left\{ \exists t : \left| \frac{S(k-1)}{\sqrt{n}} - W_n(t) \right| > \varepsilon \right\}.$$

Writing $S(k) = B(T_k) = \sqrt{n}W_n(T_k/n)$, we get

$$A_n \subset \left\{ \exists t : \left| W_n(\frac{T_k}{n}) - W_n(t) \right| > \varepsilon \right\} \bigcup \left\{ \exists t : \left| W_n(\frac{T_{k-1}}{n}) - W_n(t) \right| > \varepsilon \right\}.$$

Given $\delta \in (0, 1)$, the event on the right implies that either

$$\{ \exists t : |T_k/n - t| \vee |T_{k-1}/n - t| \ge \delta \} \tag{7.4.4}$$

or

$$\{\exists s, t \in [0,2] : |s-t| < \delta, |W_n(s) - W_n(t)| > \varepsilon\}.$$

Since each W_n is a standard Brownian motion, by choosing δ small, the probability of the later event can be made arbitrarily small.

To conclude the proof, all we have to show is that for each δ, the probability of (7.4.4) converges to 0 as $n \to \infty$. In fact, we will show that this event only happens for finitely many n a.s. Since we chose k so that t is in the interval $[(k-1)/n, k/n]$, the absolute differences in (7.4.4) are bounded above by the maximum of these distances when we let $t = (k-1)/n$ and k/n. This implies that (7.4.4) is a subset of the union of the events

$$\left\{\sup_{0 \le k \le n} \frac{|T_k - k + c|}{n} > \delta\right\} \tag{7.4.5}$$

for $c = -1, 0, 1$. Note the deterministic fact that if a real sequence $\{a_n\}$ satisfies $\lim a_n/n \to 1$, then $\sup_{0 \le k \le n} |a_k - k|/n \to 0$. Since T_n is a sum of i.i.d. mean 1 random variables, the Law of Large Numbers enables us to apply this to $a_n = T_n + c$, and conclude that (7.4.5) happens only finitely many times, as desired. $\qquad\square$

7.5 Harmonic functions and Brownian motion in $\mathbb{R}^d$

Definition 7.5.1 Let $D \subset \mathbb{R}^d$ be a domain (a connected open set). A function $u : D \to \mathbb{R}$ is **harmonic** if it is measurable, locally bounded (i.e., bounded on closed balls in D), and for any ball $B = B(x, r) \subset D$,

$$u(x) = \frac{1}{\mathcal{L}_d(B)} \int_B u(y)\, dy.$$

If u is harmonic in D, then it is continuous in D: if $x_n \to x$ then

$$u(y)\mathbf{1}_{B(x_n,r)}(y) \xrightarrow[n\to\infty]{a.e.} u(y)\mathbf{1}_{B(x,r)}(y),$$

thus, by the Dominated Convergence Theorem, $u(x_n) \to u(x)$.

Theorem 7.5.2 *Let u be measurable and locally bounded in D. Then, u is harmonic in D if and only if*

$$u(x) = \frac{1}{\sigma_{d-1}(S(x,r))} \int_{S(x,r)} u(y)\, d\sigma_{d-1}(y), \tag{7.5.1}$$

where $S(x,r) = \{y : |y-x| = r\}$, and σ_{d-1} is the $(d-1)$-dimensional Hausdorff measure.

Proof   Assume u is harmonic. Define

$$\int_{S(x,r)} u(y) \, d\sigma_{d-1}(y) = \Psi(r) r^{d-1}.$$

We will show that Ψ is constant. Indeed, for any $R > 0$,

$$R^d \mathcal{L}_d(B(x,1)) u(x) = \mathcal{L}_d(B(x,R)) u(x)$$
$$= \int_{B(x,R)} u(y) \, dy = \int_0^R \Psi(r) r^{d-1} \, dr.$$

Differentiate with respect to R to obtain:

$$d\mathcal{L}_d(B(x,1)) u(x) = \Psi(R),$$

and so $\Psi(R)$ is constant. From the identity $d\mathcal{L}_d(B(x,r))/r = \sigma_{d-1}(S(x,r))$, it follows that (7.5.1) holds.

For the other direction, note that (7.5.1) implies that

$$u(x) = \mathcal{L}_d(B(x,r))^{-1} \int_{B(x,r)} u(y) \, dy$$

by Fubini's Theorem.   □

An equivalent definition for harmonicity states that u is harmonic if u is continuous, twice differentiable and

$$\Delta u \equiv \sum_i \frac{\partial^2 u}{(\partial x_i)^2} = 0.$$

Definition 7.5.3

$$G(x,y) = \int_0^\infty p(x,y,t) \, dt, \ x,y \in \mathbb{R}^d$$

is **Green's function** in $\mathbb{R}^d$, where $p(x,y,t)$ is the Brownian transition density function, $p(x,y,t) = (2\pi t)^{-d/2} \exp\left(-\frac{|x-y|^2}{2t}\right)$.

Proposition 7.5.4   *Green's function G satisfies:*

(1) $G(x,y)$ *is finite if and only if* $x \neq y$ *and* $d > 2$.
(2) $G(x,y) = G(y,x) = G(y-x,0)$.
(3) $G(x,0) = c_d |x|^{2-d}$ *where* $c_d = \Gamma(d/2 - 1)/(2\pi^{d/2})$, $d > 2$ *and* $x \neq 0$.

Proof   Facts (1) and (2) are immediate. For (3), note that

$$G(x,0) = \int_0^\infty (2\pi t)^{-d/2} \exp\left(-\frac{|x|^2}{2t}\right) dt.$$

Substituting $s = \frac{|x|^2}{2t}$, we obtain:

$$G(x,0) = \int_0^\infty (\frac{\pi|x|^2}{s})^{-d/2} e^{-s} \frac{|x|^2}{2s^2} ds = |x|^{2-d} \frac{\pi^{-d/2}}{2} \int_0^\infty e^{-s} s^{\frac{d}{2}-2} ds.$$

(The integral is known as $\Gamma(\frac{d}{2} - 1)$.)   □

One probabilistic meaning of G is given in the following proposition:

Proposition 7.5.5   *Define* $F_r(x) = \int_{B(0,r)} G(x,z)\, dz$. *Then*

$$F_r(x) = \mathbb{E}_x \int_0^\infty \mathbf{1}_{W_t \in B(0,r)}\, dt. \tag{7.5.2}$$

In words: $F_r(x)$ *is the expected time the Brownian motion started at* x *spends in* $B(0,r)$.

Proof   By Fubini's Theorem and the definition of the Brownian transition density function p, we have

$$F_r(x) = \int_0^\infty \int_{B(0,r)} p(x,z,t)\, dz dt = \int_0^\infty \mathbb{P}_x(W_t \in B(0,r))\, dt.$$

Applying Fubini another time,

$$F_r(x) = \mathbb{E}_x \int_0^\infty \mathbf{1}_{W_t \in B(0,r)}\, dt, \tag{7.5.3}$$

as needed.   □

Theorem 7.5.6   *For* $d \geq 3$: $x \mapsto G(x,0)$ *is harmonic on* $\mathbb{R}^d \setminus \{0\}$.

Proof   We prove that $F_\varepsilon(x)$ is harmonic in $\mathbb{R}^d \setminus B(0,\varepsilon)$, i.e.,

$$F_\varepsilon(x) = \frac{1}{\mathscr{L}_d(B(x,r))} \int_{B(x,r)} F_\varepsilon(y)\, dy \tag{7.5.4}$$

for $0 < r < |x| - \varepsilon$. The theorem will follow from (7.5.4), since using the continuity of G gives:

$$\begin{aligned}
G(x,0) &= \lim_{\varepsilon \to 0} \frac{F_\varepsilon(x)}{\mathscr{L}_d(B(0,\varepsilon))} \\
&= \lim_{\varepsilon \to 0} \frac{1}{\mathscr{L}_d(B(0,r))} \int_{B(x,r)} \frac{F_\varepsilon(y)}{\mathscr{L}_d(B(0,\varepsilon))}\, dy \\
&= \frac{1}{\mathscr{L}_d(B(x,r))} \int_{B(x,r)} G(y,0)\, dy,
\end{aligned}$$

where the last equality follows from the bounded convergence theorem.

Denote by $\nu_{d-1} = \sigma_{d-1}/\|\sigma_{d-1}\|$ the rotation-invariant probability measure on the unit sphere in $\mathbb{R}^d$. Fix $x \neq 0$ in $\mathbb{R}^d$, let $0 < r < |x|$ and let $\varepsilon < |x| - r$.

For Brownian motion W denote $\tau = \min\{t : |W(t) - x| = r\}$. Since W spends no time in $B(0,\varepsilon)$ before time τ, we can write $F_\varepsilon(x)$ as

$$\mathbb{E}_x \int_\tau^\infty \mathbf{1}_{W_t \in B(0,\varepsilon)} \, dt = \mathbb{E}_x \mathbb{E}_x \left[\int_\tau^\infty \mathbf{1}_{W_t \in B(0,\varepsilon)} \, dt \,\big|\, W_\tau \right].$$

By the strong Markov property and since W_τ is uniform on the sphere of radius r about x by rotational symmetry, we conclude:

$$F_\varepsilon(x) = \mathbb{E}_x F_\varepsilon(W_\tau) = \int_{S(0,1)} F_\varepsilon(x + ry) \, d\nu_{d-1}(y).$$

Hence (7.5.4) follows from Theorem 7.5.2. This proves Theorem 7.5.6. $\square$

The above proof of Theorem 7.5.6 is a probabilistic one. One could also prove this result by showing that $\Delta_x G(x,0) = 0$.

We have therefore proved that $x \mapsto \frac{1}{|x|^{d-2}}$ is harmonic in $\mathbb{R}^d \setminus \{0\}$, $d \geq 3$. For $d \geq 3$, the time Brownian motion spends in the ball $B(0,R)$ around the origin has expectation $F_R(0)$ by (7.5.2). By the definition of $F_R(0)$, this expectation can be written as

$$\int_{B(0,R)} G(0,x) \, dx = c_d \int_{B(0,R)} |x|^{2-d} \, dx = \tilde{c}_d \int_0^R r^{d-1} r^{2-d} \, dr = c_d' R^2,$$

in particular, it is finite. Next we will show that Brownian motion in $\mathbb{R}^d$, $d \geq 3$, is transient.

Proposition 7.5.7   *For $d \geq 3$ and $|x| > r$,*

$$h_r(x) \stackrel{\text{def}}{=} \mathbb{P}_x \left(\exists t \geq 0 : W_t \in B(0,r) \right) = \left(\frac{r}{|x|} \right)^{d-2}.$$

Proof   Recall we defined

$$F_r(x) = \int_{B(0,r)} G(x,z) \, dz = \int_{B(0,r)} G(x-z,0) \, dz. \tag{7.5.5}$$

Since G is harmonic, from (7.5.5) we have

$$F_r(x) = \mathscr{L}_d(B(0,r)) G(x,0) = \mathscr{L}_d(B(0,r)) c_d |x|^{2-d}.$$

In particular, $F_r(x)$ depends only on $|x|$. We define $\tilde{F}_r(|x|) = F_r(x)$.

Suppose $|x| > r$. Since $F_r(x)$ is the expected time spent in $B(0,r)$ starting from x, it must equal the probability of hitting $S(0,r)$ starting from x, times the expected time spent in $B(0,r)$ starting from the hitting point of this sphere. Therefore $F_r(x) = h_r(x)\tilde{F}_r(r)$. This implies $h_r(x) = (r/|x|)^{d-2}$. $\square$

Proposition 7.5.8   *Brownian motion W in dimension $d \geq 3$ is transient, i.e.,* $\lim_{t \to \infty} |W(t)| = \infty$.

Proof   We use the fact that $\limsup_{t\to\infty}|W(t)| = \infty$ almost surely. Therefore, for any $0 < r < R$,

$$
\begin{aligned}
\mathbb{P}(W \text{ visits } B(0,r) &\text{ for arbitrarily large t}) \\
&\leq \mathbb{P}(W \text{ visits } B(0,r) \text{ after hitting } S(0,R)) \\
&= \left(\frac{r}{R}\right)^{d-2},
\end{aligned}
$$

which goes to 0 as $R \to \infty$. The proposition follows.   □

We are now also able to calculate the probability that a Brownian motion starting between two spheres will hit the smaller one before hitting the larger one.

Proposition 7.5.9   *Define*

$$
a = \mathbb{P}_x(\text{Brownian motion } W \text{ hits } S(0,r) \text{ before } S(0,R)),
$$

where $r < |x| < R$*. Then*

$$
a = \frac{(r/|x|)^{d-2} - (r/R)^{d-2}}{1 - (r/R)^{d-2}}. \tag{7.5.6}
$$

Proof   It follows from Proposition 7.5.7 and the strong Markov property that

$$
\begin{aligned}
a = \mathbb{P}_x(W &\text{ hits } S(0,r)) \\
&- \mathbb{P}_x(W \text{ hits } S(0,R) \text{ first and then hits } S(0,r)) \\
&= \left(\frac{r}{|x|}\right)^{d-2} - (1-a)\left(\frac{r}{R}\right)^{d-2}. \tag{7.5.7}
\end{aligned}
$$

Solving (7.5.7), we get (7.5.6).   □

Since a is fixed under scaling, the visits of a Brownian path to $S(0,e^k)$, $k \in \mathbb{Z}$ form a discrete random walk with constant probability to move up (k increasing) or down (k decreasing). The probability to move down is

$$
\frac{e^{2-d} - e^{4-2d}}{1 - e^{4-2d}}.
$$

It is easy to see that this probability is less than $1/2$ (this also follows from the fact that the Brownian motion is transient, and therefore the random walk should have an upward drift).

7.6 The maximum principle for harmonic functions

Proposition 7.6.1 (Maximum Principle)   *Suppose that u is harmonic in D, a connected, open subset of $\mathbb{R}^d$.*

(i) *If u attains its maximum in D, then u is a constant.*

(ii) *If u is continuous on $\bar{D}$ and D is bounded, then $\max_{\bar{D}} u = \max_{\partial D} u$.*

(iii) *Assume that D is bounded, u_1 and u_2 are two harmonic functions on D that are continuous on $\bar{D}$. If u_1 and u_2 take the same values on ∂D, then they are identical on D.*

Proof   (i) Set $M = \sup_D u$. Note that $V = \{x \in D : u(x) = M\}$ is relatively closed in D. Since D is open, for any $x \in V$, there is a ball $B(x,r) \subset D$. By the mean-value property of u,

$$u(x) = \frac{1}{\mathscr{L}_d(B(x,r))} \int_{B(x,r)} u(y)\, dy \le M.$$

Equality holds if and only if $u(y) = M$ almost everywhere on $B(x,r)$, or, by continuity, $B(x,r) \subset V$. This means that V is also open. Since D is connected and $V \ne \emptyset$ we get that $V = D$. Therefore, u is constant on D.

(ii) $\sup_{\bar{D}} u$ is attained on $\overline{D}$ since u is continuous and $\overline{D}$ is closed and bounded. The conclusion now follows from (i).

(iii) Consider $u_1 - u_2$. It follows from (ii) that

$$\sup_{\bar{D}}(u_1 - u_2) = \sup_{\partial D}(u_1 - u_2) = 0.$$

Similarly $\sup_{\bar{D}}(u_2 - u_1) = 0$. So $u_1 = u_2$ on $\overline{D}$. $\square$

A function u on $\{r < |x| < R\} \subset \mathbb{R}^d$ is called **radial** if $u(x) = u(y)$ whenever $|x| = |y|$.

Corollary 7.6.2   *Suppose that u is a radial harmonic function in the annulus $D := \{r < |x| < R\} \subset \mathbb{R}^d$, and u is continuous on $\bar{D}$.*

(i) *If $d \ge 3$, there exist constants a and b such that $u(x) = a + b|x|^{2-d}$.*

(ii) *If $d = 2$, there exist constants a and b such that $u(x) = a + b\log|x|$.*

Proof   For $d \ge 3$, choose a and b such that

$$a + br^{2-d} = \tilde{u}(r),$$

and

$$a + bR^{2-d} = \tilde{u}(R).$$

Notice that the radial harmonic function $u(x) = \tilde{u}(|x|)$ and the harmonic function $x \mapsto a + b|x|^{2-d}$ agree on ∂D. They also agree on D by Proposition 7.6.1. So $u(x) = a + b|x|^{2-d}$. Similarly, we can show that $u(x) = a + b\log|x|$ in the case $d = 2$. $\qquad\square$

7.7 The Dirichlet problem

Definition 7.7.1 Let $D \subset \mathbb{R}^d$ be a domain. We say that D satisfies the **Poincaré cone condition** if for each point $x \in \partial D$ there exists a cone $C_x(\alpha, h)$ of height $h(x)$ and angle $\alpha(x)$ such that $C_x(\alpha, h) \subset D^c$ and $C_x(\alpha, h)$ is based at x.

Proposition 7.7.2 (Dirichlet Problem) *Suppose $D \subset \mathbb{R}^d$ is a bounded domain with boundary ∂D, such that D satisfies the Poincaré cone condition, and f is a continuous function on ∂D. Then there exists a unique function u that is harmonic on D, continuous on $\overline{D}$ and satisfies $u(x) = f(x)$ for all $x \in \partial D$.*

Proof The uniqueness claim follows from Proposition 7.6.1. To prove existence, let W be a Brownian motion in $\mathbb{R}^d$ and define

$$u(x) = \mathbb{E}_x f(W_{\tau_{\partial D}}), \quad \text{where} \quad \tau_A = \inf\{t \geq 0 : W_t \in A\}$$

for any Borel set $A \subset \mathbb{R}^d$. For a ball $B(x, r) \subset D$, the strong Markov property implies that

$$u(x) = \mathbb{E}_x[\mathbb{E}_x[f(W_{\tau_{\partial D}})|\mathscr{F}_{\tau_{S(x,r)}}]] = \mathbb{E}_x[u(W_{\tau_{S(x,r)}})] = \int_{S(x,r)} u(y) d\mu_r,$$

where μ_r is the uniform distribution on the sphere $S(x, r)$. Therefore, u has the mean value property and so it is harmonic on D (by Theorem 7.5.2).

It remains to be shown that the Poincaré cone condition implies

$$\lim_{x \to z, x \in D} u(x) = f(z) \text{ for all } z \in \partial D.$$

Fix $z \in \partial D$, then there is a cone with height $h > 0$ and angle $\alpha > 0$ in D^c based at z. Let

$$\phi = \sup_{x \in B(0, \frac{1}{2})} \mathbb{P}_x[\tau_{S(0,1)} < \tau_{C_0(\alpha,1)}].$$

Then $\phi < 1$. Note that if $x \in B(0, 2^{-k})$ then by the strong Markov property:

$$\mathbb{P}_x[\tau_{S(0,1)} < \tau_{C_0(\alpha,1)}] \leq \prod_{i=0}^{k-1} \sup_{x \in B(0, 2^{-k+i})} \mathbb{P}_x[\tau_{S(0,2^{-k+i+1})} < \tau_{C_0(\alpha, 2^{-k+i+1})}] = \phi^k.$$

Therefore, for any positive integer k, we have

$$\mathbb{P}_x[\tau_{S(z,h)} < \tau_{C_z(\alpha,h)}] \leq \phi^k$$

for all x with $|x - z| < 2^{-k}h$.

Given $\varepsilon > 0$, there is a $0 < \delta \leq h$ such that $|f(y) - f(z)| < \varepsilon$ for all $y \in \partial D$ with $|y - z| < \delta$. For all $x \in \bar{D}$ with $|z - x| < 2^{-k}\delta$,

$$|u(x) - u(z)| = |\mathbb{E}_x f(W_{\tau_{\partial D}}) - f(z)| \leq \mathbb{E}_x |f(W_{\tau_{\partial D}}) - f(z)|. \tag{7.7.1}$$

If the Brownian motion hits the cone $C_z(\alpha, \delta)$, which is outside the domain D, before it hits the sphere $S(z, \delta)$, then $|z - W\tau_{\partial D}| < \delta$, and $f(W_{\tau_{\partial D}})$ is close to $f(z)$. The complement has small probability. More precisely, (7.7.1) is bounded above by

$$2\|f\|_\infty \mathbb{P}_x\{\tau_{S(z,\delta)} < \tau_{C_z(\alpha,\delta)}\} + \varepsilon \mathbb{P}_x\{\tau_{\partial D} < \tau_{S(z,\delta)}\} \leq 2\|f\|_\infty \phi^k + \varepsilon.$$

Hence u is continuous on $\bar{D}$. ☐

7.8 Polar points and recurrence

Given $x \in \mathbb{R}^2, 1 \leq |x| \leq R$, we know that

$$\mathbb{P}_x[\tau_{S(0,R)} < \tau_{S(0,1)}] = a + b\log|x|.$$

The left-hand side is clearly a function of $|x|$, and it is a harmonic function of x for $1 < |x| < R$ by averaging over a small sphere surrounding x. Setting $|x| = 1$ implies $a = 0$, and $|x| = R$ implies $b = \frac{1}{\log R}$. It follows that

$$\mathbb{P}_x[\tau_{S(0,R)} < \tau_{S(0,1)}] = \frac{\log|x|}{\log R}.$$

By scaling, for $0 < r < R$ and $r \leq |x| \leq R$,

$$\mathbb{P}_x[\tau_{S(0,R)} < \tau_{S(0,r)}] = \frac{\log \frac{|x|}{r}}{\log \frac{R}{r}}. \tag{7.8.1}$$

Definition 7.8.1 A set A is **polar** for a Markov process X if for all x we have

$$\mathbb{P}_x[X_t \in A \text{ for some } t > 0] = 0.$$

The image of (δ, ∞) under Brownian motion W is the random set

$$W(\delta, \infty) \stackrel{\text{def}}{=} \bigcup_{\delta < t < \infty} \{W_t\}.$$

Proposition 7.8.2 *Points are polar for a planar Brownian motion W, that is, for all $z \in \mathbb{R}^2$ we have $\mathbb{P}_0\{z \in W(0,\infty)\} = 0$.*

Proof Take $z \neq 0$ and $0 < \varepsilon < |z| < R$,

$$\mathbb{P}_0\{\tau_{S(z,R)} < \tau_{S(z,\varepsilon)}\} = \frac{\log \frac{|z|}{\varepsilon}}{\log \frac{R}{\varepsilon}}.$$

Let $\varepsilon \to 0+$,

$$\mathbb{P}_0\{\tau_{S(z,R)} < \tau_{\{z\}}\} = \lim_{\varepsilon \to 0+} \mathbb{P}_0\{\tau_{S(z,R)} < \tau_{S(z,\varepsilon)}\} = 1,$$

and then

$$\mathbb{P}_0\{\tau_{S(z,R)} < \tau_{\{z\}} \text{ for all integers } R > |z|\} = 1.$$

It follows that

$$\mathbb{P}_0\{z \in W(0,\infty)\} = \mathbb{P}_0\{\tau_{\{z\}} < \infty\} = 0.$$

Let $f(z) = \mathbb{P}_z(0 \in W(0,\infty))$. Given $\delta > 0$, by the Markov property

$$\mathbb{P}_0\{0 \in W(\delta,\infty)\} = \mathbb{E}_0[f(W_\delta)] = 0.$$

Finally, $f(0) = \mathbb{P}(\bigcup_{n=1}^\infty \{0 \in W(\frac{1}{n},\infty)\}) = 0$. Hence any fixed single point is a polar set for a planar Brownian motion. $\qquad\square$

Corollary 7.8.3 *Almost surely, a Brownian path has zero area.*

Proof The expected area $\mathbb{E}_0[\mathscr{L}_2(W(0,\infty))]$ of planar Brownian motion can be written as

$$\mathbb{E}_0\Big[\int_{\mathbb{R}^2} I_{\{z \in W(0,\infty)\}} \, dz\Big] = \int_{\mathbb{R}^2} \mathbb{P}_0\{z \in W(0,\infty)\} \, dz = 0,$$

where the first equality is by Fubini's Theorem, the second from the previous theorem. So almost surely, the image of a planar Brownian motion is a set with zero Lebesgue measure. $\qquad\square$

This was previously proven as Theorem 6.8.2 using a different method.

Proposition 7.8.4 *Planar Brownian motion W is neighborhood recurrent. In other words,*

$$\mathbb{P}_0\{W(0,\infty) \text{ is dense in } \mathbb{R}^2\} = 1.$$

Proof Note that $\limsup_{t \to \infty} |W_t| = \infty$, so for all $z \in \mathbb{R}^2$ and $\varepsilon > 0$,

$$\mathbb{P}_0\{\tau_{B(z,\varepsilon)} = \infty\} = \lim_{R \to \infty} \mathbb{P}_0\{\tau_{S(z,R)} < \tau_{B(z,\varepsilon)}\} = 0.$$

Summing over all rational z and ε completes the proof. $\qquad\square$

7.9 Conformal invariance *

If $V, U \subset \mathbb{C}$ are open, a map $f : V \to U$ is called *conformal* if it is holomorphic and 1-to-1. Conformal maps preserve angles. This property also holds for anti-conformal maps, i.e., complex conjugates of conformal maps.

If u is harmonic on U and $f : V \to U$ is conformal, then $u \circ f$ is harmonic on V. In fact, conformal and anti-conformal maps are the only homeomorphisms with this property (Exercise 7.6). Since the hitting distribution of Brownian motion solves the Dirichlet problem on both domains, it is easy to verify that, assuming f extends continuously to ∂V, it maps the Brownian hitting distribution on ∂V (known as harmonic measure) to the harmonic measure on ∂U. Does f take individual Brownian paths in V to Brownian paths in U? This is not quite correct: if $f(z) = 2z$, then $f(B(t))$ leaves a disk of radius 2 in the same expected time that $B(t)$ leaves a disk of radius 1, so $f(B(t))$ is "too fast" to be Brownian motion. However, it is Brownian motion up to a time change.

What does this mean? Suppose $f : V \to U$ is conformal and suppose $0 \in V$. For a Brownian path started at 0 let τ be the first hitting time on ∂V. For $0 \leq t < \tau$, define

$$\varphi(t) = \int_0^t |f'(B(t))|^2 \, dt. \tag{7.9.1}$$

Why does the integral makes sense? For $0 \leq t < \tau$, we know that $B([0,t])$ is a compact subset of V and that $|f'|$ is a continuous function that is bounded above and below on this compact set. Thus the integrand above is a bounded continuous function. Therefore $\varphi(t)$ is continuous and strictly increasing, and so $\varphi^{-1}(t)$ is well defined.

Theorem 7.9.1   *Suppose that $V, U \subset \mathbb{C}$ are open sets with $0 \in V$, the map $f : V \to U$ is conformal and φ is defined by (7.9.1). If $B(\cdot)$ is Brownian motion in V and τ is the exit time from V, then $\{X(t) : 0 \leq t \leq \phi(\tau)\}$ defined by $X(t) := f(B(\varphi^{-1}(t)))$ is a Brownian motion in U started at $f(0)$ and stopped at ∂U.*

Proof   Let $Y(t)$ be Brownian motion in U started at $f(0)$. The idea of the proof is to show that both $X(t)$ and $Y(t)$ are limits of discrete random walks that depend on a parameter ε, and that as $\varepsilon \to 0$, the two random walks get closer and closer, and hence have the same limiting distribution.

Fix a small $\varepsilon > 0$, and starting at $f(0)$, sample $Y(t)$ every time it moves distance ε from the previous sample point. We stop when a sample point lands within 2ε of ∂U. Since $Y(t)$ is almost surely continuous, it is almost surely the limit of the linear interpolation of these sampled values. Because Brownian motion is rotationally invariant, the increment between samples is uniformly distributed on a circle of radius ε. Thus $Y(t)$ is the limit of the following discrete process: starting at $z_0 = f(0)$, choose z_1 uniformly on $|z - z_0| = \varepsilon$. In general, z_{n+1} is chosen uniformly on $|z - z_n| = \varepsilon$.

Now sample $X(t)$ starting at z_0, each time it first moves distance ε from the previous sample. We claim that, as above, z_{n+1} is uniformly distributed on an ε-circle around z_n. Note that if $D = D(z_n, \varepsilon) \subset U$, then the probability that $X(t)$ first hits ∂D in a set $E \subset \partial D$ is the same as the probability that $B(t)$ started at $w_n = f^{-1}(z_n)$ first hits $\partial f^{-1}(D)$ in $F = f^{-1}(E)$. This probability is the solution at w_n of the Dirichlet problem on $f^{-1}(D)$ with boundary data $\mathbf{1}_F$. Since f is conformal, this value is the same as the solution at z_n of the Dirichlet problem on D with boundary data $\mathbf{1}_E$, which is just the normalized angle measure of E. Thus the hitting distribution of $X(t)$ on ∂D starting from z_n is the uniform distribution, just as it is for usual Brownian motion, only the time needed for $f(B(t))$ to hit ∂D may be different.

How different? The time $T_n - T_{n-1}$ between the samples z_{n-1} and z_n for the Brownian motion Y are i.i.d. random variables with expectation $\varepsilon^2/2$ and variance $O(\varepsilon^4)$, see Exercise 7.5. So taking $t_n := n\varepsilon^2/2$, the time T_n for $Y(t)$ to reach z_n satisfies $\mathbb{E}[|T_n - t_n|^2] = O(n\varepsilon^4) = O(t_n \varepsilon^2)$. Moreover, by Kolmogorov's maximal inequality (Exercise 1.60), we have

$$\mathbb{P}[\max_{1 \leq k \leq n} |T_k - t_k| \geq h] \leq O(n\varepsilon^4/h^2).$$

This remains true even if we condition on the sequence $\{z_k\}$.

Next we do the same calculation for the process $X(t)$. Note that we may pick the points $\{z_n\}$ to be the same for the two processes X and Y without altering their distributions – this amounts to a *coupling* of X and Y. The exit time for $X(t)$ from $D = D(z_n, \varepsilon)$ is same as the exit time for $B(\varphi^{-1}(t))$ from $f^{-1}(D)$ starting at the point $p = f^{-1}(z_n)$. Since f is conformal, f is close to linear on a neighborhood of p with estimates that only depend on the distance of p from ∂V. Thus for any $\delta > 0$, we can choose ε so small (uniformly for all p in any compact subset of V) that

$$D\left(p, \frac{\varepsilon}{(1+\delta)|f'(p)|}\right) \subset f^{-1}(D) \subset D\left(p, \frac{\varepsilon(1+\delta)}{|f'(p)|}\right).$$

Therefore the expected exit time for the Brownian motion $B(\cdot)$ from $f^{-1}(D)$ starting at p is bounded above and below by the expected exit times for these two disks, which are

$$\frac{1}{2}\left(\frac{\varepsilon}{(1+\delta)|f'(p)|}\right)^2 \quad \text{and} \quad \frac{1}{2}\left(\frac{\varepsilon(1+\delta)}{|f'(p)|}\right)^2$$

respectively. As long as $B(s)$ is inside Ω,

$$\frac{|f'(p)|^2}{(1+\delta)^2} \leq |f'(B(s))|^2 \leq |f'(p)|^2(1+\delta)^2,$$

if ε is small enough. Therefore $\varphi(s)$ has derivative between these two bounds during this time and so φ^{-1} has derivative bounded between the reciprocals, i.e.,

$$\frac{1}{(1+\delta)^2|f'(p)|^2} \leq \frac{d}{ds}\varphi^{-1} \leq \frac{(1+\delta)^2}{|f'(p)|^2}.$$

Thus the expected exit time of $B \circ \varphi^{-1}$ from $f^{-1}(D)$ is between

$$\frac{\varepsilon^2}{2(1+\delta)^4} \quad \text{and} \quad \frac{\varepsilon^2(1+\delta)^4}{2}.$$

The bounds are uniform as long as ε is small enough and z_n is in a compact subset of U. Let S_n denote the time it takes $X(\cdot)$ to reach z_n. The random variables $S_n - S_{n-1}$ are not i.i.d., but they are independent given the sequence $\{z_k\}$, and have variances $O(\varepsilon^4)$, so

$$\mathbb{P}[\max_{1 \leq k \leq n} |S_k - s_k| \geq h \,|\, \{z_k\}_{k=1}^n] = O(n\varepsilon^4/h^2),$$

where $s_n = \mathbb{E}(S_n \,|\, \{z_k\}_{k=1}^n)$. We have already proved that

$$(1+\delta)^{-4} \leq s_n/t_n \leq (1+\delta)^4,$$

so it follows that for $n \leq 2C\varepsilon^{-2}$

$$\mathbb{P}[\max_{1 \leq k \leq n} |S_k - T_k| \geq 5C\delta + 2h \,|\, \{z_k\}_{k=1}^n] \to 0$$

as $\varepsilon \to 0$. Using the uniform continuity of Brownian motion on $[0,C]$, given any $\eta > 0$ we can choose δ and h sufficiently small so that

$$\mathbb{P}[\max_{t \leq C} |X(t) - Y(t)| \geq \eta \,|\, \{z_k\}_{k=1}^n] \to 0$$

as $\varepsilon \to 0$. Since η can be taken arbitrarily small, this coupling implies that $X(\cdot)$ and $Y(\cdot)$ indeed have the same distribution until the first exit from U. $\qquad \square$

The following elegant result is due to Markowsky (2011).

Lemma 7.9.2 *Suppose that $f(z) = \sum_{n=0}^\infty a_n z^n$ is conformal in $\mathbb{D}$. Then the expected time for Brownian motion to leave $\Omega = f(\mathbb{D})$ starting at $f(0)$ is $\frac{1}{2}\sum_{n=1}^\infty |a_n|^2$.*

We give two proofs, the first relies on Lemma 7.3.1 and the second on an exercise.

Proof 1   We may assume that $f(0) = 0$ since translating the domain and starting point does not change either the expected exit time or the infinite sum (it doesn't include a_0). We use the identity

$$2\mathbb{E}[\tau] = \mathbb{E}[|B_\tau|^2],$$

where τ is a stopping time for a 2-dimensional Brownian motion B (Lemma 7.3.1). We apply it in the case when B starts at $p = f(0)$ and is stopped when it hits $\partial\Omega$. Then the expectation on the right side above is

$$\int_{\partial\Omega} |z|^2 \, d\omega_p(z),$$

where ω_p is harmonic measure on $\partial\Omega$ with respect to p, i.e., the hitting distribution of Brownian motion started at p. By the conformal invariance of Brownian motion, we get

$$\mathbb{E}[|B_\tau|^2] = \frac{1}{2\pi} \int_{\partial\mathbb{D}} |f(z)|^2 \, d\theta = \sum_{n=1}^\infty |a_n|^2. \qquad \square$$

Proof 2   By definition, the expected exit time from Ω for Brownian motion started at $w = f(0)$ is (writing $z = x + iy$)

$$\iint_\Omega G_\Omega(z,w) dxdy.$$

By the conformal invariance of Green's functions (Exercise 7.12), this is the same as

$$\iint_\mathbb{D} G_\mathbb{D}(z,0)|f'(z)|^2 dxdy.$$

The Green's function for the disk is $G_\mathbb{D}(z,0) = \frac{1}{\pi} \log |z|^{-1}$ (Exercise 7.13), so this formula becomes

$$\frac{1}{\pi} \iint_\mathbb{D} |f'(z)|^2 \log \frac{1}{|z|} dxdy.$$

This can be evaluated using the identities (writing $z = e^{i\theta}$),

$$\int_0^{2\pi} \Big| \sum c_n z^n \Big|^2 d\theta = \int_0^{2\pi} \Big(\sum c_n z^n \Big) \overline{\Big(\sum c_n z^n \Big)} \, d\theta = 2\pi \sum |c_n|^2,$$

and

$$\int_0^x t^m \log t \, dt = x^{m+1} \left(\frac{\log x}{m+1} - \frac{1}{(m+1)^2} \right), \qquad m \neq -1,$$

as follows:

$$\frac{1}{\pi} \iint_{\mathbb{D}} \log \frac{1}{|z|} |f'(z)|^2 \, dx \, dy = \frac{1}{\pi} \int_0^{2\pi} \int_0^1 \log \frac{1}{r} |f'(re^{i\theta})|^2 r \, dr \, d\theta$$

$$= 2 \sum_{n=1}^{\infty} n^2 |a_n|^2 \int_0^1 r^{2n-1} \log \frac{1}{r} \, dr$$

$$= 2 \sum_{n=1}^{\infty} n^2 |a_n|^2 \left[-r^{2n} \left(\frac{\log r}{2n} - \frac{1}{(2n)^2} \right) \right]_0^1$$

$$= 2 \sum_{n=1}^{\infty} n^2 |a_n|^2 \frac{1}{4n^2}$$

$$= \frac{1}{2} \sum_{n=1}^{\infty} |a_n|^2. \qquad \square$$

Lemma 7.9.2 can be used to derive a number of formulas. For example, the expected time for a 1-dimensional Brownian motion started at zero to leave $[-1, 1]$ is 1 (this is calculated at the beginning of Section 7.3) and is the same as the time for a 2-dimensional path to leave the infinite strip $S = \{x + iy : |y| < 1\}$. This strip is the image of the unit disk under the conformal map

$$f(z) = \frac{2}{\pi} \log \frac{1+z}{1-z}$$

since the linear fractional map $(1+z)/(1-z)$ maps the disk to the right half-plane and the logarithm carries the half-plane to the strip $\{|y| \leq \pi/2\}$. Since

$$f(z) = \frac{2}{\pi} [\log(1+z) - \log(1-z)] = \frac{4}{\pi} \left(z + \frac{1}{3} z^3 + \frac{1}{5} z^5 + \cdots \right),$$

the expected time a Brownian motion spends in S is

$$1 = \frac{1}{2} \left(\frac{4}{\pi} \right)^2 \left(1 + \frac{1}{9} + \frac{1}{25} + \cdots \right),$$

so

$$\frac{\pi^2}{8} = \left(1 + \frac{1}{9} + \frac{1}{25} + \cdots \right).$$

From this it is easy to derive the more famous identity

$$\frac{\pi^2}{6} = 1 + \frac{1}{4} + \frac{1}{9} + \frac{1}{16} + \cdots = \zeta(2).$$

Corollary 7.9.3 *Among all simply connected domains with area π and containing 0, Brownian motion started at 0 has the largest expected exit time for the unit disk.*

Proof If $f : \mathbb{D} \to \Omega$ is conformal, then

$$
\begin{aligned}
\pi = \text{area}(\Omega) &= \iint_{\mathbb{D}} |f'(z)|^2 dx dy \\
&= \int_0^{2\pi} \int_0^1 |\sum_{n=1}^{\infty} n a_n r^{n-1} e^{i(n-1)\theta}|^2 r \, dr \, d\theta \\
&= 2\pi \int_0^1 \sum_{n=1}^{\infty} n^2 |a_n|^2 r^{2n-1} \, dr \\
&= \pi \sum_{n=1}^{\infty} n |a_n|^2 \\
&\geq \pi \sum_{n=1}^{\infty} |a_n|^2.
\end{aligned}
$$

By Lemma 7.9.2 the expected exit time is $\leq \frac{1}{2}$ with equality if and only if $|a_1| = 1, a_n = 0$ for $n \geq 2$, so the disk is optimal. $\qquad\square$

7.10 Capacity and harmonic functions

The central question of this section is the following: which sets $\Lambda \subset \mathbb{R}^d$ does Brownian motion hit with positive probability? This is related to the following question: for which $\Lambda \subset \mathbb{R}^d$ are there bounded, non-constant harmonic functions on $\mathbb{R}^d \setminus \Lambda$?

Consider the simplest case first. When Λ is the empty set, the answer to the first question is trivial, whereas the answer to the second one is provided by Liouville's Theorem. We will give a probabilistic proof of this theorem.

Theorem 7.10.1 *For $d \geq 1$ any bounded harmonic function on $\mathbb{R}^d$ is constant.*

Proof Let $u : \mathbb{R}^d \to [-M, M]$ be a harmonic function, x, y two distinct points in $\mathbb{R}^d$, and H the hyperplane so that the reflection in H takes x to y.

Let W_t be Brownian motion started at x, and $\overline{W}_t$ its reflection in H. Let $\tau_H = \min\{t : W_t \in H\}$. Note that

$$
\{W_t\}_{t \geq \tau_H} \overset{d}{=} \{\overline{W}_t\}_{t \geq \tau_H}. \tag{7.10.1}
$$

If v_x is the distribution of $|W_t - x|$, then harmonicity implies that for any $r > 0$,

$$
\mathbb{E}_x u(W_t) = \int \mathbb{E}_x \big(u(W_t) \,\big|\, |W_t - x| = r \big) dv_x(r) = \mathbb{E}_x (u(x)) = u(x),
$$

since the conditional expectation above is just the average of u on a sphere about x of radius $r = |W(t) - x|$. Decomposing the above into $t < \tau_H$ and $t \geq \tau_H$ we get

$$u(x) = \mathbb{E}_x u(W_t)\mathbf{1}_{t < \tau_H} + \mathbb{E}_x u(W_t)\mathbf{1}_{t \geq \tau_H}.$$

A similar equality holds for $u(y)$. Now using (7.10.1):

$$|u(x) - u(y)| = |\mathbb{E}_x u(W_t)\mathbf{1}_{t < \tau_H} - \mathbb{E}_x u(\overline{W}_t)\mathbf{1}_{t < \tau_H}|$$
$$\leq 2M\mathbb{P}(t < \tau_H) \to 0$$

as $t \to \infty$. Thus $u(x) = u(y)$, and since x and y were chosen arbitrarily, u must be constant. $\qquad\square$

A stronger result is also true.

Theorem 7.10.2 *For $d \geq 1$, any positive harmonic function on $\mathbb{R}^d$ is constant.*

Proof Let $x, y \in \mathbb{R}^d$, $a = |x - y|$. Suppose u is a positive harmonic function. Then $u(x)$ can be written as

$$\frac{1}{\mathcal{L}_d B_R(x)} \int_{B_R(x)} u(z)\,dz \leq \frac{\mathcal{L}_d B_{R+a}(y)}{\mathcal{L}_d B_R(x)} \frac{1}{\mathcal{L}_d B_{R+a}(y)} \int_{B_{R+a}(y)} u(z)\,dz$$
$$= \frac{(R+a)^d}{R^d} u(y).$$

This converges to $u(y)$ as $R \to \infty$, so $u(x) \leq u(y)$, and by symmetry, $u(x) = u(y)$ for all x, y. Hence u is constant. $\qquad\square$

Nevanlinna (1936) proved that for $d \geq 3$ there exist non-constant bounded harmonic functions on $\mathbb{R}^d \setminus \Lambda$ if and only if $\mathrm{Cap}_G(\Lambda) > 0$. Here G denotes the Green's function $G(x, y) = c|x - y|^{2-d}$. By Nevanlinna's result and Theorem 3.4.2, $\dim \Lambda > d - 2$ implies existence of such functions, and $\dim \Lambda < d - 2$ implies non-existence. Kakutani (1944) showed that there exist such functions if and only if $\mathbb{P}(W \text{ hits } \Lambda) > 0$. Note that the Green's function is translation invariant, while the hitting probability of a set is invariant under scaling. It is therefore better to estimate hitting probabilities by a capacity function with respect to a scale-invariant modification of the Green's kernel, called the **Martin kernel**:

$$K(x, y) = \frac{G(x, y)}{G(0, y)} = \frac{|y|^{d-2}}{|x - y|^{d-2}}$$

for $x \neq y$ in $\mathbb{R}^d$, and $K(x, x) = \infty$. The following theorem shows that capacity is indeed a good estimate of the hitting probability.

Theorem 7.10.3 *Let Λ be any closed set in $\mathbb{R}^d$, $d \geq 3$. Then*

$$\frac{1}{2}\text{Cap}_K(\Lambda) \leq \mathbb{P}_0(\exists t > 0 \; : \; W_t \in \Lambda) \leq \text{Cap}_K(\Lambda). \tag{7.10.2}$$

Here

$$\text{Cap}_K(\Lambda) = \left[\inf_{\mu(\Lambda)=1} \int_\Lambda \int_\Lambda K(x,y)\, d\mu(x) d\mu(y) \right]^{-1}.$$

Proof To bound the probability of ever hitting Λ from above, consider the stopping time $\tau = \min\{t \; : \; W_t \in \Lambda\}$. The distribution of W_τ on the event $\tau < \infty$ is a possibly defective distribution ν satisfying

$$\nu(\Lambda) = \mathbb{P}_0(\tau < \infty) = \mathbb{P}(\exists t > 0 \; : \; W_t \in \Lambda). \tag{7.10.3}$$

Now recall the standard formula from Proposition 7.5.7: when $0 < \varepsilon < |y|$,

$$\mathbb{P}_0(\exists t > 0 : |W_t - y| < \varepsilon) = \left(\frac{\varepsilon}{|y|}\right)^{d-2}. \tag{7.10.4}$$

By a first entrance decomposition, the probability in (7.10.4) is at least

$$\mathbb{P}(|W_\tau - y| > \varepsilon \text{ and } \exists t > \tau : |W_t - y| < \varepsilon) = \int_{x:|x-y|>\varepsilon} \frac{\varepsilon^{d-2} d\nu(x)}{|x-y|^{d-2}}.$$

Dividing by ε^{d-2} and letting $\varepsilon \to 0$ we obtain

$$\int_\Lambda \frac{d\nu(x)}{|x-y|^{d-2}} \leq \frac{1}{|y|^{d-2}},$$

i.e., $\int_\Lambda K(x,y)\, d\nu(x) \leq 1$ for all $y \in \Lambda$. Therefore, if

$$\mathscr{E}_K(\nu) = \int_\Lambda \int_\Lambda \frac{|y|^{d-2} d\nu(x) d\nu(y)}{|x-y|^{d-2}},$$

then $\mathscr{E}_K(\nu) \leq \nu(\Lambda)$ and thus if we use $\frac{\nu}{\nu(\Lambda)}$ as a probability measure we get

$$\text{Cap}_K(\Lambda) \geq [\mathscr{E}_K(\nu/\nu(\Lambda))]^{-1} \geq \nu(\Lambda),$$

which by (7.10.3) yields the upper bound on the probability of hitting Λ.

To obtain a lower bound for this probability, a second moment estimate is used. It is easily seen that the Martin capacity of Λ is the supremum of the capacities of its compact subsets, so we may assume that Λ is itself compact. For $\varepsilon > 0$ and $y \in \mathbb{R}^d$ let $B_\varepsilon(y)$ denote the Euclidean ball of radius ε about y and let $h_\varepsilon(|y|)$ denote the probability that the standard Brownian path hits this ball; that is, $(\varepsilon/|y|)^{d-2}$ if $|y| > \varepsilon$, and 1 otherwise.

Given a probability measure μ on Λ, and $\varepsilon > 0$, consider the random variable

$$Z_\varepsilon = \int_\Lambda \mathbf{1}_{\{\exists t > 0: W_t \in B_\varepsilon(y)\}} h_\varepsilon(|y|)^{-1} d\mu(y).$$

Clearly $\mathbb{E}Z_\varepsilon = 1$. By symmetry, the second moment of Z_ε can be written as

$$\mathbb{E}Z_\varepsilon^2 = 2\mathbb{E}\int_\Lambda \int_\Lambda \mathbf{1}_{\{\exists t > 0: W_t \in B_\varepsilon(x), \exists s > t: W_s \in B_\varepsilon(y)\}} \frac{d\mu(x)d\mu(y)}{h_\varepsilon(|x|)h_\varepsilon(|y|)}$$

$$\leq 2\mathbb{E}\int_\Lambda \int_\Lambda \mathbf{1}_{\{\exists t > 0: W_t \in B_\varepsilon(x)\}} \frac{h_\varepsilon(|y - x| - \varepsilon)}{h_\varepsilon(|x|)h_\varepsilon(|y|)} d\mu(x)d\mu(y)$$

$$= 2\int_\Lambda \int_\Lambda \frac{h_\varepsilon(|y - x| - \varepsilon)}{h_\varepsilon(|y|)} d\mu(x)d\mu(y).$$

The last integrand is bounded by 1 if $|y| \leq \varepsilon$. On the other hand, if $|y| > \varepsilon$ and $|y - x| \leq 2\varepsilon$ then $h_\varepsilon(|y-x| - \varepsilon) = 1 \leq 2^{d-2}h_\varepsilon(|y-x|)$, so that the integrand on the right-hand side of the equation above is at most $2^{d-2}K(x,y)$. Thus

$$\mathbb{E}Z_\varepsilon^2 \leq 2\mu(B_\varepsilon(0)) + 2^{d-1}\int_\Lambda \int_\Lambda \mathbf{1}_{|y-x| \leq 2\varepsilon}K(x,y)\, d\mu(x)d\mu(y)$$

$$+ 2\int_\Lambda \int_\Lambda \mathbf{1}_{|y-x| > 2\varepsilon}\left(\frac{|y|}{|y - x| - \varepsilon}\right)^{d-2} d\mu(x)d\mu(y). \quad (7.10.5)$$

Since the kernel is infinite on the diagonal, any measure with finite energy must have no atoms. Restricting attention to such measures μ, we see that the first two summands in (7.10.5) drop out as $\varepsilon \to 0$ by dominated convergence. Thus by the Dominated Convergence Theorem,

$$\lim_{\varepsilon \downarrow 0} \mathbb{E}Z_\varepsilon^2 \leq 2\mathscr{E}_K(\mu). \quad (7.10.6)$$

The hitting probability $\mathbb{P}(\exists t > 0, y \in \Lambda : W_t \in B_\varepsilon(y))$ is at least

$$\mathbb{P}(Z_\varepsilon > 0) \geq \frac{(\mathbb{E}Z_\varepsilon)^2}{\mathbb{E}Z_\varepsilon^2} = (\mathbb{E}Z_\varepsilon^2)^{-1}.$$

Transience of Brownian motion implies that if the Brownian path visits every ε-neighborhood of the compact set Λ then it almost surely intersects Λ itself. Therefore, by (7.10.6),

$$\mathbb{P}(\exists t > 0 : W_t \in \Lambda) \geq \lim_{\varepsilon \downarrow 0}(\mathbb{E}Z_\varepsilon^2)^{-1} \geq \frac{1}{2\mathscr{E}_K(\mu)}.$$

Since this is true for all probability measures μ on Λ, we get the desired conclusion:

$$\mathbb{P}(\exists t > 0 : W_t \in \Lambda) \geq \frac{1}{2}\text{Cap}_K(\Lambda). \qquad \square$$

The right-hand inequality in (7.10.2) can be an equality; a sphere centered at the origin has hitting probability and capacity both equal to 1. To prove that the constant 1/2 on the left cannot be increased consider the spherical shell

$$\Lambda_R = \{x \in \mathbb{R}^d : 1 \le |x| \le R\}.$$

We claim that $\lim_{R \to \infty} \mathrm{Cap}_K(\Lambda_R) = 2$. Indeed by Theorem 7.10.3, the Martin capacity of *any* compact set is at most 2, while lower bounds tending to 2 for the capacity of Λ_R are established by computing the energy of the probability measure supported on Λ_R, with density a constant multiple of $|x|^{1-d}$ there. The details are left as Exercise 7.4.

7.11 Notes

McKean's theorem (Theorem 7.1.2) is stated for closed sets, but is true for all analytic sets (which include all Borel sets). The proof is the same as given in the text, but we use Frostman's theorem for analytic sets, which is proven in Appendix B. A proof of McKean's theorem using Fourier transforms is given in Chapter 12 of Mattila (2015). This proof also shows that $B_d(A) \subset \mathbb{R}^d$ is a Salem set (see the notes of Chapter 3) for any Borel $A \subset \mathbb{R}$, giving examples of all possible dimensions.

The LIL for Brownian motion was proved by Khinchin (1933); for i.i.d. random variables with finite variance it is due to Hartman and Wintner (1941). A proof for Dubins' stopping rule (described in Section 7.3.2) is given in Dudley (2002) and in Mörters and Peres (2010). The idea of using Skorokhod embedding to prove Donsker's theorem and the Hartman–Wintner LIL is due to Strassen (1964). The connection between Brownian motion and the Dirichlet problem was discovered by Kakutani (1944). Indeed, in 1991 Kakutani told the second author that in the early 1940s he met with Ito and Yosida to select a topic to collaborate on, as they were isolated from the rest of the world due to the war; they chose Brownian motion. Ultimately the three of them worked on this topic separately, with seminal results: Brownian potential theory, stochastic calculus and semigroup theory. Conformal invariance of Brownian motion paths was discovered by Paul Lévy, who sketched the proof in Lévy (1948). The proof we give in Section 7.9 follows the methods used by Kakutani and Lévy; a proof using stochastic calculus can be found in Bass (1995) or Mörters and Peres (2010).

Instead of using a disk of a fixed radius ε in the proof of Theorem 7.9.1, we could have sampled Brownian motion in Ω using disks of the form $D(z_n, \lambda r_n)$

where $r_n = \text{dist}(z_n, \partial\Omega)$ and $0 < \lambda < 1$. See Figure 7.11.1. As before, this discrete walk always has the same hitting distribution on $\partial\Omega$ and is well defined up until the hitting time on $\partial\Omega$, and converges to Brownian motion as $\lambda \to 0$ (but now we have to use a more difficult distortion estimate for conformal maps to control the shapes of disk preimages under f. This process is called the "Walk on Spheres" in Binder and Braverman (2009) and (2012), where the method is credited to Muller (1956) (however, the first author learned the method in class from Shizuo Kakutani, and thinks of it as "Kakutani's Algorithm"). Kakutani is also credited with summarizing Propositions 7.5.8 and 7.8.4 by saying "A drunk man will find his way home, but a drunk bird may get lost forever."

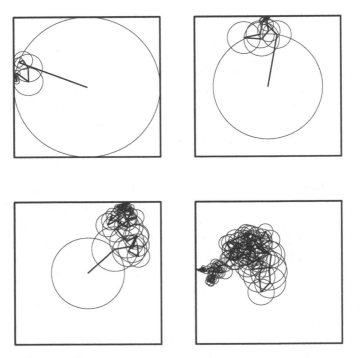

Figure 7.11.1 The random walk where we step $\lambda \, \text{dist}(z, \partial\Omega)$ in a random direction. Here the domain is a square and we show sample paths for $\lambda = 1, .75, .5, .25$. The hitting distribution on the boundary is the same as for Brownian motion and the paths converge to Brownian motion as $\lambda \to 0$.

Let W be a Brownian motion in $\mathbb{R}^3$. The orthogonal projection of W onto a 2-plane is a 2-dimensional Brownian motion in that plane, hence almost surely recurrent. Hence W hits every cylinder perpendicular to the plane. Is there any infinite cylinder avoided by W? In fact, an avoided cylinder does exist almost surely; this is due to Adelman et al. (1998).

Theorem 7.10.3 is from Benjamini et al. (1995). The result as stated does not cover the case $d = 2$. In the plane, a set is hit by Brownian motion if and only if it has positive logarithmic capacity. This is due to Kakutani (1944), building on work of Nevanlinna. There are several proofs of this; one approach is to deduce it from the 3-dimensional case and this is sketched in Exercises 7.21 and 7.22.

7.12 Exercises

Exercise 7.1 Prove Theorem 7.1.5 (dimension doubling) for two-dimensional Brownian motion.

• **Exercise 7.2** Let $\mathbf{C}_{1/2}$ be the "middle halves" Cantor set (Example 1.1.3). Show that $\dim B(A) = 2 \dim A$ for any $A \subset \mathbf{C}_{1/2}$, where B is 1-dimensional Brownian motion.

• **Exercise 7.3** Suppose B, W are independent, one-dimensional Brownian motions and let Z be the zero set of W. Show that $\dim B(A) = 2 \dim A$ for any $A \subset Z$.

Exercise 7.4 Consider the spherical shell

$$\Lambda_R = \{x \in \mathbb{R}^d \ : \ 1 \le |x| \le R\}.$$

Show that $\lim_{R \to \infty} \mathrm{Cap}_K(\Lambda_R) = 2$. Here K is the Martin kernel as in Theorem 7.10.3.

• **Exercise 7.5** Show that the exit time for Brownian motion in a ball in $\mathbb{R}^d$, $d \ge 1$ has finite variance. More generally, if the expected exit time from a domain is at most α, independent of the starting point, show that the variance of the exit time is at most $2\alpha^2$.

• **Exercise 7.6** Suppose $f : \Omega \to \Omega'$ is a homeomorphism between planar domains and $u \circ f$ is harmonic on Ω for every harmonic function u on Ω'. Show that f is conformal or anti-conformal.

Exercise 7.7 A holomorphic map is B-**proper** if $f(B(t))$ exits $\Omega = f(\mathbb{D})$ almost surely where $B(t)$ is Brownian motion on $\mathbb{D}$ run until it exits $\mathbb{D}$ (e.g., f would not be B-proper if some boundary set of positive measure maps to the interior of Ω). Show that Lemma 7.9.2 holds for such maps. See Lemma 2 of Markowsky (2011).

Exercise 7.8   Use conformal invariance of Brownian motion to prove Liouville's Theorem: any bounded holomorphic function on the plane is constant.

Exercise 7.9   Show that among all planar domains of area π the expected exit time is largest for the unit disk.

Exercise 7.10   Compute the expected exit time of a 2-dimensional Brownian motion from $\mathbb{D}$ if it starts at $a \in \mathbb{D}$.

• **Exercise 7.11**   Show that the expected exit time of Brownian motion starting at 1 from the l $\{z = re^{i\theta} : r > 0, |\theta| < \theta_0\}$ is finite if and only if $\theta_0 < \pi/4$.

• **Exercise 7.12**   If $f : V \to U$ is conformal and G_V, G_U are Green's functions for these domains then show that

$$G_V(x,y) = G_U(f(x), f(y)).$$

• **Exercise 7.13**   Prove that Green's function for Brownian motion stopped when it leaves the unit disk, $\mathbb{D}$, is

$$G_{\mathbb{D}}(x,y) = \frac{1}{\pi} \log \left| \frac{1 - \bar{x}y}{x - y} \right|.$$

In particular $G(x,0) = (1/\pi) \log |x|^{-1}$.

• **Exercise 7.14**   Prove Brownian motion is not conformally invariant in $\mathbb{R}^d$, $d \geq 3$. What step of the proof of Theorem 7.9.1 breaks down in higher dimensions?

• **Exercise 7.15**   Prove Kakutani's walk converges exponentially. More precisely, suppose Ω is simply connected and $z_0 \in \Omega$. Fix $0 < \lambda < 1$ and iteratively define a random walk in Ω by setting $r_n = \text{dist}(z_n, \partial\Omega)$ and choosing z_{n+1} uniformly on the circle $\{|z - z_n| = \lambda r_n\}$. Prove that there $0 < a, b < 1$ so that $\mathbb{P}(r_n \geq a^n) \leq b^n$.

Exercise 7.16   Let $A \subset [0,1]$ be a compact set with $\dim(A) > 1/2$ and let Z be the zero set of a linear Brownian motion. Then $\dim(A \cap Z) > 0$ with positive probability. See Kaufman (1972).

• **Exercise 7.17**   Let B_t be one-dimensional Brownian motion. Assume f is a real-valued function so that (f, B) almost surely doubles the Hausdorff dimension of every compact set in $\mathbb{R}$. If $\alpha > 0$ then there is no set $A \subset [0,1]$ with dimension $> 1/2$ so that f restricted to A is α-Hölder. See Balka and Peres (2014).

• **Exercise 7.18** Let B_t be linear Brownian motion and assume $\alpha > 1/2$. Then almost surely there is no set A with $\dim(A) > 1/2$ so that B is α-Hölder on A. See Balka and Peres (2014).

Exercise 7.19 Let B_t be standard linear Brownian motion. Then for every $A \subset [0,1]$ we almost surely have

$$\underline{\dim}_{\mathcal{H}}(B(A)) \geq \frac{2\underline{\dim}_{\mathcal{H}}A}{1 + \underline{\dim}_{\mathcal{H}}A}, \quad \text{and} \quad \overline{\dim}_{\mathcal{H}}(B(A)) \geq \frac{2\overline{\dim}_{\mathcal{H}}A}{1 + \overline{\dim}_{\mathcal{H}}A}.$$

This is from Charmoy et al. (2014). Analogous results for packing dimension were obtained in Talagrand and Xiao (1996).

Exercise 7.20 Show equality in Exercise 7.19 can be attained (use Exercise 1.3).

Exercise 7.21 Show that a set $K \subset \mathbb{R}^2$ has positive logarithmic capacity (see the discussion following Theorem 3.4.2) if and only if $K \times [0,1] \subset \mathbb{R}^3$ has positive capacity for the kernel $|x|^{-1}$.

Exercise 7.22 Use Exercise 7.21 to show that a set $K \subset \mathbb{R}^2$ is hit by 2-dimensional Brownian motion if and only if it has positive logarithmic capacity.

Exercise 7.23 Show that a set $K \subset \mathbb{R}^2$ is hit by 2-dimensional Brownian motion with positive probability if and only if it is hit almost surely.

Exercise 7.24 Suppose **C** is the middle thirds Cantor set and we start a 2-dimensional Brownian motion at a point $z \in \mathbb{R}$ that is distance $r < 1$ from **C**. Since the Cantor set has positive capacity the Brownian motion will almost surely hit the Cantor set at a point w. Show that

$$\mathbb{P}(|w - z| > n) = O(n^{\log 2/\log 3}),$$

where the probability measure is the counting measure

A more general result relating distances and Minkowski dimension is given in Batakis, Levitz and Zinsmeister (2011).

8

Random walks, Markov chains and capacity

Given a random process on a space X, what is the probability that the process eventually hits some subset $A \subset X$? What is the probability that it hits A infinitely often? In this chapter, we will consider these problems for discrete Markov processes and show that the answers are given in terms of capacities with respect to kernels built from the Green's function of the process. We give applications to the simple random walk on $\mathbb{Z}^d$ and deduce an elegant result of Russell Lyons concerning percolation on trees.

8.1 Frostman's theory for discrete sets

In this section we discuss some ways of measuring the size of a discrete set, usually a subset of $\mathbb{N}$ or $\mathbb{Z}^d$. The discussion is very similar to the one on capacity in $\mathbb{R}^d$ in Chapter 3.

Definition 8.1.1 Let Λ be a set and β a σ-algebra of subsets of Λ. Given a measurable function $F : \Lambda \times \Lambda \to [0, \infty]$ with respect to the product σ-algebra, and a finite positive measure μ on (Λ, β), the F-**energy** of μ is

$$\mathscr{E}_F(\mu) = \int_\Lambda \int_\Lambda F(x, y) \, d\mu(x) \, d\mu(y).$$

The **capacity** of Λ in the kernel F is

$$\mathrm{Cap}_F(\Lambda) = \left[\inf_\mu \mathscr{E}_F(\mu) \right]^{-1}$$

where the infimum is over probability measures μ on (Λ, β), and by convention, $\infty^{-1} = 0$.

If Λ is contained in Euclidean space, we always take β to be the Borel

σ-algebra; if Λ is countable, we take β to be the σ-algebra of all subsets. When Λ is countable we also define the **asymptotic capacity** of Λ in the kernel F:

$$\mathrm{Cap}_F^\infty(\Lambda) = \inf_{\{\Lambda_0 \text{ finite}\}} \mathrm{Cap}_F(\Lambda \setminus \Lambda_0). \qquad (8.1.1)$$

For a subset $A \subset \mathbb{Z}^d$ we also have an analog of Hausdorff dimension: a cube in $\mathbb{Z}^d$ is a set of the form $[a_1, a_1 + n] \times \cdots \times [a_d, a_d + n]$, where $n \geq 1$. Given such a cube Q we let $d(Q)$ be the distance of the farthest point in Q from 0 and let $|Q|$ be the diameter of Q. Both distance and diameters in $\mathbb{Z}^d$ will be measured with respect to some norm on $\mathbb{Z}^d$, which we denote $|\cdot|$. This could be $\|\cdot\|_2, \|\cdot\|_\infty$, or any other norm; the particular choice will not be important since any two norms on a finite-dimensional space are comparable (certain constants in some argument will depend on the dimension d and the choice of norm, but this will not affect any of our results).

Definition 8.1.2   If $A \subset \mathbb{Z}^d$, define the **discrete Hausdorff content**

$$\mathscr{H}_D^\alpha(A) = \inf \sum_j \Big(\frac{|Q_j|}{d(Q_j)} \Big)^\alpha,$$

where the infimum is over all coverings of A by cubes. The **discrete Hausdorff dimension** of A, $\dim_D(A)$, is the infimum of the $\alpha > 0$ such that $\mathscr{H}_D^\alpha(A) < \infty$. (See Barlow and Taylor, 1992.)

As for subsets of $\mathbb{R}^d$, it suffices in the definition to consider only coverings by dyadic cubes, i.e., cubes of the form

$$[a_1 2^n, (a_1 + 1) 2^n] \times \cdots \times [a_d 2^n, (a_d + 1) 2^n].$$

These cubes are nested, and so may be considered as the vertices of an infinite tree; for each dyadic cube, its parent is the dyadic cube of twice the side-length that contains it. However, unlike the case of dyadic subcubes of $[0,1]^d$, this tree has leaves, but no root (i.e., unit cubes of $\mathbb{Z}^d$ form the leaves of the tree and every cube has a parent). However, the ideas of a flow and cut-set still make sense, a flow still corresponds to a measure on the boundary of the tree ($\partial T = \mathbb{Z}^d$) and cut-sets correspond to coverings by dyadic cubes. (See Section 3.1.)

As in Section 3.4, dimension and capacity are closely related. For $\alpha > 0$, let

$$F_\alpha(x,y) = \frac{|y|^\alpha}{1 + |x - y|^\alpha}$$

denote the Riesz kernels; (as above, $|\cdot|$ denotes our fixed choice of norm on $\mathbb{Z}^d$). The capacity and asymptotic capacity associated to these kernels are denoted Cap_α and $\mathrm{Cap}_\alpha^\infty$, respectively.

Theorem 8.1.3   *Suppose $A \subset \mathbb{Z}^d$. Then*

(1) *For all $\alpha > \beta > 0$, we have $\mathrm{Cap}_\beta(A) \geq C_{\alpha,\beta}\mathcal{H}_D^\alpha(A)$, where $C_{\alpha,\beta}$ is a positive number that depends only on α and β.*

(2) *If $\mathrm{Cap}_\alpha^\infty(A) > 0$ then $\mathcal{H}_D^\alpha(A) = \infty$.*

(3) *$\dim_D(A) = \inf\{\alpha : \mathrm{Cap}_\alpha^\infty(A) = 0\}$.*

Proof   The proof is similar to that of Theorem 3.4.2, which the reader may wish to review before continuing. Dyadic cubes are nested, so the dyadic cubes that intersect A may be considered as a tree. If we define the conductance of an edge connecting cubes $Q' \subset Q$ to be $(|Q|/d(Q))^\alpha$, then a legal flow on the tree is exactly a measure on A such that $\mu(Q) \leq (|Q|/d(Q))^\alpha$, for all cubes Q. The norm of a flow is the mass of the measure.

A cut-set is a covering of A by dyadic cubes $\{Q_j\}$ and the norm of the flow through the cut-set is at most the sum $\sum_j(|Q_j|/d(Q_j))^\alpha$. Thus $\mathcal{H}_D^\alpha(A)$ is the infimum of all such cut-sets sums. By the MaxFlow–MinCut Theorem (Theorem 3.1.5 and Corollary 3.1.6) there is a positive measure μ on A so that $\|\mu\| = \mathcal{H}_D^\alpha(A)$ and $\mu(Q_j) \leq (|Q_j|/d(Q_j))^\alpha$ for all dyadic cubes Q_j. Given y, we want to show that if $\beta < \alpha$ then

$$\int F_\beta(x,y)\,d\mu(x) \leq C,$$

where C is a positive constant depending only on β, α and $\|\mu\|$. Write

$$\int F_\beta(x,y)\,d\mu(x) = \int_{x:|x-y|>\frac{1}{2}|y|} F_\beta(x,y)\,d\mu(x) + \int_{x:|x-y|\leq\frac{1}{2}|y|} F_\beta(x,y)\,d\mu(x).$$

If $|x-y| > \frac{1}{2}|y|$ then $|F_\beta(x,y)| \leq 2^\beta$, so the first term is bounded by $2^\beta\|\mu\|$. To bound the second term, choose N so that $2^{N-1} \leq |y| < 2^N$, and break the integral into integrals over the sets $A_0 = \{y\}$ and $A_n = \{x : 2^{n-1} \leq |x-y| < 2^n\}$ for $n = 1,\ldots,N-1$. Each A_n can be covered by a bounded number of dyadic cubes of side-length 2^n. Call these cubes $\{Q_j^n\}$. Note that $d(Q_j^n) \sim |y|$. Thus

$$\int_{x:|x-y|\leq\frac{1}{2}|y|} F_\beta(x,y)\,d\mu(x) \leq \sum_{n=0}^{N-1}\int_{A_n}\frac{|y|^\beta}{1+|x-y|^\beta}\,d\mu(x)$$

$$\leq |y|^{\beta-\alpha} + \sum_{n=1}^{N-1}\sum_j\frac{|y|^\beta}{1+2^{(n-1)\beta}}|Q_j^n|^\alpha d(Q_j^n)^{-\alpha}$$

$$\leq C_1|y|^{\beta-\alpha}\sum_{n=0}^{N-1}2^{n(\alpha-\beta)}.$$

If $\beta < \alpha$ then the exponent in the sum is positive, so the sum is bounded by a constant times $2^{N(\alpha-\beta)}$. Thus the integral is bounded by

$$C_2|y|^{\beta-\alpha}2^{N(\alpha-\beta)} \leq C_3|y|^{\beta-\alpha}|y|^{\alpha-\beta} \leq C_3.$$

Thus

$$\mathcal{E}_\beta(\mu) = \int\int F_\beta(x,y)\,d\mu(y)\,d\mu(x),$$

is uniformly bounded. By normalizing μ so that it be a probability measure we deduce that $\mathrm{Cap}_\beta(A) \geq C_{\alpha,\beta}\mathscr{H}_D^\alpha(A)$. This proves (1).

To prove (2), suppose that A has positive α-capacity, i.e., there is a probability measure μ on A such that

$$\mathcal{E}_\alpha(\mu) = \int\int F_\alpha(x,y)\,d\mu(y)\,d\mu(x) = M.$$

Then by Markov's inequality the set $B \subset A$ of ys such that

$$\int F_\alpha(x,y)\,d\mu(x) < 2M,$$

satisfies $\mu(B) \geq \frac{1}{2}$.

Next assume that A satisfies $\mathscr{H}_D^\alpha(A) < 2^{-\alpha}$ and take a covering $\{Q_j\}$ of B such that $\sum_j |Q_j|^\alpha d(Q_j)^{-\alpha} \leq 2^{-\alpha}$. In particular every cube in the covering satisfies $d(Q_j) \geq 2|Q_j|$.

Suppose y is a point of B contained in a cube Q_j. If x is another point of Q_j, then

$$F_\alpha(x,y) = \frac{|y|^\alpha}{1+|x-y|^\alpha} \geq \frac{d(Q_j)^\alpha}{2^\alpha(1+|Q_j|^\alpha)} \geq \frac{1}{2^{\alpha+1}}d(Q_j)^\alpha|Q_j|^{-\alpha}.$$

Thus

$$d(Q_j)^\alpha|Q_j|^{-\alpha}\mu(Q_j) \leq 2^{\alpha+1}\int_{Q_j} F_\alpha(x,y)\,d\mu(x) < 2^{\alpha+2}M,$$

so

$$\mu(Q_j) < 2^{\alpha+2}M|Q_j|^\alpha d(Q_j)^{-\alpha}.$$

Therefore we obtain

$$\frac{1}{2} \leq \mu(B) \leq \sum_j \mu(Q_j) < 2^{\alpha+2}M\sum_j |Q_j|^\alpha d(Q_j)^{-\alpha}.$$

Taking the infimum over all coverings, we deduce

$$\mathrm{Cap}_\alpha(A) \leq \frac{1}{M} \leq 2^{\alpha+3}\mathscr{H}_D^\alpha(B) \leq 2^{\alpha+3}\mathscr{H}_D^\alpha(A).$$

To prove the second part of (2) note that if $\mathscr{H}_D^\alpha(A) < \infty$ and $\{Q_j\}$ is a cover of A with $\sum_j |Q_j|^\alpha d(Q_j)^{-\alpha} < \infty$, then setting $A_n = A \setminus \bigcup_{j=1}^n Q_j$, we get

$\mathcal{H}_D^\alpha(A_n) \to 0$. By the previous argument this implies $\text{Cap}_\alpha(A_n) \to 0$ which implies $\text{Cap}_\alpha^\infty(A) = 0$, a contradiction.

Finally, we prove (3). If $\alpha > \dim_D(A)$ then $\mathcal{H}_D^\alpha(A) < \infty$ by definition so $\text{Cap}_\alpha^\infty(A) = 0$ by part (2). Thus

$$\dim_D(A) \geq \inf\{\alpha : \text{Cap}_\alpha^\infty(A) = 0\}.$$

On the other hand, if $\alpha < \dim_D(A)$ then $\mathcal{H}_D^\alpha(A) = \infty$, and for any finite subset A_0 of A, $\mathcal{H}_D^\alpha(A \setminus A_0) = \infty > 1$. So using part (1), for any $\beta < \alpha$, we have $\text{Cap}_\beta(A \setminus A_0) > C_{\alpha,\beta} > 0$. Thus $\text{Cap}_\beta^\infty(A) > 0$ and so

$$\dim_D(A) = \inf\{\alpha : \text{Cap}_\alpha^\infty(A) = 0\}. \qquad \square$$

Example 8.1.4 $A = \{\lfloor n^a \rfloor\}$ with $a > 1$. By considering the covering of A by unit cubes we see that

$$\mathcal{H}_D^\beta(A) \leq \sum_{k \in A} k^{-\beta} \leq \sum_n (n^a)^{-\beta}.$$

Thus $\mathcal{H}_D^\beta(A) < \infty$ for $\beta > 1/a$. Thus $\dim_D(A) \leq 1/a$. To prove equality, let $A_n = A \cap [2^n, 2^{n+1})$ and consider the probability measure μ_n on A that gives all $\sim 2^{n/a}$ points in A_n equal mass. Also note that the points of A_n are separated by at least distance $2^{n(1-1/a)}$. Then

$$\begin{aligned}
\int_{A_n} F_\beta(x,y)\, d\mu(x) &\leq \sum_{k=0}^{C_1 2^{n/a}} \frac{2^{n\beta}}{1 + |k 2^{n(1-1/a)}|^\beta} C_1 2^{-n/a} \\
&\leq C_1 2^{n\beta - n/a} \sum_{k=0}^{C_1 2^{n/a}} \frac{1}{1 + |k 2^{n(1-1/a)}|^\beta} \\
&\leq C_1 2^{n\beta - n/a} \left(1 + 2^{-n\beta(1-1/a)} \sum_{k=1}^{C_1 2^{n/a}} k^{-\beta} \right) \\
&\leq C_1 2^{n\beta - n/a} (1 + C_2 2^{-n\beta(1-1/a)} 2^{n(1-\beta)/a}) \\
&\leq C_1 2^{n\beta - n/a} + C_3.
\end{aligned}$$

Thus the integral is bounded if $\beta \leq 1/a$. Hence $\mathscr{E}_\beta(\mu_n)$ is uniformly bounded and we conclude

$$\text{Cap}_\beta^\infty(A) > 0 \text{ if and only if } \beta \leq \frac{1}{a}.$$

Thus $\dim_D(A) = 1/a$ by Theorem 8.1.3.

Example 8.1.5 **Integer Cantor sets.** If $b \geq 2$ is an integer, consider a set of allowed "digits" $D \subset \{0, 1, \ldots, b-1\}$ define $A = \{\sum a_n b^n, a_n \in D, n \geq 0\}$. Let

$A_n = A \cap [b^n, b^{n+1})$ and let d be the number of elements in D. Then A_n contains d^n points each at distance $\sim b^n$ to the origin. Thus

$$\sum_{k \in A} k^{-\beta} = \sum_{j=1}^{\infty} \sum_{k \in A_j} k^{-\beta} \leq \sum_{j=1}^{\infty} d^j b^{-j\beta}.$$

This converges if $\beta > \log_b d$, and so as in the previous example we deduce $\dim_D(A) \leq \log_b d$.

To prove equality we produce a probability measure on A_n that has energy bounded independently of n. As in the previous example, it suffices to take the measure that gives equal mass to each of the d^n points. Fixing $y \in A_n$, we get

$$\int_{A_n} F_\beta(x,y) \, d\mu(x) \leq \sum_{k=0}^{n+1} \sum_{b^{k-1} < |x-y| \leq b^k} \frac{b^{n\beta} d^{-n}}{1 + |x-y|^\beta} + \sum_{|x-y| \leq 1} \frac{b^{n\beta} d^{-n}}{1 + |x-y|^\beta}$$

$$\leq \sum_{k=0}^{n+1} \sum_{b^{k-1} < |x-y| \leq b^k} \frac{b^{n\beta}}{1 + b^{(k-1)\beta}} d^{-n} + 3 b^{n\beta} d^{-n}$$

$$\leq C_1 b^{n(\beta - \log_b d)} \left(1 + \sum_{k=1}^{n+1} \sum_{b^{k-1} < |x-y| \leq b^k} \frac{1}{1 + b^{k\beta}} \right) +$$

$$\leq C_1 b^{n(\beta - \log_b d)} \left(1 + \sum_{k=1}^{n} \frac{d^k}{1 + b^{k\beta}} \right)$$

$$\leq C_1 b^{n(\beta - \log_b d)} \left(1 + \sum_{k=1}^{n} b^{k(\log_b d - \beta)} \right).$$

If $\beta < \log_b d$ the exponent in the sum is positive and the sum is bounded by a multiple of its largest term. Thus the integral is bounded by

$$C_1 b^{n(\beta - \log_b d)} b^{n(\log_b d - \beta)} \leq C_2.$$

Therefore $\mathrm{Cap}_\beta^\infty(A) > 0$ if $\beta < \log_b d$. Thus $\dim(A) = \log_b d$. Note that we have not computed $\mathrm{Cap}_\beta^\infty(A)$ when $\beta = \log_b d$; see Exercise 8.15.

Example 8.1.6 Let $A = \bigcup_n [2^n, 2^n + 2^{\alpha n}]$, where $0 < \alpha < 1$. In this case

$$\sum_{k \in A} k^{-\beta} = \sum_{n=1}^{\infty} \sum_{k \in A \cap [2^n, 2^{n+1})} k^{-\beta} \sim \sum_{n=1}^{\infty} 2^{n\alpha - n\beta}$$

converges if and only if $\beta > \alpha$ which implies $\dim_D(A) \leq \alpha$. Unlike the previous examples, this is not sharp. In fact, taking the obvious covering of A by the intervals $I_n = [2^n, 2^n + 2^{\alpha n})$, we get

$$\sum_n |I_n|^\beta d(I_n)^{-\beta} \leq \sum_n 2^{n\alpha\beta} 2^{-n\beta} = \sum_n 2^{n(\alpha - 1)\beta},$$

which converges for any $\beta > 0$. Thus $\dim_D(A) = 0$.

Example 8.1.7   We can make a discrete analog of the random Cantor sets considered in Section 3.7. Break $\mathbb{N}$ into the intervals $I_n = [2^n, 2^{n+1})$ and for each n perform an independent percolation with parameter p on an n level binary tree. The level n vertices that can be connected to the root by kept edges defines a random subset on I_n. The union of these random subsets over n defines our discrete random Cantor set. (For the definition of percolation on trees see Definition 8.4.1.) We leave it as Exercise 8.8 to show that the discrete Hausdorff dimension is $1 + \log_2 p$.

8.2 Markov chains and capacity

A **Markov chain** on a countable space Y is a sequence of random variables $\{X_n\}$ with values in Y, with a transition function $p(x,y) : Y \times Y \to [0,1]$ such that for all $x_0, x_1, \ldots, x_{n-1}, x, y \in Y$,

$$\mathbb{P}(X_{n+1} = y | X_n = x, X_{n-1} = x_{n-1}, \ldots, X_0 = x_0) = p(x,y).$$

The n-step transition function is

$$p^{(n)}(x,y) = \mathbb{P}(X_n = y | X_0 = x),$$

and can be computed from p using the iteration

$$p^{(n)}(x,y) = \sum_{z \in Y} p(z,y) p^{(n-1)}(x,z).$$

Note that we must have

$$\sum_{y \in Y} p(x,y) = 1$$

for every $x \in Y$. Moreover, given any p with this property and any initial state $X_0 = \rho \in Y$ there is a corresponding Markov chain.

We say that a state y in a Markov chain is **recurrent** if the probability of returning to y given that we start at y is 1. Otherwise the state is **transient**. If every state is recurrent (transient), we say the Markov chain is **recurrent** (**transient**, respectively). We consider as transient also a chain that has a finite life time and then transitions to a cemetery state. Note that it is possible for a chain to be neither recurrent nor transient (i.e., it could have states of both types).

Example 8.2.1   Consider the probability space $[0,1]$ with Lebesgue measure. Let $X_n : [0,1] \to \{0,1\}$ be the nth binary digit of x. Then $\{X_n\}$ is a Markov chain on the two element set $X = \{0,1\}$ with $p(x,y) = \frac{1}{2}$ everywhere.

Example 8.2.2 The most familiar example is the simple random walk on the integers. Here $Y = \mathbb{Z}$ and $p(x,y) = \frac{1}{2}$ if $|x - y| = 1$ and is zero otherwise. Similarly, we define the simple random walk on $\mathbb{Z}^d$ by $p(x,y) = \frac{1}{2d}$ if $\sum_{j=1}^{d} |x_j - y_j| = 1$ and $p(x,y) = 0$ otherwise. It is a well-known result that this walk is recurrent if $d = 1, 2$ and is transient if $d \geq 3$.

Example 8.2.3 Define a transition function $p : \mathbb{N} \times \mathbb{N} \rightarrow [0,1]$ by $p(n,0) = \varepsilon_n$, $p(n,n+1) = 1 - \varepsilon_n$ and $p(n,m) = 0$ otherwise. Then it is easy to see that the corresponding Markov chain is recurrent if and only if $\sum_n \varepsilon_n = \infty$, and is transient otherwise.

Given a Markov chain, define the Green's function $G(x,y)$ as the expected number of visits to y starting at x, i.e.,

$$G(x,y) = \sum_{n=0}^{\infty} p^{(n)}(x,y) = \sum_{n=0}^{\infty} \mathbb{P}_x[X_n = y]$$

where $\mathbb{P}_x$ is the law of the chain $\{X_n : n \geq 0\}$ when $X_0 = x$. Observe that if $G(x,y) < \infty$ for all x and y in Y, then the chain must be transient (if there was a positive probability of hitting y infinitely often then the expected number of returns to y is infinite). Also note that G satisfies the following "mean value property"

$$G(x,y) = \sum_{z \in Y} G(z,y)p(x,z), \quad \text{for } y \neq x,$$

i.e., it is a discrete harmonic function away from the initial state.

Theorem 8.2.4 *Let $\{X_n\}$ be a transient Markov chain on the countable state space Y with initial state ρ and transition probabilities $p(x,y)$. For any subset Λ of Y we have*

$$\frac{1}{2}\text{Cap}_K(\Lambda) \leq \mathbb{P}_\rho[\exists n \geq 0 : X_n \in \Lambda] \leq \text{Cap}_K(\Lambda) \qquad (8.2.1)$$

and

$$\frac{1}{2}\text{Cap}_K^\infty(\Lambda) \leq \mathbb{P}_\rho[X_n \in \Lambda \text{ infinitely often}] \leq \text{Cap}_K^\infty(\Lambda) \qquad (8.2.2)$$

where K is the Martin kernel

$$K(x,y) = \frac{G(x,y)}{G(\rho,y)} \qquad (8.2.3)$$

defined using the initial state ρ.

The Martin kernel $K(x,y)$ can obviously be replaced by the symmetric kernel $\frac{1}{2}(K(x,y) + K(y,x))$ without affecting the energy of any measure or the capacity of any set.

If the Markov chain starts according to an initial measure π on the state space, rather than from a fixed initial state, the theorem may be applied by adding an abstract initial state ρ.

Proof of Theorem 8.2.4   The right-hand inequality in (8.2.1) will follow from an entrance time decomposition. Let τ be the first hitting time of Λ and let v be the hitting measure $v(x) = \mathbb{P}_\rho[X_\tau = x]$ for $x \in \Lambda$. Note that v may be defective, i.e., of total mass less than 1. In fact,

$$v(\Lambda) = \mathbb{P}_\rho[\exists n \geq 0 : X_n \in \Lambda]. \tag{8.2.4}$$

Note that we may assume $v(\Lambda) > 0$, for otherwise the desired inequality is trivial. Now for all $y \in \Lambda$:

$$\int G(x,y)\,dv(x) = \sum_{x \in \Lambda} \mathbb{P}_\rho[X_\tau = x]G(x,y) = G(\rho,y).$$

Thus $\int K(x,y)\,dv(x) = 1$ for every $y \in \Lambda$. Consequently,

$$\mathscr{E}_K\left(\frac{v}{v(\Lambda)}\right) = v(\Lambda)^{-2}\mathscr{E}_K(v) = v(\Lambda)^{-1},$$

so that $\mathrm{Cap}_K(\Lambda) \geq v(\Lambda)$. By (8.2.4), this proves the right half of (8.2.1).

To establish the left-hand inequality in (8.2.1), we use the second moment method. Given a probability measure μ on Λ, consider the random variable

$$Z = \int_\Lambda G(\rho,y)^{-1} \sum_{n=0}^\infty \mathbf{1}_{\{X_n=y\}}\,d\mu(y).$$

By Fubini's Theorem and the definition of G,

$$\mathbb{E}_\rho Z = \mathbb{E}_\rho \int_\Lambda G(\rho,y)^{-1} \sum_{n=0}^\infty \mathbf{1}_{\{X_n=y\}}\,d\mu(y)$$

$$= \int_\Lambda G(\rho,y)^{-1} \sum_{n=0}^\infty \mathbb{P}(X_n = y|X_0 = \rho)\,d\mu(y)$$

$$= \int_\Lambda G(\rho,y)^{-1}G(\rho,y)\,d\mu(y)$$

$$= 1.$$

Thus by Cauchy–Schwarz

$$1 = (\mathbb{E}_\rho Z)^2 \leq \mathbb{E}_\rho(Z^2)\mathbb{E}_\rho(\mathbf{1}_{Z>0}),$$

and hence

$$\mathbb{P}_\rho[\exists n \geq 0 : X_n \in \Lambda] \geq \mathbb{P}_\rho(Z > 0) \geq \mathbb{E}_\rho(Z^2)^{-1}.$$

Now we bound the second moment:

$$\mathbb{E}_\rho Z^2 = \mathbb{E}_\rho \int_\Lambda \int_\Lambda G(\rho,y)^{-1}G(\rho,x)^{-1} \sum_{m,n=0}^{\infty} \mathbf{1}_{\{X_m=x,X_n=y\}}\, d\mu(x)\,d\mu(y)$$

$$\leq 2\mathbb{E}_\rho \int_\Lambda \int_\Lambda G(\rho,y)^{-1}G(\rho,x)^{-1} \sum_{0\leq m\leq n<\infty} \mathbf{1}_{\{X_m=x,X_n=y\}}\, d\mu(x)\,d\mu(y).$$

(Note that this is not equality since the diagonal terms are counted twice.) For each m we have

$$\mathbb{E}_\rho \sum_{m\leq n}^{\infty} \mathbf{1}_{\{X_m=x,X_n=y\}} = \mathbb{P}_\rho[X_m=x]G(x,y).$$

Summing this over all $m \geq 0$ yields $G(\rho,x)G(x,y)$, and therefore

$$\mathbb{E}_\rho Z^2 \leq 2\int_\Lambda \int_\Lambda G(\rho,y)^{-1}G(x,y)\, d\mu(x)\,d\mu(y) = 2\mathscr{E}_K(\mu).$$

Thus

$$\mathbb{P}_\rho[\exists n \geq 0 : X_n \in \Lambda] \geq \frac{1}{2\mathscr{E}_K(\mu)}.$$

Since the left-hand side does not depend on μ, we conclude that

$$\mathbb{P}_\rho[\exists n \geq 0 : X_n \in \Lambda] \geq \frac{1}{2}\mathrm{Cap}_K(\Lambda)$$

as claimed.

To infer (8.2.2) from (8.2.1) observe that since $\{X_n\}$ is a transient chain, almost surely every state is visited only finitely often and therefore

$$\{X_n \in \Lambda \text{ infinitely often }\} = \bigcap_{\Lambda_0 \text{ finite}} \{\exists n \geq 0 : X_n \in \Lambda \setminus \Lambda_0\} \quad \text{a.s.}$$

Applying (8.2.1) and the definition (8.1.1) of asymptotic capacity yields (8.2.2). $\qquad\square$

Example 8.2.5 Perhaps the best known example of a transient Markov chain is the simple random walk on $\mathbb{Z}^3$, for which the Green's function is known to satisfy $G(x,y) \asymp |x-y|^{-1}$ (see Exercise 8.16 or Chapter 1 in Lawler (1991)). Thus a set in $\mathbb{Z}^3$ is hit infinitely often if and only if it has positive asymptotic capacity for the kernel

$$F_1(x,y) = \frac{|y|}{|x-y|}.$$

Based on Example 8.1.4, we see that a 3-dimensional simple random walk hits $\mathbb{N} \times \{0\} \times \{0\} \subset \mathbb{Z}^3$ infinitely often, but only hits $\{(\lfloor n^a \rfloor,0,0) : n \in \mathbb{N}\}$ finitely often for any $a > 1$.

8.3 Intersection equivalence and return times

This section is devoted to deriving some consequences of Theorem 8.2.4. The first involves a widely applicable equivalence relation between distributions of random sets.

Definition 8.3.1   Say that two random subsets W_1 and W_2 of a countable space are **intersection-equivalent** (or more precisely, that their **laws** are intersection-equivalent) if there exist positive finite constants C_1 and C_2, such that for every subset A of the space,

$$C_1 \le \frac{\mathbb{P}[W_1 \cap A \neq \emptyset]}{\mathbb{P}[W_2 \cap A \neq \emptyset]} \le C_2.$$

It is easy to see that if W_1 and W_2 are intersection-equivalent then

$$C_1 \le \frac{\mathbb{P}[\#(W_1 \cap A) = \infty]}{\mathbb{P}[\#(W_2 \cap A) = \infty]} \le C_2,$$

for all sets A, with the same constants C_1 and C_2. An immediate corollary of Theorem 8.2.4 is the following, one instance of which is given in Corollary 8.3.8.

Corollary 8.3.2   *Suppose Green's functions for two transient Markov chains on the same state space (with the same initial state) are bounded by constant multiples of each other. (It suffices that this bounded ratio property holds for the corresponding Martin kernels $K(x,y)$ or for the symmetric version $K(x,y) + K(y,x)$.) Then the ranges of the two chains are intersection-equivalent.*

Lamperti (1963) gave an alternative criterion for a transient Markov chain $\{X_n\}$ to visit the set Λ infinitely often. This is essentially a variant of Wiener's criteria for whether a set has positive capacity.

Theorem 8.3.3 (Lamperti's test)   *With the notation as in Theorem 8.2.4, fix $b > 1$ and let $Y(n) = \{x \in Y : b^{-n-1} < G(\rho,x) \le b^{-n}\}$. Assume that the set $\{x \in Y : G(\rho,x) > 1\}$ is finite. Also, assume that there exists a finite constant C such that for all sufficiently large m and n we have*

$$G(x,y) < Cb^{-(m+n)}, \qquad G(y,x) < Cb^{-n} \tag{8.3.1}$$

for all $x \in Y(m)$ and $y \in Y(m+n)$. Then

$$\mathbb{P}_\rho[X_n \in \Lambda \text{ infinitely often}] > 0 \tag{8.3.2}$$

$$\Longleftrightarrow \sum_{n=1}^{\infty} b^{-n} \mathrm{Cap}_G(\Lambda \cap Y(n)) = \infty.$$

Sketch of proof: Clearly

$$\sum_{n=1}^{\infty} b^{-n} \text{Cap}_G(\Lambda \cap Y(n)) = \infty \quad \text{if and only if} \quad \sum_n \text{Cap}_K(\Lambda \cap Y(n)) = \infty.$$

By Theorem 8.2.4, the series on the right above is comparable to $\sum \mathbb{P}(A_k)$, where A_k is the event $\{\exists n : X_n \in \Lambda \cap Y(k)\}$. If the series converges, then the Borel–Cantelli lemma says that almost surely, only finitely many A_k occur.

On the other hand, (8.3.1) implies that the $\{A_k\}$ are quasi-independent, i.e.,

$$\mathbb{P}(A_k \cap A_j) \leq C \cdot \mathbb{P}(A_k) \cdot \mathbb{P}(A_j)$$

(see Exercise 8.36). The Borel–Cantelli lemma for quasi-independent events then says that $\sum_k \mathbb{P}(A_k) = \infty$ implies that infinitely many of the A_k occur with positive probability. This version of the Borel–Cantelli lemma is proved in Lamperti's paper. A better proof is in Kochen and Stone (1964) and is sketched in Exercises 8.34 and 8.35. □

Lamperti's test is useful in many cases; however, condition (8.3.1) excludes some natural transient chains such as simple random walk on a binary tree. Next, we deduce from Theorem 8.2.4 a criterion for a recurrent Markov chain to visit its initial state infinitely often within a prescribed time set.

Corollary 8.3.4 *Let $\{X_n\}$ be a recurrent Markov chain on the countable state space Y, with initial state $X_0 = \rho$ and transition probabilities $p(x,y)$. For non-negative integers $m \leq n$ denote*

$$\widetilde{G}(m,n) = \mathbb{P}[X_n = \rho | X_m = \rho] = p^{(n-m)}(\rho,\rho)$$

and

$$\widetilde{K}(m,n) = \frac{\widetilde{G}(m,n)}{\widetilde{G}(0,n)}.$$

Then for any set of times $A \subseteq \mathbb{Z}^+$:

$$\frac{1}{2} \text{Cap}_{\widetilde{K}}(A) \leq \mathbb{P}_\rho[\exists n \in A : X_n = \rho] \leq \text{Cap}_{\widetilde{K}}(A) \tag{8.3.3}$$

and

$$\frac{1}{2} \text{Cap}_{\widetilde{K}}^{\infty}(A) \leq \mathbb{P}_\rho\left[\sum_{n \in A} \mathbf{1}_{\{X_n=\rho\}} = \infty\right] \leq \text{Cap}_{\widetilde{K}}^{\infty}(A). \tag{8.3.4}$$

Proof We consider the space-time chain $\{(X_n,n) : n \geq 0\}$ on the state space $Y \times \mathbb{Z}^+$. This chain is obviously transient since the second coordinate tends to infinity; let G denote its Green's function. Since $G((\rho,m),(\rho,n)) = \widetilde{G}(m,n)$

for $m \leq n$, applying Theorem 8.2.4 with the set $\Lambda = \{\rho\} \times A$ shows that (8.3.3) and (8.3.4) follow respectively from (8.2.1) and (8.2.2). $\qquad\square$

The next few examples make use of the Local Central Limit Theorem, e.g., Theorem 2.5.2 in Durrett (1996) or 7.P10 in Spitzer (1964). In fact, all we will need is the simpler consequence that if S_n is a mean zero, finite variance, aperiodic random walk on $\mathbb{N}$ then

$$\lim_{n \to \infty} \sqrt{n}\mathbb{P}(S_n = 0) = c > 0.$$

Example 8.3.5 **Does a random walk on $\mathbb{Z}$ return to 0 during A?** Let S_n be the partial sums of mean-zero finite variance i.i.d. integer random variables. By the Local Central Limit Theorem (see Spitzer, 1964),

$$\widetilde{G}(0,n) = \mathbb{P}[S_n = 0] \asymp n^{-1/2}$$

provided that the summands $S_n - S_{n-1}$ are aperiodic. Therefore

$$\mathbb{P}\Big[\sum_{n \in A} \mathbf{1}_{\{S_n = 0\}} = \infty\Big] > 0 \Longleftrightarrow \mathrm{Cap}_F^\infty(A) > 0, \tag{8.3.5}$$

with $F(m,n) = (n^{1/2}/(n-m+1)^{1/2})\mathbf{1}_{\{m \leq n\}}$. By the Hewitt–Savage zero-one law, the event in (8.3.5) must have probability 0 or 1. Consider the special case in which A consists of separated blocks of integers:

$$A = \bigcup_{n=1}^\infty [2^n, 2^n + L_n]. \tag{8.3.6}$$

A standard calculation (e.g., with Lamperti's test applied to the space-time chain) shows that in this case $S_n = 0$ for infinitely many $n \in A$ with probability 1 if and only if $\sum_n L_n^{1/2} 2^{-n/2} = \infty$. On the other hand, the expected number of returns $\sum_{n \in A} \mathbb{P}[S_n = 0]$ is infinite if and only if $\sum_n L_n 2^{-n/2} = \infty$. Thus an infinite expected number of returns in a time set does not suffice for almost sure return in the time set. When the walk is periodic, i.e.

$$r = \gcd\{n : \mathbb{P}[S_n = 0] > 0\} > 1,$$

the same criterion holds as long as A is contained in $r\mathbb{Z}^+$. Similar examples may be found in Ruzsa and Székely (1982) and Lawler (1991).

Example 8.3.6 In some cases, the criterion of Corollary 8.3.4 can be turned around and used to estimate asymptotic capacity. For instance, if $\{S_n'\}$ is an independent random walk with the same distribution as $\{S_n\}$ and A is the random set $A = \{n : S_n' = 0\}$, then the positivity of $\mathrm{Cap}_F^\infty(A)$ for

$$F(m,n) = (n^{1/2}/(n-m+1)^{1/2})\mathbf{1}_{\{m \leq n\}}$$

follows from the recurrence of the planar random walk $\{(S_n, S'_n)\}$. Therefore $\dim_D(A) \geq 1/2$. On the other hand, by the Local Central Limit Theorem,

$$\mathbb{E}(\#(A \cap [0,N])) \leq CN^{1/2},$$

for some positive constant C. This easily implies $\dim_D(A) \leq 1/2$, see Exercise 8.5. Thus the discrete Hausdorff dimension of A is almost surely $1/2$; detailed estimates of the discrete Hausdorff measure of A were obtained by Khoshnevisan (1994).

Example 8.3.7 **Random walk on $\mathbb{Z}^2$.** Now we assume that S_n are partial sums of mean-zero, finite variance i.i.d. random variables in $\mathbb{Z}^2$. We also assume that the distribution of S_1 is not supported on a line. Denote

$$r = \gcd\{n : \mathbb{P}[S_n = 0] > 0\},$$

and let $A \subseteq r\mathbb{Z}^+$. Again, by the Hewitt–Savage zero-one law,

$$\mathbb{P}[S_n = 0 \text{ for infinitely many } n \in A]$$

is zero or one. Therefore

$$\mathbb{P}[S_n = 0 \text{ for infinitely many } n \in A]$$

is one if and only if $\operatorname{Cap}_F^\infty(A) > 0$ where

$$F(m,n) = (n/(1+n-m))\mathbf{1}_{\{m \leq n\}}. \tag{8.3.7}$$

Here we applied the Local Central Limit Theorem to get that

$$\widetilde{G}(0,rn) = \mathbb{P}[S_{rn} = 0] \asymp n^{-1} \text{ as } n \to \infty.$$

For instance, if A consists of disjoint blocks

$$A = \bigcup_n [2^n, 2^n + L_n]$$

then $\operatorname{Cap}_F^\infty(A) > 0$ if and only if $\sum_n 2^{-n} L_n / \log L_n = \infty$. The expected number of returns to zero is infinite if and only if $\sum 2^{-n} L_n = \infty$.

Comparing the kernel F in (8.3.7) with the Martin kernel for simple random walk on $\mathbb{Z}^3$ leads to the next corollary.

Corollary 8.3.8 *For $d = 2, 3$, let $\{S_n^{(d)}\}$ be a truly d-dimensional random walk on the d-dimensional lattice (the linear span of its range is $\mathbb{R}^d$ almost surely), with increments of mean zero and finite variance. Assume that the walks are aperiodic, i.e., the set of positive integers n for which $\mathbb{P}[S_n^{(d)} = 0] > 0$*

has g.c.d. 1. *Then there exist positive finite constants C_1 and C_2 such that for any set of positive integers A,*

$$C_1 \le \frac{\mathbb{P}[S_n^{(2)} = 0 \text{ for some } n \in A]}{\mathbb{P}[S_n^{(3)} \in \{0\} \times \{0\} \times A \text{ for some } n]} \le C_2, \tag{8.3.8}$$

where $\{0\} \times \{0\} \times A = \{(0,0,k) : k \in A\}$. Consequently,

$$\mathbb{P}[S_n^{(2)} = 0 \text{ for infinitely many } n \in A] \tag{8.3.9}$$
$$= \mathbb{P}[S_n^{(3)} \in \{0\} \times \{0\} \times A \text{ infinitely often}].$$

Note that both sides of (8.3.9) take only the values 0 or 1. Corollary 8.3.8 follows from Corollary 8.3.2, in conjunction with Example 8.3.7 above and the asymptotics $G(0,x) \sim c/|x|$ as $|x| \to \infty$ for the random walk $S_n^{(3)}$ (cf. Spitzer, 1964). The Wiener test implies the equality (8.3.9) but not the estimate (8.3.8). To see why Corollary 8.3.8 is surprising, observe that the space-time chain $\{(S_n^{(2)}, n)\}$ travels to infinity faster than $S_n^{(3)}$, yet by Corollary 8.3.8, the same subsets of lattice points on the positive z-axis are hit infinitely often by the two processes.

8.4 Lyons' Theorem on percolation on trees

Theorem 8.2.4 yields a short proof of a fundamental result of R. Lyons concerning percolation on trees.

Definition 8.4.1 Let T be a finite rooted tree. Vertices of degree one in T (apart from the root ρ) are called **leaves,** and the set of leaves is the **boundary** ∂T of T. The set of edges on the path connecting the root to a leaf x is denoted Path(x).

Independent percolation on T is defined as follows. To each edge e of T, a parameter p_e in $[0,1]$ is attached, and e is removed with probability $1 - p_e$, retained with probability p_e, with mutual independence among edges. Say that a leaf x **survives the percolation** if all of Path(x) is retained, and say that the tree boundary ∂T survives if some leaf of T survives.

Theorem 8.4.2 (Lyons (1992)) *Let T be a finite rooted tree. With the notation above, define a kernel F on ∂T by*

$$F(x,y) = \prod_{e \in \text{Path}(x) \cap \text{Path}(y)} p_e^{-1}$$

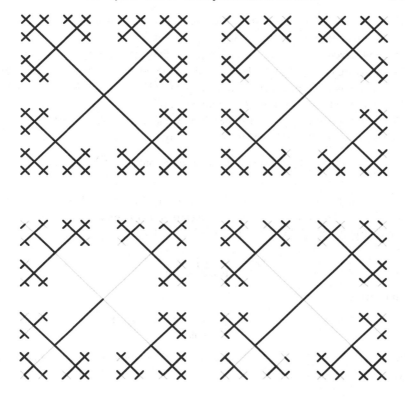

Figure 8.4.1 Percolations on a finite tree where we keep edges with probabilities $p = 1, .8, .6, .4$.

for $x \neq y$ and

$$F(x,x) = 2 \prod_{e \in \text{Path}(x)} p_e^{-1}.$$

Then

$$\text{Cap}_F(\partial T) \leq \mathbb{P}[\partial T \text{ survives the percolation}] \leq 2\text{Cap}_F(\partial T).$$

(The kernel F differs from the kernel used in Lyons (1992) on the diagonal, but this difference is unimportant in all applications).

Proof Embed T in the lower half-plane, with the root at the origin. The random set of $r \geq 0$ leaves that survive the percolation may be enumerated from left to right as $V_1, V_2, \ldots, V_r$. The key observation is that the sequence

$\rho, V_1, V_2, \ldots, V_r, \Delta, \Delta, \ldots$ is a Markov chain on the state space $\partial T \cup \{\rho, \Delta\}$ (here ρ is the root and Δ is a formal absorbing cemetery).

Indeed, *given* that $V_k = x$, all the edges on Path(x) are retained, so survival of leaves to the right of x is determined by the edges strictly to the right of Path(x), and is thus conditionally independent of $V_1, \ldots, V_{k-1}$. This verifies the Markov property, so Theorem 8.2.4 may be applied.

The transition probabilities for the Markov chain above are complicated, but it is easy to write down Green's kernel. Clearly,

$$G(\rho, y) = \mathbb{P}[y \text{ survives the percolation}] = \prod_{e \in \text{Path}(y)} p_e.$$

Also, if x is to the left of y, then $G(x, y)$ is equal to the probability that the range of the Markov chain contains y given that it contains x, which is just the probability of y surviving given that x survives. Therefore

$$G(x, y) = \prod_{e \in \text{Path}(y) \backslash \text{Path}(x)} p_e \mathbf{1}_{\{x \leq y\}}$$

and hence

$$K(x, y) = \frac{G(x, y)}{G(\rho, y)} = \prod_{e \in \text{Path}(x) \cap \text{Path}(y)} p_e^{-1} \mathbf{1}_{\{x \leq y\}}.$$

Thus $K(x, y) + K(y, x) = F(x, y)$ for all $x, y \in \partial T$, and Lyons' Theorem follows from Theorem 8.2.4. □

Lyons' Theorem extends to infinite trees. Given an infinite tree truncate it to level n and apply Lyons' Theorem to compute the probability that percolation survives for n levels. The probability it survives all levels is the limit of this (decreasing) sequence. Then use the Monotone Convergence Theorem in the definition of $\mathscr{E}_F$ to show that the capacities $\text{Cap}_F(\partial T_n)$ converge to $\text{Cap}_F(\partial T)$.

The same method of recognizing a "hidden" Markov chain may be used to prove more general results on random labeling of trees due to Evans (1992) and Lyons (1992).

8.5 Dimension of random Cantor sets (again)

Recall the random Cantor sets discussed in Section 3.7: fix a number $0 < p < 1$ and define a random set by dividing the unit cube $[0, 1]^d$ of $\mathbb{R}^d$ into b^d b-adic congruent closed cubes with disjoint interiors. Each cube is kept with probability p, and the kept cubes are further subdivided and again each subcube is kept with probability p. We shall denote such a set as $\Lambda_d(b, p)$. We denote by $Z_n = Z_n(p)$ the number of cubes kept in the nth generation. We showed in that section that if $p > b^{-d}$ then $\Lambda_d(b,p)$ is a non-empty set of dimension $d + \log_b p$ with positive probability. Here we shall use Lyons' Theorem to prove this again (and a little more). The following combines results of Hawkes (1981) and Lyons (1990).

Theorem 8.5.1 *Let α be a positive real number, d an integer ≥ 1 and b an integer ≥ 2. Set $p = b^{-\alpha}$. Then for any closed set K in $[0,1]^d$,*

$$\mathbb{P}(\Lambda_d(b,p) \cap K \neq \emptyset) \asymp \mathrm{Cap}_\alpha(K),$$

i.e., each of the two terms is bounded by a constant multiple of the other, and these constants do not depend on K.

Proof Recall the rooted tree Γ^{b^d} in which each vertex has b^d children. Let $T = T(K) \subset \Gamma^{b^d}$ be the subtree corresponding to the compact set K (the vertices correspond to b-adic cubes that hit K, and each b-adic cube is connected to its "parent"). Perform percolation on T with $p = b^{-\alpha}$. The probability that K hits $\Lambda_d(b,p)$ is then just the probability that this percolation survives. By Lyons' Theorem (Theorem 8.4.2), the probability that the percolation survives is comparable to the tree capacity of the boundary of T with respect to the kernel $F(\xi, \eta) = f(|\xi \wedge \eta|)$ where $f(n) = b^{-\alpha n}$. By Theorem 3.6.1, this capacity is comparable to the α-capacity of K, as desired. □

Theorem 8.5.2 *Let $p = b^{-\alpha} < 1$. For any closed set $K \subset [0,1]^d$,*

(1) *If $\dim(K) < \alpha$ then $K \cap \Lambda_d(b,p) = \emptyset$ almost surely.*
(2) *If $\dim(K) > \alpha$ then $K \cap \Lambda_d(b,p) \neq \emptyset$ with positive probability.*

Proof Parts (1) and (2) follow from the previous result and Frostman's Theorem (Theorem 3.4.2), which asserts $\dim(K) = \inf\{\alpha : \mathrm{Cap}_\alpha(K) = 0\}$. □

Corollary 8.5.3 *Let $p = b^{-\alpha}$.*

(1) *Let E be a closed set in $[0,1]^d$. Assume $\dim(E) > \alpha$. If $\beta < \dim(E) - \alpha$ then $\dim(E \cap \Lambda_d(b,p)) \geq \beta$ with positive probability. If $\beta > \dim(E) - \alpha$ then $\dim(E \cap \Lambda_d(b,p)) \leq \beta$ almost surely.*
(2) *Assume $d > \alpha$. Then $\dim(\Lambda_d(b,p)) = d + \log_b p$ almost surely, conditioned on $\Lambda_d(b,p)$ being non-empty.*

Proof (1) Suppose $\Lambda_d(b,p)$ and $\Lambda_d'(b,q)$ are independent and that $p = b^{-\alpha}$, $q = b^{-\beta}$. Then $\Lambda_d(b,p) \cap \Lambda_d'(b,q)$ has the same distribution as $\Lambda_d(b,pq)$. Hence, by Theorem 8.5.2 (2), if $\dim(E) > \alpha + \beta$, then with positive probability $E \cap \Lambda_d(b,p) \cap \Lambda_d'(b,q)$ is non-empty. Thinking of $K = E \cap \Lambda_d(b,p)$ as the fixed set in Theorem 8.5.2 (1), we deduce that if $\dim(E) > \alpha + \beta$, then with positive probability $\dim(E \cap \Lambda_d(b,p)) \geq \beta$. Similarly, if $\dim(E) < \alpha + \beta$ then almost surely $E \cap \Lambda_d(b,p) \cap \Lambda_d'(b,q)$ is empty, which implies $\dim(E \cap \Lambda_d(b,p)) \leq \beta$ almost surely.

(2) Recall the generating function

$$f(x) = \sum_{k=0}^{b^d} q_k x^k, \quad q_k = p^k (1-p)^{b^d - k} \binom{b^d}{k},$$

that was discussed in the proof of Theorem 3.7.4. In particular, recall that this function is increasing and concave up on $[0,1]$ and if $p > b^{-d}$ it has a unique fixed point in the open interval $(0,1)$. Moreover, we proved that this fixed point is equal to $\mathbb{P}(\Lambda_d(b,p)) = \emptyset)$. Notice that in our setting,

$$q_k = \mathbb{P}(Z_1 = k).$$

Given $\varepsilon > 0$, set $x_\varepsilon = \mathbb{P}(\dim(\Lambda_d(b,p)) \leq d + \log_b p - \varepsilon)$. Then

$$\begin{aligned}
x_\varepsilon &= \mathbb{P}(\dim(\Lambda_d(b,p)) \leq d + \log_b p - \varepsilon) \\
&= \sum_{k=1}^{b^d} \mathbb{P}(\dim(\Lambda_d(b,p)) \leq d + \log_b p - \varepsilon | Z_1 = k) \mathbb{P}(Z_1 = k) \\
&= \sum_{k=1}^{b^d} (\mathbb{P}(\dim(\Lambda_d(b,p)) \leq d + \log_b p - \varepsilon))^k \mathbb{P}(Z_1 = k) \\
&= f(\mathbb{P}(\dim(\Lambda_d(b,p)) \leq d + \log_b p - \varepsilon) = f(x_\varepsilon),
\end{aligned}$$

i.e., x_ε is a fixed point of f; by taking $E = [0,1]^d$ in item (1), we see that $0 < x_\varepsilon < 1$. Therefore x_ε equals $\mathbb{P}(\Lambda_d(b,p)) = \emptyset)$. By letting $\varepsilon \to 0$, we get

$$\mathbb{P}(\dim(\Lambda_d(b,p)) < d + \log_b p) = \mathbb{P}(\Lambda_d(b,p) = \emptyset).$$

Similarly,

$$\mathbb{P}(\dim(\Lambda_d(b,p)) \leq d + \log_b p) = 1. \qquad \square$$

We can use similar techniques to compute the dimension of projections of random sets. By Marstrand's Projection Theorem (see Corollary 3.5.2),

$$\dim(\Pi_\theta \Lambda_2(b,p)) = \max(1, \dim(\Lambda_2(b,p))),$$

for almost every direction, but does not ensure equality for any particular direction. For random sets, we can do better in the sense that we get equality for every direction for almost every set.

Theorem 8.5.4 *Let L be a linear or affine map from $\mathbb{R}^d$ to $\mathbb{R}^k$, $k \leq d$, and let $0 < p < 1$.*

(1) *If $d + \log_b p > k$ then $L(\Lambda_d(b,p))$ has positive k-dimensional Lebesgue measure almost surely on the event $\Lambda_d(b,p) \neq \emptyset$.*
(2) *If $d + \log_b p \leq k$ then $L(\Lambda_d(b,p))$ has Hausdorff dimension $d + \log_b p$ almost surely on the event $\Lambda_d(b,p) \neq \emptyset$.*

For the coordinate directions the Minkowski dimension of these projections was computed by Dekking and Grimmett (1988) and the Hausdorff dimension by Falconer (1989b).

Proof of Theorem 8.5.4 The proof relies again on the properties of the generating function

$$f(x) = \sum_{k=0}^{b^d} \mathbb{P}(Z_1 = k)x^k = \sum_{k=0}^{b^d} q_k x^k$$

recalled in the proof of Corollary 8.5.3(2).

To prove (1), note that for every point $y \in L((0,1)^d)$, the preimage L^{-1} of y has dimension $d - k$. Thus it intersects $\Lambda_d(b,p)$ with positive probability by Theorem 8.5.2(2). Therefore, by Fubini's Theorem the expected k-dimensional volume of $L(\Lambda_d(b,p))$ is positive. Moreover,

$$\mathbb{P}(\mathscr{L}^k(L(\Lambda_d(b,p))) = 0) = \sum_{j=1}^{b^d} \mathbb{P}(\mathscr{L}^k(L(\Lambda_d(b,p))) = 0 | Z_1 = j)\mathbb{P}(Z_1 = j)$$

$$= \sum_{j=1}^{b^d} \mathbb{P}(\mathscr{L}^k(L(\Lambda_d(b,p))) = 0)^j \mathbb{P}(Z_1 = j)$$

$$= f(\mathbb{P}(\mathscr{L}^k(L(\Lambda_d(b,p))) = 0)).$$

Thus $x = \mathbb{P}(\mathscr{L}^k(L(\Lambda_d(b,p))) = 0)$ is a fixed point of f; it is not equal to 1 since $\mathbb{P}(\mathscr{L}^k(L(\Lambda_d(b,p))) > 0) > 0$. Thus it must be the unique fixed point of f in $(0,1)$ and hence is equal to $\mathbb{P}(\Lambda_d(b,p) = \emptyset)$.

To prove (2), we need only to prove the lower bound (since L is Lipschitz, see Exercise 1.9). Let $\gamma < d + \log_b p$, by assumption $d + \log_b p \leq k$, hence by Corollary 8.5.3, $\Lambda_k(b,b^{-\gamma})$ has dimension $k - \gamma$ almost surely upon non-extinction. Thus by the Marstrand Product Theorem (Theorem 3.2.1) the preimage $L^{-1}(\Lambda_k(b,b^{-\gamma}))$ has dimension greater than

$$(d-k) + \dim(\Lambda_k(b,b^{-\gamma})) = d - k + k - \gamma > -\log_b p = \alpha,$$

and hence intersects $\Lambda_d(b,p)$ with positive probability, by Theorem 8.5.2(2). But $L^{-1}(\Lambda_k(b,b^{-\gamma}))$ intersects $\Lambda_d(b,p)$ if and only if $\Lambda_k(b,b^{-\gamma})$ intersects $L(\Lambda_d(b,p))$. Thus $\dim(L(\Lambda_d(b,p))) \geq \gamma$ with positive probability by Theorem 8.5.2. An argument like that in part (1) shows that $\mathbb{P}(\dim(L(\Lambda_d(b,p))) < \gamma)$ is a fixed point of the generating function f in $(0, 1)$ hence equals $\mathbb{P}(\Lambda_d(b, p) = \emptyset)$; this concludes the proof. $\square$

8.6 Brownian motion and Martin capacity

Kakutani (1944) discovered that a compact set $\Lambda \subseteq \mathbb{R}^d$ (not containing the origin) is hit with positive probability by a d-dimensional Brownian motion ($d \geq 3$) if and only if Λ has $\mathrm{Cap}_{d-2}(\Lambda) > 0$. A quantitative version of this assertion is given in Theorem 7.10.3. This theorem clearly contains the classical criterion

$$\mathbb{P}[\exists t > 0 : B_d(t) \in \Lambda] > 0 \Longleftrightarrow \mathrm{Cap}_G(\Lambda) > 0,$$

where $G(x, y) = |x - y|^{2-d}$; passing from the Green's kernel $G(x, y)$ to the Martin kernel $K(x, y) = G(x, y)/G(0, y)$ yields sharper estimates. To explain this, note that while Green's kernels, and hence the corresponding capacity, are translation invariant, the hitting probability of a set Λ by standard d-dimensional Brownian motion is not translation invariant, but is invariant under scaling. This scale-invariance is shared by the Martin capacity.

Corollary 8.6.1 *For $d \geq 3$, Brownian motion started uniformly in the unit cube is intersection equivalent in the unit cube to the random sets $\Lambda_d(2, 2^{-(d-2)})$.*

Proof This is clear since in both cases the probability of hitting a closed set $A \subset [0, 1]^d$ is equivalent to $\mathrm{Cap}_{d-2}(A)$. $\square$

Lemma 8.6.2 *Suppose that $A_1, \ldots, A_k$, $B_1, \ldots, B_k$ are independent random closed sets with A_i intersection equivalent to B_i for $i = 1, \ldots, k$. Then $\bigcap_{j=1}^k A_j$ is intersection equivalent to $\bigcap_{j=1}^k B_j$.*

Proof By induction it is enough to do $k = 2$. It suffices to show $A_1 \cap A_2$ is intersection equivalent to $B_1 \cap B_2$, and this is done by conditioning on A_2,

$$\mathbb{P}(A_1 \cap A_2 \cap \Lambda \neq \emptyset) = \mathbb{E}(\mathbb{P}(A_1 \cap A_2 \cap \Lambda \neq \emptyset)|A_2)$$
$$\asymp \mathbb{E}(\mathbb{P}(B_1 \cap A_2 \cap \Lambda \neq \emptyset)|A_2)$$
$$= \mathbb{P}(B_1 \cap A_2 \cap \Lambda \neq \emptyset),$$

and repeating the argument conditioning on B_1. $\square$

Corollary 8.6.3 *If B_1 and B_2 are independent Brownian motions in $\mathbb{R}^3$ (started uniformly in $[0, 1]^3$) and $A \subset [0, 1]^3$ is closed, then*

$$\mathbb{P}([B_1] \cap [B_2] \cap A \neq \emptyset) \asymp \mathrm{Cap}_2(A),$$

where $[B_i]$ is the range of B_i. However, the paths of three independent Brownian motions in $\mathbb{R}^3$ (started uniformly in $[0,1]^3$) almost surely do not intersect.

Proof To prove the last statement, use the preceding lemma to deduce that $[B_1] \cap [B_2] \cap [B_3]$ is intersection-equivalent in $[0,1]^3$ to $\Lambda_3(2, 1/8)$. The latter set is empty a.s. since critical branching processes are finite a.s., see Athreya and Ney (1972); moreover, the special case needed here is an easy consequence of Theorem 8.4.2. □

Next, we pass from the local to the global behavior of Brownian paths. Barlow and Taylor (1992) noted that for $d \geq 2$ the set of nearest-neighbor lattice points to a Brownian path in $\mathbb{R}^d$ is a subset of $\mathbb{Z}^d$ with dimension 2. This is a property of the path near infinity; another such property is given by

Proposition 8.6.4 *Let $B_d(t)$ denote Brownian motion in $\mathbb{R}^d$, $d \geq 3$. If $\Lambda \subseteq \mathbb{R}^d$, let Λ_1 be the cubical fattening of Λ defined by*

$$\Lambda_1 = \{x \in \mathbb{R}^d : \exists y \in \Lambda \text{ such that } |y - x|_\infty \leq 1\}.$$

Then a necessary and sufficient condition for the almost sure existence of times $t_j \uparrow \infty$ at which $B_d(t_j) \in \Lambda_1$ is that $\mathrm{Cap}_{d-2}^\infty (\Lambda_1 \cap \mathbb{Z}^d) > 0$.

The proof is very similar to the proof of Theorem 8.2.4 and is omitted.

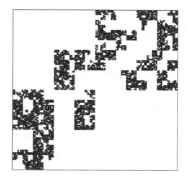

Figure 8.6.1 Brownian motion in $\mathbb{R}^2$ is intersection equivalent to random Cantor sets formed by choosing each of four subsquares with probability $p_n = (n-1)/n$ at generation n. Two examples of such sets are shown.

8.7 Notes

Frostman Theory for discrete sets was developed by Barlow and Taylor (1992) although "integer Cantor sets" and their dimensional properties are considered in many other works.

Lyons' Theorem 8.4.2 and its variants have been used in the analysis of a variety of probabilistic processes on trees, including random walks in a random environment, first-passage percolation and the Ising model. (See Lyons, 1989, 1990, 1992; Lyons and Pemantle, 1992; Benjamini and Peres, 1994; Pemantle and Peres, 1994 and Lyons and Peres, 2016.)

The idea of using intersections with random sets to determine the dimension of a set was first used by Taylor (1966) to study multiple points of stable processes.

Sections 8.2, 8.3 and 8.4 follow the presentation in Benjamini et al. (1995) and Section 8.5 is based on Peres (1996b). Lemma 8.6.2 is from Peres (1996b).

8.8 Exercises

Exercise 8.1 Consider a random walk on $\mathbb{Z}$ that at position n steps to the left with probability p_n and to the right with probability $1 - p_n$. If $p_n = \frac{1}{2} + \frac{1}{n}$, is this walk recurrent or transient?

Exercise 8.2 Consider a random walk on a binary tree so that from a vertex there is probability $0 < p < 1/2$ of moving to each of the children and probability $1 - 2p$ of moving to the parent. For what values of p is this walk recurrent?

Exercise 8.3 Construct a rooted tree so that the number of vertices of distance n from the root is $\geq 2^n$, but such that the nearest neighbor equal probability walk is recurrent.

Exercise 8.4 Consider a random walk on the integers that at time n steps to the left or right by distance $\lfloor \log n \rfloor$ with equal probability. Does this walk return to the origin infinitely often? What if the step is size $\lfloor \sqrt{n} \rfloor$ at time n?

Exercise 8.5 Show that for any random set $A \subset \mathbb{N}$ satisfying $\mathbb{E}(\#(A \cap [0,n]))$ $\leq Cn^\beta$, for some $C > 0$ and $0 < \beta < 1$, we have $\dim_D(A) \leq \beta$ almost surely.

Exercise 8.6 Show for $A \subset \mathbb{N}$, $\dim(A) \leq \limsup_{n \to \infty} \log \#(A \cap [0,n]) / \log n$.

Exercise 8.7 Construct a set A with $\dim(A) = 1$, but

$$\liminf_{n \to \infty} \log \#(A \cap [0,n])/\log n = 0.$$

Exercise 8.8 Prove that the discrete random Cantor set in Example 8.1.7 has discrete Hausdorff dimension $1 + \log_2 p$.

Exercise 8.9 Characterize the gauge functions on $\mathbb{N}$ that give $\mathrm{Cap}_F^\infty(\mathbb{N}) > 0$.

Exercise 8.10 Show that the set $A = \{\lfloor n \log n \rfloor\}_{n \geq 2} \subset \mathbb{N}$ satisfies $\mathrm{Cap}_1^\infty(A) > 0$.

Exercise 8.11 Prove the same for the prime numbers in $\mathbb{N}$. This was proved by Erdős. See the adjacent papers Erdős (1961) and McKean (1961). This requires more than just the prime number theorem. Use the estimates that the interval $I = [2^n, 2^{n+1}]$ contains at least $c_1 2^n/n$ primes (the prime number theorem) and any subinterval of length m on I contains at most $c_2 m/\log m$ primes (Selberg's sieve estimate, Selberg (1952)).

Exercise 8.12 Use Exercise 8.11 to prove that almost surely, a simple random walk on $\mathbb{Z}$ returns to the origin at infinitely many times of the form $2p$, where p is a prime.

Exercise 8.13 Show that the previous exercise also holds for a simple random walk on $\mathbb{Z}^2$.

Exercise 8.14 Show that a simple random walk in $\mathbb{Z}^3$ hits infinitely many points $(p,0,0)$ where p is a prime. See Kochen and Stone (1964), Erdős (1961) and McKean (1961).

• **Exercise 8.15** Show that the β-capacity of the integer Cantor set in Example 8.1.5 is positive for $\beta = \log_b d$.

Exercise 8.16 Prove that Green's function of the simple random walk in $\mathbb{Z}^3$ satisfies $G(x,y) \asymp |x-y|^{-1}$.

Exercise 8.17 Does a 2-dimensional (simple) random walk on $\mathbb{Z}^2$ return infinitely often to 0 during times $\{n^2\}$ almost surely?

Exercise 8.18 Does a random walk in $\mathbb{Z}^3$ hit the set

$$\{(\lfloor \log n \rfloor, \lfloor \log^2 n \rfloor, \lfloor \log \log n \rfloor) : n \in \mathbb{Z}\}$$

infinitely often almost surely?

Exercise 8.19 Does a random walk in $\mathbb{Z}^2$ occupy position $(\lfloor \sqrt{n} \rfloor, 0)$ at time n for infinitely many n almost surely?

Exercise 8.20 Suppose S_n and S'_m are independent simple random walks in Z_d. Show that $\mathbb{P}(\exists \text{ infinitely many } n : S_n = S'_n) = 1$ if $d = 1$ or 2 and equals 0 if $d \geq 3$.

Exercise 8.21 If S_n is the standard random walk on $\mathbb{Z}^2$, what is the dimension of the zero set $\{n : S_n = 0\}$ almost surely?

Exercise 8.22 Suppose S_n is the standard random walk on $\mathbb{Z}^2$ and $A \subset \mathbb{N}$ has dimension α. What can we say about the dimension of $\{S_n : n \in A\}$ almost surely?

Exercise 8.23 Suppose S_n is the standard random walk on $\mathbb{Z}^2$ and $A \subset \mathbb{Z}^2$ has dimension β. What can we say about the dimension of $\{n : S_n \in A\}$ almost surely?

Exercise 8.24 Estimate the probability (within a factor of 2) that Brownian motion in $\mathbb{R}^2$ started at the origin hits a set E in the unit disk before it hits the unit circle. (Hint: the Green's function in the unit disk is $\log \left| \frac{z-w}{1-\bar{w}z} \right|$.)

Exercise 8.25 Choose a random subset S of $\mathbb{N}$ by independently putting n in S with probability $1/n$. What is the dimension of S almost surely?

Exercise 8.26 Let S be the set of integers whose decimal representations do not contain 1 or 3. What is the dimension of S?

Exercise 8.27 Suppose S_1 and S_2 are subsets of $\mathbb{N}$ with $\dim(S_1) = \dim(S_2) = 1$. Is there necessarily an n so that $\dim((S_1 + n) \cap S_2) = 1$?

Exercise 8.28 Given a set $S \subset \mathbb{N}$, consider a random walk in $\mathbb{Z}^2$ that steps left or right with equal probability for times $n \in S$ and steps up or down (again with equal probability) if $n \notin S$. If $S = \{n^2\}$ does this walk return to the origin infinitely often almost surely? What if $S = \{n\}$?

Exercise 8.29 Given a rooted tree T we say percolation occurs for $0 < p < 1$ if percolation occurs with positive probability when each edge is given the parameter p. The critical index p_c for T is the infimum of all p such that percolation occurs.

Show that for the infinite binary tree, the critical index is $1/2$.

Exercise 8.30 What is the critical index for the infinite k-tree?

Exercise 8.31 Suppose T is the rooted tree such that vertices of distance n from the root each have $(n \mod 4) + 1$ children. What is the critical index p_c for T?

Exercise 8.32 A random tree T is constructed by giving each vertex 2 children with probability $1/2$ and 4 children with probability $1/2$. What is the critical index for T almost surely?

Exercise 8.33 Let G be the group generated by a, b with the relations $a^3 = b^3 = e$. Is the simple random walk on G (with these generators) recurrent?

• **Exercise 8.34** Suppose $\{X_n\}_1^\infty$ are random variables such that $\mathbb{E}X_n > 0$ and $0 < \mathbb{E}(X_n^2) < \infty$. Suppose also that $\limsup_{n \to \infty} (\mathbb{E}X_n)^2 / \mathbb{E}(X_n^2) > 0$. Show that
  1. $\mathbb{P}(\limsup X_n / EX_n \geq 1) > 0$.
  2. $\mathbb{P}(\liminf X_n / EX_n \leq 1) > 0$.
  3. $\mathbb{P}(\limsup X_n / EX_n > 0) \geq \limsup (EX_n)^2 / EX_n^2$.
This is due to Kochen and Stone (1964).

• **Exercise 8.35** Suppose $\{E_n\}$ are events and $\mathbb{P}(E_j \cap E_k) \leq C\mathbb{P}(E_j)\mathbb{P}(E_k)$ for some $C < \infty$ that is independent of j and k. This is called **quasi-independence**. Show that if $\sum_n \mathbb{P}(E_n) = \infty$, then infinitely many of the events occur with positive probability. Give an example to show that we cannot replace "positive probability" with "almost surely".

This result is in Lamperti (1963) and Kochen and Stone (1964), but has also been discovered independently by others; e.g., Sullivan (1982) stated and used it in the context of approximation by rational numbers and geodesics in hyperbolic manifolds.

• **Exercise 8.36** In Lamperti's test (Theorem 8.3.3), let A_k be the event that the process X_n hits $\Lambda \cap Y(k)$. Prove that (8.3.1) implies these events are quasi-independent, completing the proof of Theorem 8.3.3.

9

Besicovitch–Kakeya sets

A Besicovitch set $K \subset \mathbb{R}^d$ is a set of zero d-measure that contains a line segment in every direction. These are also called Kakeya sets or Besicovitch–Kakeya sets, although we use the term Kakeya set to mean a set where a line segment can continuously move so as to return to its original position in the opposite orientation. In this chapter we give several deterministic and one random construction of Besicovitch sets for $d = 2$, leaving the construction of a Kakeya set with small area as an exercise. We also discuss an application of Besicovitch sets to Fourier analysis: Fefferman's disk multiplier theorem. One of the constructions of Besicovitch sets involves projections of self-similar planar Cantor sets in random directions; this will lead us to consider self-similar sets that do not satisfy the open set condition (OSC) from the theory of self-similar sets in Chapter 2. In that chapter we proved that a self-similar set has positive measure in its similarity dimension if OSC holds; here we will prove the measure must be zero if OSC does not hold.

9.1 Existence and dimension

We start with the following result of Besicovitch (1919, 1928):

Theorem 9.1.1   *There is a set $K \subset \mathbb{R}^2$ that has zero area and contains a unit line segment in every direction.*

Proof   Let $\{a_k\}_0^\infty$ be a dense sequence in $[0,1]$, chosen with $a_0 = 0$, and so that $\varepsilon(k) = |a_{k+1} - a_k| \searrow 0$ and so that $[a_k - \varepsilon(k), a_k + \varepsilon(k)]$ covers every point of $[0,1]$ infinitely often. Set

$$g(t) = t - \lfloor t \rfloor, \quad f_k(t) = \sum_{m=1}^{k} \frac{a_{m-1} - a_m}{2^m} g(2^m t), \quad f(t) = \lim_{k \to \infty} f_k(t).$$

270

We claim that the closure of $K = \{(a, f(t) + at) : a, t \in [0, 1]\}$ has zero area. By telescoping series $f_k'(t) = -a_k$ on each component of $U = [0, 1] \setminus 2^{-k}\mathbb{N}$. Fix $a \in [0, 1]$ and choose k so that $|a - a_k| \leq \varepsilon(k)$. Since

$$f(t) + at = (f(t) - f_k(t)) + (f_k(t) + a_k t) + (a - a_k)t,$$

and

$$|f(t) - f_k(t)| \leq \sum_{m=k+1}^{\infty} \frac{|a_{m-1} - a_m|}{2^m} g(2^m t) \leq \varepsilon(k) \sum_{m=k+1}^{\infty} 2^{-m} = \varepsilon(k)2^{-k},$$

each of the 2^k components I of U is mapped to a set of diameter less than

$$2\varepsilon(k)2^{-k} + 0 + \varepsilon(k)|I| \leq 3\varepsilon(k)2^{-k}$$

under $f(t) + at$. This proves that every vertical slice $\{t : (a, t) \in \overline{K}\}$ has length zero, so by Fubini's Theorem, area$(\overline{K}) = 0$.

Fixing t and varying a shows K contains unit segments of all slopes in $[0, 1]$, so a union of four rotations of $\overline{K}$ proves the theorem. $\square$

We can modify the proof slightly to remove the use of Fubini's Theorem, replacing it by an explicit covering of $\overline{K}$ by small squares. This approach has the advantage of proving that the closure of K also has zero area, and by choosing the $\{a_k\}$ more carefully, we can get an explicit estimate on the size of our set, which we will show later is optimal (Theorem 9.1.3). For a compact set K, let $K(\varepsilon) = \{z : \text{dist}(z, K) < \varepsilon\}$.

Lemma 9.1.2 *There is a Besicovitch set K so that $0 < \delta < 1$ implies*

$$\text{area}(K(\delta)) = O\left(\frac{1}{\log(1/\delta)}\right).$$

Proof Repeat the previous proof using the sequence

$$\{a_k\}_{k=0}^{\infty} = \left\{0, 0, 1, \frac{1}{2}, 0, \frac{1}{4}, \frac{2}{4}, \frac{3}{4}, 1, \frac{7}{8}, \frac{6}{8}, \frac{5}{8}, \dots\right\},$$

i.e., set $a_0 = 0$ and if $k \in [2^n, 2^{n+1})$, then take $a_k = k2^{-n} - 1$ if n is even, and $a_k = 2 - k2^{-n}$ if n is odd. This just traverses the dyadic rationals $\{j2^{-n}\}_0^{2^n}$, reversing the order each time. The estimates in the proof of Theorem 9.1.1 now hold with $\varepsilon(k) = 2^{-n}$ for $k \in [2^n, 2^{n+1})$. See Figures 9.1.1 and 9.1.2 for the sets resulting from this choice of $\{a_k\}$.

For each $a \in [0, 1]$ we can choose $k \in [2^n, 2^{n+1}]$ so that $f_k(t) + at$ is piecewise linear with slopes of absolute value $\leq 2^{-n}$ on 2^k connected components of $U = [0, 1] \setminus 2^{-k}\mathbb{N}$. Hence this function maps each such interval I into an interval of length at most $2^{-n}|I| = 2^{-n-k}$. The estimate $|f(t) - f_k(t)| \leq \varepsilon(k)2^{-k} = 2^{-n-k}$

Besicovitch–Kakeya sets

Figure 9.1.1 The graph of f from the proof of Lemma 9.1.2.

Figure 9.1.2 The set constructed in Lemma 9.1.2.

from before implies that $f(t) + at$ maps each such I into an interval of length $\leq 3 \cdot 2^{-n-k}$ (but the image may not be an interval since f is not continuous).

To show that $\overline{K}$ has zero area, instead of taking a fixed, as in the previous paragraph, we allow a to vary over an interval J of length 2^{-k-n}. Then each component I of U still maps into a fixed interval of length $4 \cdot 2^{-n-k}$, independent of which $a \in J$ we choose. Thus

$$K_J := \{(a, f(t) + at) : a \in J, t \in [0, 1]\}$$

is covered by 2^k squares of side length $O(2^{-n-k})$ and hence has area $O(2^{-2n-k})$. Since $[0,1]$ is covered by 2^{n+k} such intervals J, the whole of K is covered by a finite union of closed squares whose total area is $O(2^{-n})$. The closure of K is also contained in this finite union of squares. Thus $\overline{K}$ can be covered by squares of size $\delta_n = 2^{-n-2^{n+1}}$ and total area $O(2^{-n})$. This implies that

$$\text{area}(K(\delta_n)) = O\left(\frac{1}{\log(1/\delta_n)}\right).$$

Since $\log(1/\delta_{n+1}) = O(\log(1/\delta_n))$, we get the estimate for all $\delta \in (0,1)$ (with a slightly larger constant). $\square$

Lemma 9.1.2 is optimal (first proved in Córdoba (1993)):

Theorem 9.1.3 *If $K \subset \mathbb{R}^2$ contains a unit line segment in all directions, then for all ε sufficiently small,*

$$\text{area}(K(\varepsilon)) \geq \frac{1-\varepsilon}{1+\log(1/\varepsilon)}.$$

Proof Let $\varepsilon > 0$ and $n = \lfloor \varepsilon^{-1} \rfloor$ (so $1 - \varepsilon \leq n\varepsilon \leq 1$ for ε small). Choose unit line segments $\{\ell_i\}_1^n$ in K so that the angle between ℓ_{i-1} and ℓ_i is $\frac{\pi}{n}$ (indices considered mod n). For $i = 1,\ldots,n$ let Ψ_i be the indicator function of $\ell_i(\varepsilon)$ and let $\Psi = \sum_{i=1}^n \Psi_i$. See Figure 9.1.3.

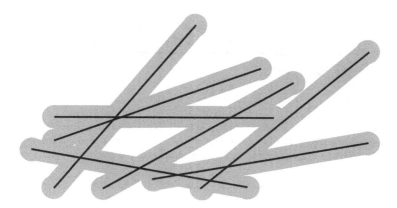

Figure 9.1.3 Neighborhoods of the angle-separated lines.

By Cauchy–Schwarz

$$\left(\int_{\mathbb{R}^2} \Psi(x)\,dx\right)^2 \leq \left(\int_{\Psi>0} 1\,dx\right)\left(\int_{\mathbb{R}^2} \Psi^2(x)\,dx\right),$$

so

$$\text{area}(K(\varepsilon)) \ge \text{area}(\{\Psi > 0\}) \ge \frac{(\int_{\mathbb{R}^2} \Psi(x)\,dx)^2}{\int_{\mathbb{R}^2} \Psi^2(x)\,dx}. \qquad (9.1.1)$$

By the definition of Ψ,

$$\int_{\mathbb{R}^2} \Psi(x)\,dx = \sum_{i=1}^{n} \text{area}(\ell_i(\varepsilon)) \ge 2\varepsilon n.$$

Since $\Psi_i^2 = \Psi_i$ for all i, we have

$$\int_{\mathbb{R}^2} \Psi^2(x)\,dx = \int_{\mathbb{R}^2} \Psi(x)\,dx + \sum_{i=1}^{n} \sum_{k=1}^{n-1} \text{area}(\ell_i(\varepsilon) \cap \ell_{i+k}(\varepsilon)).$$

The angle between the lines ℓ_i and ℓ_{i+k} is $k\pi/n$. A simple calculation (see Figure 9.1.4) shows that if $k\pi/n \le \pi/2$, then

$$\text{area}(\ell_i(\varepsilon) \cap \ell_{i+k}(\varepsilon)) \le \frac{4\varepsilon^2}{\sin(k\pi/n)} \le \frac{2\varepsilon^2 n}{k} \le \frac{2\varepsilon}{k}$$

(we use $\sin(x) \ge 2x/\pi$ on $[0, \frac{\pi}{2}]$) with a similar estimate for $k\pi/n > \pi/2$.

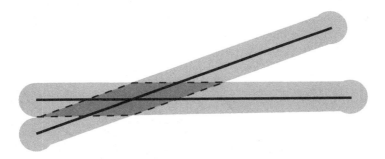

Figure 9.1.4 The intersection of two neighborhoods is contained in the intersection of two strips. The intersection of two strips of width w and angle θ is a parallelogram of area $w^2/\sin(\theta)$, which gives the estimate used in the text.

Hence (indices are mod n),

$$\sum_{i=1}^{n} \sum_{k=1}^{n-1} \text{area}(\ell_i(\varepsilon) \cap \ell_{i+k}(\varepsilon)) \le 4\varepsilon n \log n.$$

Thus by (9.1.1),

$$\text{area}(\{\Psi > 0\}) \ge \frac{(\int_{\mathbb{R}^2} \Psi(x)\,dx)^2}{\int_{\mathbb{R}^2} \Psi(x)\,dx + 4\varepsilon n \log n} \ge \frac{4\varepsilon^2 n^2}{2\varepsilon n + 4\varepsilon n \log n},$$

and then some simple arithmetic gives

$$\text{area}(\{\Psi > 0\}) \geq \frac{2\varepsilon n}{1 + 2\varepsilon n \log n} \geq \frac{1 - \varepsilon}{1 + \log(1/\varepsilon)} \qquad \square$$

The fact that the upper Minkowski dimension of a Besicovitch set must be 2 follows easily from Theorem 9.1.3 and the remarks at the beginning of Section 2.6. Computing the Hausdorff dimension requires a different argument.

Theorem 9.1.4 *If $K \subset \mathbb{R}^2$ is a compact set containing a segment in a compact set E of directions, then $\dim(K) \geq 1 + \dim(E)$. In particular, if K contains a segment in every direction, then $\dim(K) = 2$.*

Proof Let $\beta = \dim(E)$. For any interval $I \subset \mathbb{R}$, let

$$S_I = \{(a,b) : \forall x \in I \quad (x, b - ax) \in K\}.$$

By assumption $E \subset \bigcup_I \Pi_0 S_I$ where the union is over all intervals with rational endpoints and Π_0 is projection onto the first coordinate. Since this is a countable collection of intervals, for any $\varepsilon > 0$ there is such an interval I so that

$$\dim(\Pi_0 S_I) > \dim(E) - \varepsilon = \beta - \varepsilon$$

(the set of slopes of segments in K and the set of their angles have the same dimension since they are related by a smooth map, namely the tangent). Therefore, by the Marstrand Projection Theorem (Corollary 3.5.2), the projection of S_I onto almost every direction θ has dimension $> \beta - \varepsilon$. Writing $t = \tan \theta$, note that the (non-orthogonal) projection of the set S_I along lines of slope t onto the vertical axis is the same as the set $\{b - at : (a,b) \in S_I\}$; this set has the same dimension as the orthogonal projection of S_I in direction θ since the two projections differ only by a linear map. Hence for a.e. t,

$$\dim(\{b - at : (a,b) \in S_I\}) \geq \beta - \varepsilon.$$

Since $(a,b) \in S_I$ implies $(t, b - at) \in K$ for any $t \in I$, we can deduce

$$\{b - at : (a,b) \in S_I\} \subset \{b - at : (t, b - at) \in K\},$$

and hence for almost every t,

$$\dim(\{b - at : (t, b - at) \in K\}) \geq \beta - \varepsilon.$$

These sets are just vertical slices of K above t, so Marstrand's Slicing Theorem (Theorem 1.6.1) yields $\dim(K) \geq 1 + \beta - \varepsilon$. $\qquad \square$

9.2 Splitting triangles

In this section we give an alternative construction of Besicovitch sets that is closer to Besicovitch's original construction (as simplified by various authors).

We say T is a **standard triangle** if it has base $[a,b] \subset \mathbb{R}$ and its third vertex c is in the upper half-plane. We let $d = (a+b)/2$ denote the midpoint of the base. Given a standard triangle T of height k, a k-split of T divides it into two standard triangles T_1, T_2 with bases $[a,d]$ and $[d,b]$ respectively; then apply the Euclidean similarity $z \to a + \frac{k+1}{k}(z-a)$ to T_1 and $z \to b + \frac{k+1}{k}(z-b)$ to T_2. This rescales each triangle, fixing a in T_1 and fixing b in T_2. This gives two standard triangles of height $k+1$ whose bases are overlapping subintervals of $[a,b]$, the base of T. We inductively define collections $\{\mathscr{T}_k\}$ of standard triangles of height k by letting $\mathscr{T}_1$ be an isosceles triangle with base 1 and height 1, and obtaining $\mathscr{T}_{k+1}$ by applying a k-split to each triangle in $\mathscr{T}_k$. See Figure 9.2.1.

Figure 9.2.1 A triangle is cut in two by bisecting the base and then each triangle is expanded by a factor of $(1 + \frac{1}{k})$ at stage k. Here we show $k = 1$ and $k = 2$.

Each of the 2^k triangles in $\mathscr{T}_k$ has height k and base $k2^{-k}$, so each has area $k^2 2^{-k-1}$. If T_1, T_2 are the triangles that result from a k-split of $T \in \mathscr{T}_k$, then $(T_1 \cup T_2) \setminus T$ is a triangular region between heights $k-1$ and $k+1$ and has area $2\,\text{area}(T)/k^2 = 2^{-k}$. See Figure 9.2.2. Since there are 2^k triangles in $\mathscr{T}_k$, the total new area is at most 1. Let X_k be the union of the triangles in $\mathscr{T}_k$. Then $\text{area}(X_k) \le k$. Finally rescale X_k to have height 1 and so that its base is centered at the center of X_1's base. The rescaling decreases area by a factor of k^{-2}, so the new area is at most $1/k$. However, X_k still contains unit line segments in all the same directions as X_0 did. See Figure 9.2.3 to see several generations of the construction.

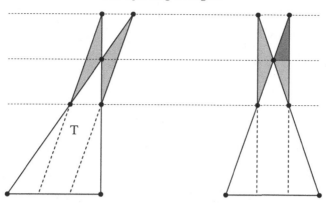

Figure 9.2.2 Computing the new area created by the splitting/expansion step.

To obtain a set of zero area, we want to pass to a limit, but this is a problem since the sets X_k are not nested, so we need to modify the construction slightly. The key observation is that when we split and rescale a triangle T to obtain T_1, T_2, we have $T \subset T_1 \cup T_2 \subset W$, where W is a trapezoid with the same base as T and sides parallel to those of T (the left side of W is parallel to the right side of T and conversely). If we split the base $[a, b]$ of a standard triangle T into n equal intervals and then apply a k-split to each resulting triangle separately, and then rescale to unit height, the resulting triangles will all be in a $(|b - a|/n)$-neighborhood of T; we can make this as small as we wish by taking n large. By subdividing each standard triangle in $\mathscr{T}_k$ before performing the k-split, we can ensure that every X_k stays within any open neighborhood of the original triangle T that we want, and still have as small an area as we wish. See Figure 9.2.4.

So start with a standard isosceles triangle Y_1 and choose an open neighborhood U_1 of Y_1 so that the $\mathrm{area}(\overline{U_1}) \le 2\,\mathrm{area}(Y_1)$. Using the argument above, find a union of triangles $Y_2 \subset U_1$ having total area $< 1/8$ and containing unit segments in all the same directions as Y_1. Choose an open neighborhood U_2 of Y_2 having area $\le 2\,\mathrm{area}(Y_2) \le 1/4$. In general, we construct a union of triangles $Y_k \subset U_{k-1}$ containing unit segments in the same directions as in Y_1 and choose an open neighborhood U_k of Y_k having $\mathrm{area}(\overline{U_k}) \le 2^{-k}$. Then $F = \bigcap_n \overline{U_n}$ is a closed set of measure zero. If we fix a direction and let $L_n \subset Y_n \subset U_n$ be a unit segment in this direction, then we can pass to a subsequence that converges to a unit segment in F, and which clearly has the same direction. Thus F contains unit segments in the same interval of directions as Y_1 and so a finite number of rotations of F is a Besicovitch set.

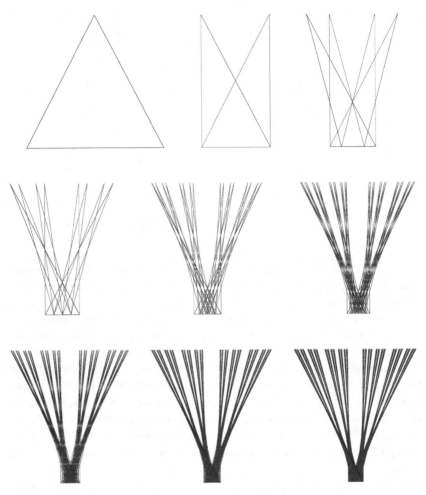

Figure 9.2.3 The first nine steps of the k-split construction.

9.3 Fefferman's Disk Multiplier Theorem *

A surprising and influential application of Besicovitch sets was given in 1971 by Charles Fefferman, solving a famous open problem about Fourier transforms. We first introduce some terminology, then state his result and sketch the proof. Given a function $f : \mathbb{R}^d \to \mathbb{R}$, its Fourier transform on $\mathbb{R}^d$ is defined

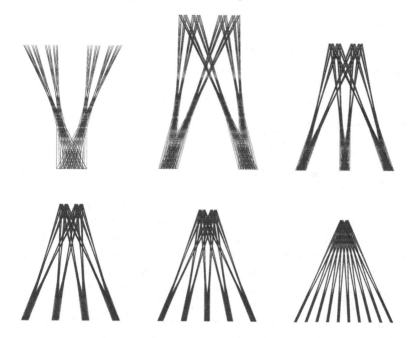

Figure 9.2.4 The construction when we subdivide the first triangle into n subtriangles. Shown are $n = 1$ (no subdivision), $2,\ldots,5$ and 10. For large n these are inside any open neighborhood of the original triangle.

as

$$\mathscr{F}f(y) = \hat{f}(y) = \int_{\mathbb{R}^d} f(x)e^{-2\pi i x \cdot y}\,dx.$$

The integral is clearly well defined when f is integrable, and the Plancherel Theorem (e.g., Theorem 8.29 of Folland (1999)) says if $f \in L^1 \cap L^2$ (which is dense in L^2), then $\|\hat{f}\|_2 = \|f\|_2$. Thus the Fourier transform extends to an isometry of L^2 to itself. The inverse is given by

$$f(x) = \mathscr{F}^{-1}\hat{f}(x) = \int_{\mathbb{R}^d} \hat{f}(y)e^{2\pi i x \cdot y}\,dy$$

(see Theorem 8.26 of Folland (1999)). Among the basic properties we will need are

$$\mathscr{F}(f_x)(y) = e^{2\pi i x \cdot y}\mathscr{F}f(y) \tag{9.3.1}$$

$$\mathscr{F}(f)_u(y) = \mathscr{F}(e^{-2\pi i x \cdot u}f(x))(y) \tag{9.3.2}$$

where $f_u(x) = f(u+x)$ denotes the translation of a function. If T is a linear, invertible map and $S = (T^*)^{-1}$ (so $Sy \cdot x = y \cdot T^{-1}x$) is the inverse of its transpose, then we have

$$\mathscr{F}(f \circ T)(y) = |\det(T)|^{-1} \mathscr{F} f(S(y)).$$

When T is a rotation ($TT^* = I$) this gives

$$\mathscr{F}(f \circ T)(y) = \mathscr{F} f(T(y)), \tag{9.3.3}$$

and when $T(x) = \lambda x$ is a dilation we have

$$\mathscr{F}(f \circ T)(y) = \lambda^{-d} \mathscr{F} f(y/\lambda). \tag{9.3.4}$$

Each of these follows from a simple manipulation of the definitions. See Theorem 8.22 of Folland (1999).

An important class of bounded, linear operators on L^2 are the multiplier operators (in this section L^p always refers to $L^p(\mathbb{R}^d, dx)$). These are defined as the composition of the Fourier transform, followed by multiplication by a bounded function m, followed by the inverse Fourier transform, or more concisely, $Mf = \mathscr{F}^{-1}(m\mathscr{F}(f))$. Because the Fourier transform is an isometry on L^2, such a multiplication operator is L^2 bounded with norm $\|m\|_\infty$. One of the most basic examples is to take $m = \mathbf{1}_E$ to be the indicator function of a set. The disk multiplier, $M_\mathbb{D}$, corresponds to multiplying $\hat{f}$ by $\mathbf{1}_\mathbb{D}$ (recall that $\mathbb{D} = \{x \in \mathbb{R}^2 : |x| < 1\}$ is the open unit disk). Although $M_\mathbb{D}$ is clearly bounded on L^2, boundedness on L^p was a famous open problem until it was settled by Fefferman (1971):

Theorem 9.3.1 *The disk multiplier is not a bounded operator from $L^p(\mathbb{R}^d)$ to itself if $d \geq 2$ and $p \neq 2$.*

For $d = 1$, the disk multiplier corresponds to multiplying by the indicator of an interval. This operator is well known to be bounded from L^p to itself for $1 < p < \infty$ (it follows from the L^p boundedness of the Hilbert transform, e.g., Chapter XVI, Volume II of Zygmund (1959)). We shall describe Fefferman's proof in the case $d = 2$ and $p > 2$; all the other cases can be deduced from this one (see Exercises 9.46–9.50). The argument breaks in four steps:

1. A geometric lemma related to Besicovitch sets.
2. A vector-valued estimate derived from randomization.
3. An application of Fatou's Lemma.
4. An explicit computation of the Hilbert transform of $\mathbf{1}_{[-1,1]}$.

We start with the geometric lemma.

Lemma 9.3.2 *For any positive integer k there are 2^k disjoint, congruent rectangles $\{R_j\}$, each of area 2^{-k}, so that translating each by less than twice its*

diameter, parallel to its long side, gives a collection $\{\tilde{R}_j\}$ of overlapping rect-
angles whose union has area $O(1/k)$.

Proof As in the last section, we iteratively define collections $\{\mathscr{T}_k\}$ of standard triangles, but using a slightly different rule this time. As before, $\mathscr{T}_1$ consists of a single standard isosceles triangle with base $[0, 1]$ on the real line. At
the kth stage, $\mathscr{T}_k$ is a collection of 2^k triangles of height k whose bases have
disjoint interiors and whose closures cover $[0, 1]$. For each triangle in the collection, lengthen the two non-base edges by a factor of $\frac{k+1}{k}$ and connect the
new endpoints to the midpoint of the base. This gives 2^{k+1} triangles of height
$(k + 1)$ and base length 2^{-k-1}. The two new triangles defined here are subsets
of the two triangles defined by the splitting rule in the previous section so the
new area formed is less than it was before, i.e., is ≤ 1. Thus the area of the new
X_k, the union of the kth generation triangles, is $\leq k$. See Figure 9.3.1.

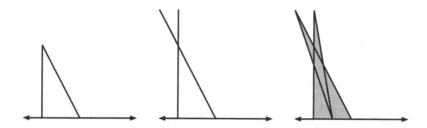

Figure 9.3.1 Define triangles by growing/splitting.

Rescale the triangles so that each has height 1 above the real axis. Extend
each triangle in $\mathscr{T}_k$ to an infinite sector by extending the sides meeting at the top
vertex and consider the part of the sector below the real line. See Figure 9.3.2.
The regions corresponding to the two triangles in $\mathscr{T}_k$ obtained by splitting a
triangle $T \in \mathscr{T}_k$ are disjoint subsets of the region corresponding to T. This
means all the 2^k regions corresponding to triangles in $\mathscr{T}_k$ are pairwise disjoint.
It is then a simple matter to choose a rectangle R_j in each such region below
the real line, and a translate $\tilde{R}_j$ of R_j in the corresponding triangle above the
real line that satisfy the lemma. Rescaling so each R_j has area 2^{-k} finishes the
proof. Figure 9.3.3 shows six generations of the construction.   □

Rather than construct a counterexample to L^p boundedness of the disk multiplier directly, it is more convenient to deduce a consequence of L^p boundedness and contradict this. Recall that the L^p norm of a function is given by

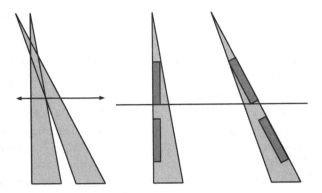

Figure 9.3.2 Extending the triangles below the real line gives disjoint quadrilaterals. For each extended triangle we can find sub-rectangles $R_j, \tilde{R}_j$ in the lower and upper parts of the triangle that are translates of each other by a vector parallel to and less than twice the length of their long sides.

$\|f\|^p = (\int |f|^p)^{1/p})$ and an operator T is L^p bounded if there is a $C < \infty$ so that $\|Tf\|_p \leq C\|f\|_p$.

The first step is to deduce a vector-valued version of L^p boundedness. This is a general result, not specific to multiplier operators:

Lemma 9.3.3 *If $M_{\mathbb{D}}$ is L^p bounded, then there is constant $C < \infty$ so that for any finite collection of real-valued functions $\{f_1, \ldots f_n\}$,*

$$\left\| \left(\sum |M_{\mathbb{D}} f_j|^2 \right)^{1/2} \right\|_p \leq C \left\| \left(\sum |f_j|^2 \right)^{1/2} \right\|_p.$$

Proof We follow the proof of Theorem XV.2.9 in Volume 2 of Zygmund (1959). Let $f = (f_1, \ldots f_n)$ take values in $\mathbb{R}^n$ and let a be a unit vector in $\mathbb{R}^n$. Then $f \cdot a$ is real valued, and by linearity

$$M_{\mathbb{D}}(f \cdot a) = (M_{\mathbb{D}} f) \cdot a \equiv (M_{\mathbb{D}} f_1, \ldots, M_{\mathbb{D}} f_n) \cdot a,$$

so (since $M_{\mathbb{D}}$ is L^p bounded),

$$\|M_{\mathbb{D}} f \cdot a\|_p^p \leq C^p \|f \cdot a\|_p^p.$$

Now integrate both sides with respect to a over the unit sphere and interchange this integral with the integral defining the L^p norm. The inner integral on both sides is of the form

$$\int_{|a|=1} |v \cdot a|^p \, da,$$

for $v = M_{\mathbb{D}} f$ and $v = f$ respectively, but this integral depends only on the norm

Figure 9.3.3 Six generations of Fefferman's triangle iteration. The top half is a union overlapping triangles with area tending to zero; the bottom half is a union of disjoint quadrilaterals with area bounded away from zero.

of v, not its direction, so equals $c|v|^p$ for some constant c independent of v. Thus c cancels from both sides, leaving the desired inequality. □

The result and proof hold for any bounded operator on L^p. We refer to this result as "randomization" since we can interpret this as an expected value of the L^p norm of a function when dotted with a random unit vector. If we computed the expectation of the dot product with a random vector of ± 1's instead of unit vectors, we would obtain Khinchin's inequality. See Exercise 9.39.

Next we want to observe that the L^p boundedness of one multiplication operator implies the boundedness of a family of similar ones.

Lemma 9.3.4 *Suppose $1 < p < \infty$, $E \subset \mathbb{R}^n$ is measurable, M_E is L^p bounded, $r > 0$ and $x \in \mathbb{R}^n$. If $\{f_1, \ldots, f_n\}$ are in L^2, then*

$$\left\| \left(\sum |M_{rE+x} f_j|^2 \right)^{1/2} \right\|_p \le C \left\| \left(\sum |f_j|^2 \right)^{1/2} \right\|_p,$$

for some constant C independent of r and x.

Proof $M_{rE+x}f$ can be written as a composition of five maps: D_r corresponding to dilation of $\hat{f}$ by r, T_x corresponding to translation of $\hat{f}$ by x, M_E, T_{-x} and $D_{1/r}$. T_x and T_{-x} act by multiplying f by a function of modulus 1 and $D_r, D_{1/r}$ act by dilating f by inverse amounts and multiplying by inverse factors. By multiplying by the appropriate scalar factors we can assume each of these maps preserve the L^p norm. Thus

$$M_{rE+x}f = D_{1/r} \circ T_{-x} \circ M_E \circ T_x \circ D_r f,$$

and using Lemma 9.3.3,

$$\left\| \left(\sum |M_{rE+rx} f_j|^2 \right)^{1/2} \right\|_p = \left\| \left(\sum |M_E \circ T_x \circ D_r f_j|^2 \right)^{1/2} \right\|_p$$

$$\leq C \left\| \left(\sum |T_x \circ D_r f_j|^2 \right)^{1/2} \right\|_p$$

$$= C \left\| \left(\sum |f_j|^2 \right)^{1/2} \right\|_p. \qquad \square$$

The next step is to note that for a unit vector u, the disk of radius $|r|$ centered at ru tends to the half-plane $H_u = \{y : y \cdot u > 0\}$. Thus we might hope that the corresponding half-plane multiplier operator M_u is a limit of disk multipliers. We will apply Fatou's Lemma to show this is indeed the case.

Lemma 9.3.5 *Suppose the disk multiplier is L^p bounded for some $p \in (2, \infty)$, $\{u_1, \dots, u_n\}$ is a collection of unit vectors in $\mathbb{R}^2$ and $\{f_1, \dots, f_n\}$ are in L^2. Let $M_j = M_{\mathbb{H}(u_j)}$ be the multiplier operator that corresponds to indicator of the half-plane $\mathbb{H}(u_j) = \{y : y \cdot u_j > 0\}$. Then*

$$\left\| \left(\sum |M_j f_j|^2 \right)^{1/2} \right\|_p \leq C \left\| \left(\sum |f_j|^2 \right)^{1/2} \right\|_p.$$

Proof We apply the previous lemma when $E = \mathbb{D}$. If we fix a unit vector u and take $r \nearrow \infty$ then the disks $r\mathbb{D} + ru$ fill up the half-plane $H_u = \{y : y \cdot u > 0\}$, so for a smooth function f of compact support we see that

$$M_u f(y) = \lim_{r \to \infty} M_{r\mathbb{D}+ru} f(y).$$

Thus Fatou's Lemma implies

$$\left\| \left(\sum |M_j f_j|^2 \right)^{1/2} \right\|_p \leq \liminf_{r \to \infty} \left\| \left(\sum |M_{r\mathbb{D}+ru_j} f_j|^2 \right)^{1/2} \right\|_p$$

and the right-hand side is $\leq C \left\| \left(\sum |f_j|^2 \right)^{1/2} \right\|_p$ by Lemma 9.3.4 (we replace the translated disks by the single disk $r\mathbb{D}$ to give E and replace the functions f_j by translates; then the lemma can be applied). $\qquad \square$

We will apply these estimates to the indicator functions for the rectangles $\{R_j\}$ in Lemma 9.3.2. Let $f_j = \mathbf{1}_{R_j}$ and let $M_j = M_{\mathbb{H}(u_j)}$ be the half-plane multiplier as above, i.e., u_j is parallel to the long side of R_j.

Lemma 9.3.6 $|M_j f_j| > c > 0$ on $\tilde{R}_j$ for some absolute $c > 0$.

Proof This is a direct calculation. Because of the invariance properties of the Fourier transform, i.e., (9.3.1)–(9.3.4), it suffices to check this estimate when $R_j = R = [-1, 1] \times [0, \varepsilon]$ and $u = (1, 0)$. In this case

$$f(x_1, x_2) = \mathbf{1}_R(x_1, x_2) = \mathbf{1}_{[-1,1]}(x_1) \cdot \mathbf{1}_{[0,\varepsilon]}(x_2) = f_1(x_1) f_2(x_2)$$

(recall $f_j = \mathbf{1}_{R_j}$). Hence $\hat{f}(y_1, y_2) = \hat{f}_1(y_1) \hat{f}_2(y_2)$, so

$$M_u f(x_1, x_2) = [M_u f_1(x_1)] f_2(x_2).$$

Thus we just have to compute the Fourier multiplier in one dimension that corresponds to multiplying $\hat{f}$ by $\mathbf{1}_{y>0}$. To evaluate this, we use a standard trick of approximating $\exp(2\pi i y z)$ by $\exp(2\pi i (y(z + is)))$ where $s > 0$; this function has exponential decay as $y \to \infty$ so the integral $\int_0^\infty \hat{f}_1(y) e^{2\pi i y(z+is)} \, dy$ is well defined and converges to $M_u f_1$ in the L^2 norm by Plancherel's Theorem. Thus for $z \in \tilde{R}_j$ we get (using Fubini's Theorem),

$$M_u f_1(z) = \int_0^\infty \left[\int_{-1}^1 e^{-2\pi i y x} dx \right] e^{2\pi i y z} dy$$

$$= \lim_{s \to 0} \int_0^\infty \left[\int_{-1}^1 e^{-2\pi i y x} dx \right] e^{2\pi i y(z+is)} \, dy$$

$$= \lim_{s \to 0} \int_{-1}^1 \int_0^\infty e^{-2\pi i y x} e^{2\pi i y(z+is)} \, dy \, dx$$

$$= \lim_{s \to 0} \int_{-1}^1 \int_0^\infty e^{-2\pi i y(x-z-is)} \, dy \, dx$$

$$= \lim_{s \to 0} \frac{1}{2\pi i} \int_{-1}^1 \frac{1}{x - z - is} \, dx$$

$$= \frac{1}{2\pi i} \int_{-1}^1 \frac{1}{x - z} \, dx.$$

(The limit exists since $z \in \tilde{R}_j$ implies the integrand converge uniformly.) The lemma follows from this since $|x - z| \simeq 1$ for all $x \in [-1, 1]$ and $z \in \tilde{R}_j$. $\square$

Corollary 9.3.7 If $E = \bigcup_j \tilde{R}_j$ then $\int_E (\sum_j |M_j f_j|^2) \, dx$ is bounded uniformly away from zero.

Proof   The previous lemma implies

$$\int_E \left(\sum_j |M_j f_j|^2 \right) dx = \sum_j \int_E \left(\sum_j |M_j f_j|^2 \right) dx$$

$$\geq c^2 \sum_j \text{area}(\tilde{R}_j)$$

$$= c^2 \sum_j \text{area}(R_j)$$

$$= c^2. \qquad \qquad \square$$

We can now prove Fefferman's Theorem:

Proof of Theorem 9.3.1   If the disk multiplier, $M_{\mathbb{D}}$, were L^p bounded, then Hölder's inequality with exponents $p/2$ and its conjugate $q = p/(p-2)$ and Lemma 9.3.5 would imply (recall $E = \cup \tilde{R}_j$))

$$\int_E \left(\sum_j |M_j f_j|^2 \right) dx \leq \left(\int_E 1^q \, dx \right)^{1/q} \left(\int_E \left(\sum_j |M_j f_j|^2 \right)^{p/2} dx \right)^{2/p}$$

$$= C \, \text{area}(E)^{1/q} \left\| \left(\sum_j |M_j f_j|^2 \right)^{1/2} \right\|_{L^p(E)}^2$$

$$\leq C \, \text{area}(E)^{1/q} \left\| \left(\sum_j |f_j|^2 \right)^{1/2} \right\|_{L^p(E)}^2$$

$$\leq C \, \text{area}(E)^{1/q} \, \text{area} \left(\bigcup_j R_j \right)^{2/p} \to 0.$$

Here we have used that E, the union of the $\tilde{R}_j$s, has area tending to zero by construction. This contradicts the previous estimate and proves that the disk multiplier must be unbounded. $\qquad \square$

9.4 Random Besicovitch sets

We started the chapter by constructing a Besicovitch set of the form

$$K = \{(a, f(t) + at) : a, t \in [0, 1]\},$$

where f was chosen so that $t \to f(t) + at$ maps $[0, 1]$ to zero Lebesgue measure for every $a \in [0, 1]$. In this section we define a random family of functions f that have this property, thus giving a random family of Besicovitch sets. We will take $f(t) = X_t$ to be the Cauchy process $[0, \infty) \to \mathbb{R}$. Essentially, X_t is the y-coordinate of a 2-dimensional Brownian motion the first time the x-coordinate equals t.

More precisely, suppose B_1, B_2 are two independent standard Brownian motions, and let $\tau_t := \inf\{s \geq 0 : B_1(s) > t\}$. Then $X_t = B_2(\tau_t)$ is the Cauchy process. An example of this discontinuous process is shown in Figure 9.4.1. The Cauchy process is a Lévy process where the increments $X_{s+t} - X_s$ have the same law as tX_1, and X_1 has the Cauchy density $(\pi(1+x^2))^{-1}$. See Bertoin (1996) or (Mörters and Peres, 2010, Theorem 2.37). Note that the Cauchy process is not continuous, but it is a.s. continuous from the right. To prove this, it suffices to verify a.s. right-continuity of τ, since $X = B_2 \circ \tau$. By continuity of B_1, the set $\{s : B_1(s) > t\}$ is open and equals the union of open sets $\bigcup_{n=1}^{\infty}\{B_1 > t + \frac{1}{n}\}$, so $\tau(t) = \lim_n \tau(t + \frac{1}{n})$. Because τ is increasing with t this means $\tau(t) = \lim_{\varepsilon \searrow 0} \tau(t + \varepsilon)$, which is right continuity. However, τ need not be left continuous, e.g., τ has a jump at t if B_1 has a local maximum at time $\min\{s : B_1(s) = t\}$. At such t, almost surely X will also fail to be left continuous.

Figure 9.4.1 A Cauchy sample path. The left shows a Brownian motion in light gray and the black shows the graph of corresponding Cauchy process (the picture is drawn by changing the color from gray to black each time the x-coordinate of the Brownian path reaches a new maximum). For clarity, the graph of the Cauchy sample is redrawn on the right (slightly rescaled) with the gray points removed.

Next we show that $t \to X_t + at$ maps $[0, 1]$ to zero length for a.e. choice of a.

Lemma 9.4.1   *There exists a constant c so that for all $a \in [0, 1]$ the r-neighborhood of $\{X_s + as : s \in [0, t]\}$ has expected total length at most*

$$\frac{ct}{\log(t/r)} + 2r.$$

Consequently, $\{X_s + as : s \in [0, t]\}$ has zero length a.s.

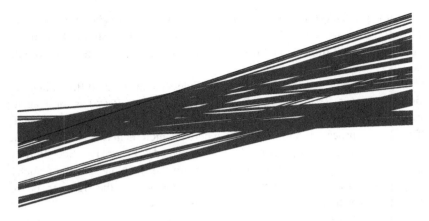

Figure 9.4.2 A Besicovitch set K formed from a Cauchy process.

Proof Let

$$\tau(x,r) = \inf\{s \geq 0 : X_s + as \in (x - r, x + r)\}$$

be the first time that $X_s + as$ takes a value in $(x - r, x + r)$. Let

$$Y(t,r) = \bigcup_{0 \leq s \leq t} (X_s + as - r, X_s + as + r)$$

be the r-neighborhood of the range up to time t. Using Fubini's Theorem, the expected length of this neighborhood is

$$\mathbb{E}[\mathscr{L}_1(Y(t,r))] = \int_{\mathbb{R}} \mathbb{P}(\tau(x,r) \leq t)\,dx$$
$$= 2r + \int_{\mathbb{R} \setminus (-r,r)} \mathbb{P}(\tau(x,r) \leq t)\,dx.$$

For $|x| \geq r$ we define

$$Z_x = \int_0^t \mathbf{1}_{|X_s + as - x| < r}\,ds \quad \text{and} \quad \tilde{Z}_x = \int_0^{2t} \mathbf{1}_{|X_s + as - x| < r}\,ds.$$

By the right continuity of the Cauchy process we deduce that up to zero probability events, we have $\{\tau(x,r) \leq t\} = \{Z_x > 0\}$. So it follows that

$$\mathbb{P}(\tau(x,r) \leq t) = \mathbb{P}(Z_x > 0) \leq \frac{\mathbb{E}\,\tilde{Z}_x}{\mathbb{E}[\tilde{Z}_x \mid Z_x > 0]}. \tag{9.4.1}$$

For the numerator we have

$$\mathbb{E}\,\tilde{Z}_x = \int_0^{2t} \int_{x-r}^{x+r} p_s(0,y)\,dy\,ds = \int_0^{2t} \int_{-r}^{r} p_s(0,x+y)\,dy\,ds,$$

where $p_s(0,y)$ is the transition density at time s of the process $(X_u + au)$. To simplify notation let $\tau = \tau(x,r)$ (drop the x and r). For the conditional expectation appearing in the denominator in (9.4.1) we have

$$\mathbb{E}\big[\tilde{Z}_x \mid Z_x > 0\big] = \mathbb{E}\left[\int_\tau^{2t} \mathbf{1}_{|X_s + as - x| < r}\, ds \;\Big|\; \tau \leq t\right]$$

$$\geq \min_{y:|y-x|<r} \mathbb{E}\int_0^t \mathbf{1}_{|X_s + as + y - x| < r}\, ds,$$

where in the last step we used the strong Markov property of X and the fact that $X_\tau + a\tau = y \in [x - r, x + r]$. We now bound from below the expectation appearing in the minimum above. If $r < t$,

$$\mathbb{E}\int_0^t \mathbf{1}_{|X_s + as + y - x| < r}\, ds = \int_0^t \int_{\frac{x}{s}-\frac{y}{s}-a-\frac{r}{s}}^{\frac{x}{s}-\frac{y}{s}-a+\frac{r}{s}} \frac{1}{\pi(1+z^2)}\, dz\, ds$$

$$\geq \int_r^t \frac{2r}{(1+16)\pi s}\, ds$$

$$= c_1 r \log \frac{t}{r}.$$

The inequality follows from the observation that when $s \geq r$ and z is in the interval of integration, then $|z| \leq 1 + 3 = 4$, since $a \in [0,1]$. Hence

$$\mathbb{E}\big[\tilde{Z}_x \mid Z_x > 0\big] \geq c_1 r \log(t/r)$$

for all x. Thus, using Fubini's Theorem and the fact p_s is a probability density,

$$\int_{\mathbb{R}\setminus(-r,r)} \mathbb{P}(Z_x > 0)\, dx \leq \frac{\int_{\mathbb{R}\setminus(-r,r)} \int_0^{2t} \int_{-r}^r p_s(0, x+y)\, dy\, ds\, dx}{c_1 r \log(t/r)}$$

$$\leq \frac{\int_0^{2t} \int_{-r}^r \int_{\mathbb{R}\setminus(-r,r)} p_s(0, x+y)\, dx\, dy\, ds}{c_1 r \log(t/r)}$$

$$\leq \frac{4rt}{c_1 r \log(t/r)} = \frac{c_2 t}{\log(t/r)}$$

and this completes the proof of the lemma. $\square$

Lemma 9.4.1 proves that almost every Cauchy sample path gives a Besicovitch set, i.e.,

$$K = \{(a, X_t + at) : a, t \in [0,1]\}$$

has zero area. Since X_s is not continuous, the set K need not be compact, but the calculation implies that the expected area of a δ-neighborhood $K(\delta)$ of the set is $O(1/\log(1/\delta))$ and since the neighborhoods are nested, we can deduce that for almost every set $\text{area}(K(\delta)) \to 0$, which implies the closure of K also has zero area a.s.

The expected area of $K(\delta)$ is optimal in these examples, but the proof above does not show that any of the sets K is an optimal Besicovitch set; it is possible each example is non-optimal for some sequence of δs tending to zero. However, this is not really the case as shown by Babichenko, Peres, Peretz, Sousi and Winkler:

Theorem 9.4.2 (Babichenko et al., 2014)   *The set K is almost surely an optimal Besicovitch set, i.e. there exist positive constants c_1, c_2 such that as $\delta \to 0$ we have*

$$\frac{c_1}{|\log \delta|} \le \operatorname{area}(K(\delta)) \le \frac{c_2}{|\log \delta|} \quad a.s.$$

This requires some estimates of higher moments. Since we have already given a (deterministic) example of an optimal Besicovitch set, we will not give the proof of this result here. See Babichenko et al. (2014).

9.5 Projections of self-similar Cantor sets

Let $K \subset [0, 1]$ denote the middle $\frac{1}{4}$-Cantor set formed by removing the center half at each stage of the construction. Let $K^2 = K \times K \subset \mathbb{R}^2$. Then K^2 is a self-similar Cantor set of dimension 1 and is referred to as the "four corner Cantor set". See Figure 9.5.1.

Figure 9.5.1 First three generations of the four corner Cantor set.

The vertical and horizontal projections of the four corner Cantor set obviously have zero length, so the following result of Besicovitch (1938b) implies its projection in almost every direction has zero length. See Figures 9.5.2 and 9.7.1.

Proposition 9.5.1   *Let F be a compact subset of $\mathbb{R}^2$ with $0 < \mathscr{H}^1(F) < \infty$. If two distinct projections of F have zero length, then almost every projection of F has zero length.*

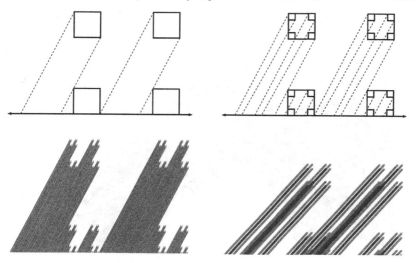

Figure 9.5.2 Induction proves that the nth generation squares project onto an interval along lines of slope 2, hence so does the Cantor set itself (generations 1, 2 and 4 are shown). The bottom right shows the projection along an irrational slope.

The proof of Proposition 9.5.1 is quite complicated, but we will give a simpler proof below that applies to a class of self-similar sets F that includes the set K^2 described above. First, we build a Besicovitch set using the result about a.e. projection of the $\frac{1}{4}$-Cantor set. This construction is due to Kahane (1969).

Consider the unit square $Q = [0, 1]^2$ and place a copy K_1 of K on the bottom edge of the square and a copy K_2 of $\frac{1}{2}K$ along the top edge. Let E be the union of all line segments with one endpoint in K_1 and one endpoint in K_2. See Figure 9.5.3. We claim that

Theorem 9.5.2 *E has zero area and contains unit segments along an interval of angles.*

Proof The horizontal slice of E at height t is simply the convex combination $E_t = (1 - t)K + \frac{t}{2}K$. This is the projection onto $\mathbb{R}$ of K^2 along the lines given by $(1 - t)x + \frac{t}{2}y = c$ whose slopes vary from ∞ to 0 as t goes from 0 to 1. Almost all of these projections have zero length, so almost every horizontal slice of E has zero length, proving E has zero area.

For each $x \in K$ and $y \in \frac{1}{2}K$, the set E contains the segment connecting $(x, 0)$ to $(y, 1)$, and this segment has slope $1/(y - x)$. This will cover an interval of slopes if $K - \frac{1}{2}K = \{x - y : x \in K, y \in \frac{1}{2}K\}$ covers an interval. But this set is the projection of K^2 onto $\mathbb{R}$ along lines of slope 2 and it is clear from a picture (Figure 9.5.2) that this particular projection covers an interval. See Figure 9.5.3.   □

Figure 9.5.3 Kahane's Besicovitch set (see Theorem 9.5.2).

Now we get to the special case of Besicovitch's Projection Theorem mentioned earlier:

Theorem 9.5.3   *Suppose that $m \geq 3$ and that*

$$\Lambda = \left\{ \sum_{n=0}^{\infty} a_n m^{-n} : a_n \in \{\mathbf{b}_1, \mathbf{b}_2, \ldots, \mathbf{b}_m\} \right\}, \tag{9.5.1}$$

where $\mathbf{b}_1, \mathbf{b}_2, \ldots, \mathbf{b}_m$ are distinct vectors in $\mathbb{R}^2$ such that the "pieces" $\Lambda_i = \mathbf{b}_i + m^{-1}\Lambda$, for $i = 1, \ldots m$, are pairwise disjoint. Then $\mathcal{H}^1(\Pi_\theta(\Lambda)) = 0$ for almost every θ in $[0, \pi)$.

We will closely follow the proof of Theorem 9.5.3 given by Peres, Simon and Solomyak (2003).We start with

Lemma 9.5.4   *Let $F \subset \mathbb{R}$ be compact with positive Lebesgue measure. For any $\delta > 0$ there is an interval J such that $\mathscr{L}_1(F \cap J) \geq (1 - \delta)|J|$.*

Proof   This follows immediately from Lebesgue's Theorem on points of density (e.g., Theorem 3.22 in Folland (1999)). For a more elementary proof, see Exercise 9.5.   □

In the next three lemmas we consider the set

$$K = K(\{d_1, \ldots, d_m\}) = \left\{ \sum_{n=0}^{\infty} a_n m^{-n} : a_n \in \{d_1, \ldots, d_m\} \right\}, \tag{9.5.2}$$

where $d_1, d_2, \ldots, d_m$ are real numbers (not necessarily distinct). Thus, K is a self-similar subset of $\mathbb{R}$, but of the special form in which the contraction ratio r is equal to m^{-1}, the reciprocal of the number of digits. In the proof of Theorem 9.5.3 we will have $K = \Pi_\theta(\Lambda)$ for a fixed θ; then $d_i = \Pi_\theta(\mathbf{b}_i)$.

Lemma 9.5.5 *For K as in (9.5.2), let $K_i = d_i + m^{-1}K$. Then $K = \bigcup_{i=1}^m K_i$ and $\mathcal{H}^1(K_i \cap K_j) = 0$ for $i \neq j$.*

Proof The first statement follows directly from the definition of K. The second statement is an easy consequence of self-similarity. The set K is a union of m pieces, each of which is a translate of $m^{-1}K$. Since $\mathcal{H}^1(K) = m \cdot \mathcal{H}^1(m^{-1}K)$, the pieces have to be pairwise disjoint in measure. □

Thus the pairwise intersections $K_i \cap K_j$ cannot be "large". However, at least one of them must be non-empty.

Lemma 9.5.6 *Let K and K_i be as in Lemma 9.5.5. There exist indices $i \neq j$ such that $K_i \cap K_j \neq \emptyset$.*

Proof Let $F(\varepsilon) = \{x : \text{dist}(x, F) < \varepsilon\}$ denote the neighborhood of radius ε of a set F. We will assume that the sets K_k are pairwise disjoint and derive a contradiction. Since they are compact, the distance between any two of them is positive. Thus we can find $\varepsilon > 0$ so that $K_i(\varepsilon) \cap K_j(\varepsilon) = \emptyset$ whenever $i \neq j$. Then we have

$$K(\varepsilon) = \bigcup_{k=1}^m K_k(\varepsilon).$$

In particular, $K(\varepsilon)$ has m times the measure of each $K_k(\varepsilon)$ (since all m such sets have equal measure). However, $K_k(\varepsilon) = d_k + m^{-1}K(m\varepsilon)$, so $K(m\varepsilon)$ has m times the measure of each $K_k(\varepsilon)$. Hence $K(\varepsilon)$ and $K(m\varepsilon)$ have equal measure. But both sets are open and the former is a strict subset of the latter, giving the desired contradiction. □

Before stating the next lemma we need to introduce some notation. Recall that the self-similar set K has a representation

$$K = \bigcup_{i=1}^m K_i = \bigcup_{i=1}^m (d_i + m^{-1}K).$$

Substituting this formula into each term in its right-hand side, we get

$$K = \bigcup_{i,j=1}^m K_{ij}, \qquad K_{ij} = d_i + m^{-1}d_j + m^{-2}K.$$

The sets K_i and K_{ij} are called the **cylinder sets** of K of orders 1 and 2, respectively. This operation can be iterated. For each positive integer ℓ the set K is the union of m^ℓ pieces, called **cylinders of order** ℓ, each of which is a translate of $m^{-\ell}K$. Let $\mathscr{A} = \{1, 2, \ldots, m\}$ and $\mathscr{A}^\ell = \{u = u_1 \ldots u_\ell : u_i \in \mathscr{A}\}$. Then

$$K = \bigcup_{u \in \mathscr{A}^\ell} K_u, \qquad K_u = K_{u_1 \ldots u_l} = \sum_{n=1}^{\ell} d_{u_n} m^{-n+1} + m^{-\ell}K.$$

Repeating the proof of Lemma 9.5.5 for this decomposition shows that

$$\mathscr{H}^1(K_u \cap K_v) = 0 \tag{9.5.3}$$

for different u and v in $\mathscr{A}^\ell$.

We want to understand when K has zero length. There is an easy sufficient condition: $\mathscr{H}^1(K) = 0$ if two cylinders of K coincide; i.e., if $K_u = K_v$ for some distinct u and v in $\mathscr{A}^\ell$. This can be seen in many ways; for instance, $\mathscr{H}^1(K) = m^\ell \mathscr{H}^1(K_u) = m^\ell \mathscr{H}^1(K_u \cap K_v) = 0$ by (9.5.3). This condition is too strong to be necessary, however: for the planar self-similar set Λ in (9.5.1), there are just countably many θ in $[0, \pi)$ for which $\Lambda^\theta = \Pi_\theta(\Lambda)$ has two coinciding cylinders. Thus we would like to know what happens if some cylinders "almost" coincide.

Definition 9.5.7 Two cylinders K_u and K_v are **ε-relatively close** if u and v belong to $\mathscr{A}^\ell$ for some ℓ and $K_u = K_v + x$ for some x with $|x| \le \varepsilon \cdot |K_u|$.

Note that $|K_u| = |K_v| = m^{-\ell}|K|$ for all u and v in $\mathscr{A}^\ell$. Let

$$d_u = \sum_{n=1}^{\ell} d_{u_n} m^{-n+1},$$

so that $K_u = d_u + m^{-\ell}K$. Then K_u and K_v are ε-relatively close whenever

$$|d_u - d_v| \le \varepsilon m^{-\ell}|K|. \tag{9.5.4}$$

Lemma 9.5.8 *If for every $\varepsilon > 0$ there exist an index ℓ and distinct u and v in $\mathscr{A}^\ell$ such that K_u and K_v are ε-relatively close, then $\mathscr{H}^1(K) = 0$.*

This lemma is a very special case of a theorem by Bandt and Graf (1992) that we will give in Section 9.6. The converse holds as well.

Proof Suppose, to the contrary, that $\mathscr{H}^1(K) > 0$. Then by Lemma 9.5.4 we can find an interval J such that $\mathscr{H}^1(J \cap K) \ge 0.9|J|$. Let $\varepsilon = |J|/(2|K|)$. By assumption, there exists an index ℓ in $\mathbb{N}$ and distinct elements u and v in $\mathscr{A}^\ell$ such that the cylinders K_u and K_v are ε-relatively close. If $J_u = d_u + m^{-\ell}J$ and $J_v = d_v + m^{-\ell}J$, then

$$J_u = J_v + (d_u - d_v)$$

and

$$|d_u - d_v| \leq \varepsilon m^{-\ell} |K| = 0.5 |J_u|.$$

This means that J_u and J_v have a large overlap – at least half of J_u lies in J_v. Since J was chosen to ensure that at least 90 percent of its length belongs to the set K, this property carries over to J_u and K_u. To be more precise,

$$\begin{aligned} \mathscr{H}^1(J_u \cap K_u) &= \mathscr{H}^1((d_u + m^{-\ell}J) \cap (d_u + m^{-\ell}K)) \\ &= m^{-\ell} \mathscr{H}^1(J \cap K) \\ &\geq 0.9 m^{-\ell} \mathscr{H}^1(J) \\ &= 0.9 |J_u|. \end{aligned}$$

Similarly,

$$\mathscr{H}^1(J_v \cap K_v) \geq 0.9 |J_v|.$$

Since at least 90 percent of J_v is in K_v and at least 50 percent of J_u is in J_v, we find that at least 40 percent of J_u is in K_v. But at least 90 percent of J_u is in K_u, so at least 30 percent of J_u is in $K_u \cap K_v$. This is in contradiction with (9.5.3), and the lemma is proved. $\qquad\square$

Proof of Theorem 9.5.3   Recall that Λ is a planar Cantor set given by (9.5.1). If we write $\Lambda^\theta = \Pi_\theta(\Lambda)$, then

$$\Lambda^\theta = \bigcup_{i=1}^{m} (\Pi_\theta(\mathbf{b}_i) + m^{-1}\Lambda^\theta),$$

so all the foregoing discussion (in particular, Lemma 9.5.8) applies to Λ^θ. Let $\mathscr{V}_\varepsilon^\ell$ be the set of $\theta \in [0, \pi)$ such that there exist u and v in $\mathscr{A}^\ell$ with Λ_u^θ and Λ_v^θ that are ε-relatively close. Then let $\mathscr{V}_\varepsilon = \bigcup_\ell \mathscr{V}_\varepsilon^\ell$. Note that if θ lies in $\bigcap_{\varepsilon>0} \mathscr{V}_\varepsilon$, then Λ^θ has zero length by Lemma 9.5.8. The proposition will be proved if we are able to show that $\bigcap_{\varepsilon>0} \mathscr{V}_\varepsilon$ has full measure in $[0, \pi)$. But $\bigcap_{\varepsilon>0} \mathscr{V}_\varepsilon = \bigcap_{n=1}^{\infty} \mathscr{V}_{1/n}$, and by DeMorgan's law,

$$[0, \pi) \setminus \bigcap_{n=1}^{\infty} \mathscr{V}_{1/n} = \bigcup_{n=1}^{\infty} ([0, \pi) \setminus \mathscr{V}_{1/n}).$$

Thus it is enough to prove that $\mathscr{V}_\varepsilon$ has full measure for each *fixed* ε.

We will do this by proving the complement of $\mathscr{V}_\varepsilon$ is porous, that is, every subinterval of $[0, \pi)$ has at least a fixed percentage of its length (depending on ε but not on the subinterval's size) lying in $\mathscr{V}_\varepsilon$. This follows if we can find positive constants C_1 and C_2 such that for any θ in $(0, \pi)$ and any ℓ in $\mathbb{N}$ there is θ_0 satisfying

$$|\theta - \theta_0| \leq C_1 m^{-\ell}, \quad (\theta_0 - C_2 \varepsilon m^{-\ell}, \theta_0 + C_2 \varepsilon m^{-\ell}) \subset \mathscr{V}_\varepsilon. \qquad (9.5.5)$$

(There is a minor technical issue when the interval in (9.5.5) is not contained in $[0, \pi)$, but it is easy to handle since either the left or right half of the interval will be in $(0, \pi)$.)

We fix θ in $(0, \pi)$ and ℓ in $\mathbb{N}$. Appealing to Lemma 9.5.6, we choose i and j with $i \neq j$ for which $\Lambda_i^{\theta} \cap \Lambda_j^{\theta} \neq \emptyset$. Since

$$\Lambda_i^{\theta} = \bigcup_{u \in \mathscr{A}^{\ell},\, u_1 = i} \Lambda_u^{\theta} \quad \text{and} \quad \Lambda_j^{\theta} = \bigcup_{v \in \mathscr{A}^{\ell},\, v_1 = j} \Lambda_v^{\theta},$$

there exist u and v in $\mathscr{A}^{\ell}$, with $u_1 = i$ and $v_1 = j$, such that $\Lambda_u^{\theta} \cap \Lambda_v^{\theta} \neq \emptyset$. This means that there are points $y_u \in \Lambda_u$ and $z_v \in \Lambda_v$ such that $\Pi_{\theta}(y_u) = \Pi_{\theta}(z_v)$. Denote by z_u the point in Λ_u corresponding to z_v, i.e. $\Lambda_v - z_v = \Lambda_u - z_u$, and let θ_0 be the angle such that $\Pi_{\theta_0}(z_u) = \Pi_{\theta_0}(z_v)$, whence $\Lambda_u^{\theta_0} = \Lambda_v^{\theta_0}$. Then $|\theta - \theta_0|$ is the angle at z_v for the triangle with vertices z_u, z_v, y_u (see Figure 9.5.4), and therefore $|z_u - y_u| \geq |y_u - z_v| \sin |\theta - \theta_0|$. This implies

$$\sin |\theta - \theta_0| \leq \frac{|\Lambda_u|}{\mathrm{dist}(\Lambda_u, \Lambda_v)}.$$

Note that $|\Lambda_u| = m^{-\ell}|\Lambda|$ and $\mathrm{dist}(\Lambda_u, \Lambda_v) \geq \mathrm{dist}(\Lambda_i, \Lambda_j) \geq \delta > 0$ for some

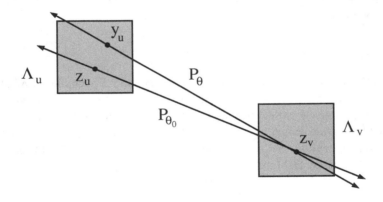

Figure 9.5.4 Finding θ_0.

$\delta = \delta(\Lambda) > 0$ by the hypothesis of pairwise disjointness in Proposition 9.5.3. Thus the first condition in (9.5.5) holds with the constant $C_1 = \pi |\Lambda| / (2\delta)$ (here we are using the fact that $x \leq (\pi/2) \sin x$ on $[0, \pi/2]$). By the choice of θ_0, we have θ_0 in $\mathscr{V}_0$, which is a subset of $\mathscr{V}_{\varepsilon}$. By the Cauchy–Schwarz inequality we have that

$$|(\Pi_{\alpha} - \Pi_{\theta_0})(\mathbf{z} - \mathbf{w})| \leq |\alpha - \theta_0| \cdot |\mathbf{z} - \mathbf{w}|$$

for any two vectors $\mathbf{z}$ and $\mathbf{w}$ and any $\alpha \in [0, \pi)$. Consequently, the projected set Λ_u^α can be obtained from Λ_v^α by the translation $\Pi_\alpha \mathbf{z}_u - \Pi_\alpha \mathbf{z}_v$ and this has length (since $\Pi_{\theta_0}(\mathbf{z}_u - \mathbf{z}_v) = 0$)

$$|\Pi_\alpha \mathbf{z}_u - \Pi_\alpha \mathbf{z}_v| = |(\Pi_\alpha - \Pi_{\theta_0})(\mathbf{z} - \mathbf{w})| \leq |\alpha - \theta_0| \cdot |\Lambda|.$$

On the other hand, the diameter of Λ_u^α is at least $m^{-\ell} \cdot \text{width}(\Lambda)$, where $\text{width}(\Lambda)$ signifies the minimal width of a strip that contains Λ; it is non-zero because the assumption that the sets Λ_i are disjoint prevents Λ from being contained in a straight line (recall Lemma 9.5.6). Set $C_2 = \text{width}(\Lambda)/|\Lambda|$. Then by Definition 9.5.7 the cylinders Λ_u^α and Λ_v^α are ε-relatively close for all α in the $C_2 \varepsilon m^{-\ell}$-neighborhood of θ_0. It follows that this neighborhood lies in $\mathcal{V}_\varepsilon$, so the second condition in (9.5.5) is verified. This completes the proof. $\qquad\square$

9.6 The open set condition is necessary *

Recall from Chapter 2 that a family of maps $\{f_1, f_2, \ldots, f_\ell\}$ of the metric space X satisfies the **open set condition (OSC)** if there is a bounded, non-empty open set $V \subset X$ such that

$$f_j(V) \subset V \text{ for } 1 \leq j \leq \ell,$$

and

$$f_i(V) \cap f_j(V) = \emptyset \text{ for } i \neq j.$$

We proved in Section 2.2 that if $f_1, \ldots, f_\ell$ are contracting similitudes of Euclidean space $\mathbb{R}^d$ that satisfy the open set condition, then $0 < \mathcal{H}^\alpha(K) < \infty$ where K is the corresponding self-similar attractor and α is the similarity dimension.

In the previous section of this chapter, we took a self-similar set in $\mathbb{R}^2$ and projected it in various directions, obtaining self-similar sets in $\mathbb{R}$ with similarity dimension 1, but with zero length. Thus these projected sets, although self-similar, cannot satisfy the open set condition. Schief (1994), building on work of Bandt and Graf (1992), proved that OSC is actually equivalent to having positive measure in the self-similarity dimension.

Theorem 9.6.1 *Let $f_1, \ldots, f_\ell$ be contracting similitudes of $\mathbb{R}^d$ and let K be the corresponding attractor. Let α be the similarity dimension determined by $f_1, \ldots, f_\ell$. If $\mathcal{H}^\alpha(K) > 0$, then $\{f_1, f_2, \ldots, f_\ell\}$ satisfy the open set condition.*

Proof The proof will be broken into five claims, each with its own proof. We start by recalling some notation from Chapter 2. Let $f_1, \ldots, f_\ell$ be contracting similitudes, i.e., $d(f_j(x), f_j(y)) = r_j d(x, y)$, with $r_j < 1$. For $\sigma = (i_1, \ldots, i_n)$, write f_σ for the composition

$$f_{i_1} \circ f_{i_2} \circ \ldots \circ f_{i_n}$$

and denote

$$K_\sigma = f_\sigma(K).$$

Also, write $r_\sigma = r_{i_1} \cdot r_{i_2} \cdot \cdots \cdot r_{i_n}$. Set $r_\emptyset = 1$. Write $r_{\max}$ for $\max_{1 \le j \le \ell} r_j$, and similarly for $r_{\min}$. The length n of σ is denoted by $|\sigma|$. If $(p_1, \ldots, p_\ell)$ is a vector of probabilities, write $p_\sigma = p_{i_1} \cdot p_{i_2} \cdot \cdots \cdot p_{i_n}$. Strings σ, τ are incomparable if each is not a prefix of the other.

Let $\theta > 0$. There exist open sets $U_1, \ldots, U_n$ such that

$$U = \bigcup_{i=1}^n U_i \supset K \text{ and } \sum_{i=1}^n |U_i|^\alpha \le (1 + \theta^\alpha) \mathscr{H}^\alpha(K).$$

Let $\delta = \text{dist}(K, U^c)$.

Step 1: For incomparable strings σ, τ with $r_\tau > \theta r_\sigma$, we have

$$d_H(K_\sigma, K_\tau) \ge \delta r_\sigma.$$

Proof Otherwise, since $\text{dist}(K_\sigma, (f_\sigma(U))^c) = \delta r_\sigma$, we get

$$K_\tau \subset (K_\sigma)(\delta r_\sigma - \varepsilon) \subset f_\sigma(U),$$

for some $\varepsilon > 0$. Recall that for any two incomparable finite strings we have $\mathscr{H}^\alpha(K_\sigma \cap K_\tau) = 0$ by part (iii) of Proposition 2.1.3. We get that

$$\mathscr{H}^\alpha(K) r_\sigma^\alpha (1 + \theta^\alpha) < \mathscr{H}^\alpha(K)(r_\sigma^\alpha + r_\tau^\alpha) = \mathscr{H}^\alpha(K_\sigma) + \mathscr{H}^\alpha(K_\tau) \quad (9.6.1)$$

and the latter is bounded above by

$$\sum_{i=1}^n |f_\sigma(U_i)|^\alpha = \sum_{i=1}^n r_\sigma^\alpha |U_i|^\alpha \le \mathscr{H}^\alpha(K) r_\sigma^\alpha (1 + \theta^\alpha).$$

This contradicts $\mathscr{H}^\alpha(K) < \infty$ (Proposition 2.1.3(i)). $\square$

Step 2: For any $0 < b < 1$ define the minimal cut-set (as in Definition 2.2.3)

$$\pi_b = \{\sigma : r_\sigma \le b < r_{\sigma'}\},$$

where σ' is obtained from σ by erasing the last coordinate. Take

$$\theta = r_{\min},$$

and let $\delta = \text{dist}(K, U^c)$ as above. Note that any distinct $\sigma, \tau \in \pi_b$ are incomparable and $r_\sigma > r_{\min} b \geq r_{\min} r_\tau$; hence by Step 1, $d_H(K_\sigma, K_\tau) \geq \delta r_\sigma$.

Fix $0 < \varepsilon < 1$ such that $(1 + 2\varepsilon) r_{\max} < 1$, and denote

$$G = \{y : \text{dist}(y, K) < \varepsilon\}.$$

Write G_σ for $f_\sigma(G)$. To simplify notation, assume $|K| = 1$. (We leave the reduction to this case as an exercise for the reader.)

Next, for every string v define

$$\Gamma(v) = \{\sigma \in \pi_{|G_v|} : K_\sigma \cap G_v \neq \emptyset\},$$

and denote $\gamma = \sup_v \#\Gamma(v)$. We claim

$$\gamma < \infty. \tag{9.6.2}$$

This is due to Bandt and Graf (1992), but their argument may be simplified as follows.

Proof of (9.6.2) Let B be a closed ball of radius 4, centered in K. Observe that for a string v,

$$f_v^{-1}(K_\sigma) \subset B \text{ for all } \sigma \in \Gamma(v). \tag{9.6.3}$$

Indeed $K_\sigma \cap G_v \neq \emptyset$ implies that $\text{dist}(y, K_v) \leq r_\sigma + \varepsilon r_v$ for all $y \in K_\sigma$. The definition of $\Gamma(v)$ guarantees that $|K_\sigma| \leq |G_v|$ holds, which in turn implies that $|f_v^{-1}(K_\sigma)| \leq |G| \leq 1 + 2\varepsilon$. Thus $\text{dist}(f_v^{-1}(y), K) < 1 + 3\varepsilon$, and hence $|y| < 4$, which is (9.6.3).

Now, distinct $\sigma, \tau \in \Gamma(v)$ satisfy

$$d_H(K_\tau, K_\sigma) \geq \delta r_\sigma > \delta r_{\min} r_v.$$

Hence, $d_H(f_v^{-1}(K_\tau), f_v^{-1}(K_\sigma)) > \delta r_{\min}$ for all v.

By the Blaschke Selection Theorem, the closed bounded subsets of a totally bounded metric space form a totally bounded space in the Hausdorff metric (see Lemma A.2.4 in Appendix A). Thus, $\#\Gamma(v)$ is bounded by the maximal number of closed subsets of B that are $\delta r_{\min}$-separated in the Hausdorff metric. This proves (9.6.2). $\qquad\qquad\square$

For the remainder of the proof, we fix v that maximizes $\#\Gamma(v)$.

Step 3: We claim that

$$\Gamma(\sigma v) = \{\sigma \tau : \tau \in \Gamma(v)\} \tag{9.6.4}$$

for all strings σ, where $\sigma \tau$ denotes the concatenation of σ and τ.

Proof Let $\tau \in \Gamma(v)$, i.e.,

$$r_\tau \leq |G_v| < r_{\tau'} \text{ and } K_\tau \cap G_v \neq \emptyset.$$

Applying f_σ, it follows that

$$r_{\sigma\tau} \leq |G_{\sigma v}| < r_{\sigma\tau'} \text{ and } K_{\sigma\tau} \cap G_{\sigma v} \neq \emptyset,$$

i.e., $\sigma\tau \in \Gamma(\sigma v)$. Thus, $\tau \to \sigma\tau$ is a one-to-one mapping from $\Gamma(v)$ to $\Gamma(\sigma v)$; by maximality of $\#\Gamma(v)$ it must be onto. This proves (9.6.4). □

Step 4: $\operatorname{dist}(K_j, K_{i\sigma v}) \geq \varepsilon r_{i\sigma v}$ for all $j \neq i$ in $\{1, 2, \ldots, \ell\}$ and any string σ.

Proof For any string $j\tau$, Step 3 implies $j\tau \notin \Gamma(i\sigma v)$. Hence, by the definition of $\Gamma(i\sigma v)$, for $j\tau \in \pi_{|G_{i\sigma v}|}$, the sets $K_{j\tau}$ and $G_{i\sigma v}$ must be disjoint, so

$$\operatorname{dist}(K_{j\tau}, K_{i\sigma v}) \geq \varepsilon r_{i\sigma v}.$$

Since $\pi_{|G_{i\sigma v}|}$ is a cut-set, K_j is the union of such $K_{j\tau}$. □

Step 5: Denote $G^* = \{y : \operatorname{dist}(y, K) < \varepsilon/2\}$. Then, $V = \bigcup_\sigma G_{\sigma v}^*$ gives the OSC.

Proof Clearly V is open and $f_i(V) = \bigcup_\sigma G_{i\sigma v}^* \subset V$. Assume that $i \neq j$ but $\exists y \in f_i(V) \cap f_j(V)$. Then, $\exists \sigma, \tau$ such that $y \in G_{i\sigma v}^* \cap G_{j\tau v}^*$. There are points $y_1 \in K_{i\sigma v}$ and $y_2 \in K_{j\tau v}$ satisfying $d(y_1, y) < \frac{\varepsilon}{2} r_{i\sigma v}$ and $d(y_2, y) < \frac{\varepsilon}{2} r_{j\tau v}$. This implies (without loss of generality) that $d(y_1, y_2) < \varepsilon r_{i\sigma v}$, which contradicts Step 4. □

This completes the proof of Theorem 9.6.1. □

The following corollary is useful when studying self-similar tilings (see Kenyon, 1997).

Corollary 9.6.2 *Let $K \subset \mathbb{R}^d$ be a self-similar set of similarity dimension d. If K has positive d-dimensional Lebesgue measure, then K has non-empty interior.*

Proof Since $\mathcal{H}^d(K) > 0$, Theorem 9.6.1 implies that there is a bounded open set $V \neq \emptyset$, with the sets $f_j(V) \subset V$ pairwise disjoint for $1 \leq j \leq \ell$. The identity

$$\sum_{j=1}^{\ell} \mathcal{H}^d(f_j(V)) = \sum_{j=1}^{\ell} r_j^d \mathcal{H}^d(V) = \mathcal{H}^d(V)$$

implies that the open set $V \setminus \bigcup_{j=1}^{\ell} f_j(\bar{V})$ has zero $\mathcal{H}^d$-measure, and is therefore empty. It follows that $\bar{V} \subset \bigcup_{j=1}^{\ell} f_j(\bar{V})$ and since the opposite inclusion is

obvious,

$$\bar{V} = \bigcup_{j=1}^{\ell} f_j(\bar{V}).$$

By uniqueness of the attractor, $K = \bar{V}$.　　　　　　　　　　　□

This corollary can be applied to the projections of a self-similar set K contained in a triangle with vertices $\{P_1, P_2, P_3\}$ defined using the similitudes

$$f_j(x) = \frac{1}{3}x + \frac{2}{3}P_j,$$

for $1 \le j \le 3$. This is a variation of the Sierpiński gasket, and has dimension 1 (see Figure 9.6.1).

Figure 9.6.1 A 1-dimensional gasket.

Let Π_θ denote orthogonal projection to the line in direction θ. Then

$$\Pi_\theta \circ f_j = \tilde{f}_j \circ \Pi_\theta,$$

where

$$\tilde{f}_j(t) = \frac{1}{3}t + \frac{2}{3}\Pi_\theta(P_j).$$

Therefore, the projection $\Pi_\theta(K)$ is self-similar:

$$\Pi_\theta(K) = \bigcup_{j=1}^{3} \tilde{f}_j(\Pi_\theta(K)).$$

Since $\Pi_\theta(K)$ has similarity dimension 1 and is contained in a line, Corollary 9.6.2 implies that $\Pi_\theta(K)$ contains intervals if $\mathscr{H}^1(\Pi_\theta(K)) > 0$. Kenyon (1997) proved this implication before Schief's Theorem was available, and used it to show there are only countably many such θ. We note that a theorem of Besicovitch (1938a) implies that $\mathscr{H}^1(\Pi_\theta(K)) = 0$ for (Lebesgue) almost every θ.

9.7 Notes

In 1919 Besicovitch was motivated by a problem involving Riemann integrability (see Exercise 9.18) to construct a set of zero area in the plane that contained a line segment in every direction. Because of World War I and the Russian revolution, his original paper (Besicovitch, 1919) was not widely known and the result was republished as Besicovitch (1928). Around the same time Kakeya (1917) and Fujiwara and Kakeya (1917) were considering the problem of the smallest area plane region in which a unit length segment can be continuously turned around. This was solved by Besicovitch using a slight modification of his original construction (see Exercise 9.9). Perron (1929) gave a simpler version of Besicovitch's construction, and this was further simplified in Schoenberg (1962). Pál (1921) proved Kakeya and Fujiwara's conjecture that the equilateral triangle is the convex body with smallest area in which a needle can be continuously turned around. Besicovitch's proof that it can be done in arbitrarily small area uses a multiply connected region, but Cunningham (1971) showed that it can also be done in a simply connected region of arbitrarily small area. The same paper also considers star-shaped regions and gives upper and lower bounds for the area of the optimal region.

Tom Körner proved that "most" Besicovitch sets have measure zero in a topological sense, i.e., he shows in Körner (2003) that measure zero sets form a residual set in a certain compact collection of Besicovitch sets in the Hausdorff metric. See Exercise 9.19.

The construction of the Besicovitch set in Theorem 9.1.1 is due to the first author. However, a closely related construction was previously given by Sawyer (1987). The random construction in Section 9.4 is due to Babichenko et al. (2014). In that paper, a stronger version of Lemma 9.4.1 was established, which implies that this random construction also yields an optimal Besicovitch set.

Kahane's construction of a Besicovitch set discussed in Section 9.5 is actually a special case of a construction due to Besicovitch himself using duality (Besicovitch, 1964). Duality in this case refers to associating a line L in $\mathbb{R}^2 \setminus \{0\}$ with the point on L that is closest to 0, inverted through the unit circle. Statements about sets of lines are then translated into statements about point sets (and conversely), which are sometimes easier to deal with. See Section 7.3 in Falconer (1990). See Exercises 9.22–9.27 where this idea is used. One such application is the construction of a set of open rays in the plane whose union has zero area, but whose endpoints have positive area (Exercise 9.24). Surprisingly, similar examples occur naturally in dynamics. Schleicher (2007) used the dynamics of cosine maps to decompose the plane $\mathbb{C} = R \cup L$ where each point in L is the endpoint of a curve connecting it to ∞, the curves are disjoint and $\dim(R) = 1$. See also Karpińska (1999).

What happens to the definition of Besicovitch sets if we replace line segments by pieces of k-planes? Is there a set K of zero n-measure in $\mathbb{R}^n$ so that every k-plane in $\mathbb{R}^n$ contains a ball with a translate inside K? At present no examples are known with $k > 1$. Marstrand (1979) showed this was impossible for $k = 2$, $n = 3$ and Falconer (1980) showed it was impossible for $k > n/2$. Bourgain (1991) improved this to $2^{k-1} + k > n$ and this was further improved to $(1 + \sqrt{2})^{k-1} + k > n$ by Oberlin (2010). Oberlin also showed that for any pair (k, n), such a set (if it exists) must have dimension at least $n - (n - k)/(1 + \sqrt{2})^k$. For proofs of some of these results and a more detailed discussion, see Mattila (2015).

Other authors have considered sets containing at least one translate of every member of some family of curves, e.g., (Wisewell, 2005), or sets that contain segments in some Cantor set of directions, e.g., (Bateman and Katz, 2008).

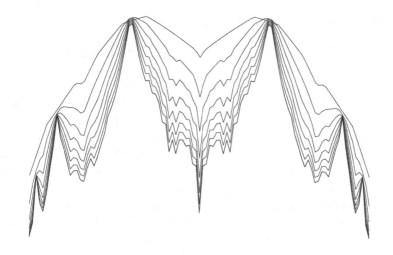

Figure 9.7.1 Projections of K^2 onto angles between 0 and $\pi/2$. Each curve represents the length of a different generation $g = 2, \ldots 8$ of the Cantor set projected. Theorem 9.5.3 implies these graphs tend to zero a.e. as $g \nearrow \infty$. Estimating the area under the nth graph is a well-known open problem.

It is an open problem to find the rate of decay of the Favard length of the nth generation of the four corner Cantor set described in Section 9.5. It is fairly easy to prove that $1/n$ is a lower bound (see Exercise 9.7) and this estimate has been improved to $(\log n)/n$ in Bateman and Volberg (2010). The best current upper bound is a power law of the form $\operatorname{Fav}(K_n) \leq C_\tau n^{-\tau}$ for any

$\tau < 1/6$, given by Nazarov et al. (2010) and improving earlier estimates in Peres and Solomyak (2002) and Mattila (1990). The situation is different for the Favard length of "random" examples constructed in Peres and Solomyak (2002), where the decay is exactly of order $1/n$ almost surely. Projections of the four corner Cantor set are also treated in Mattila (2015). This set was shown to have zero analytic capacity by Garnett (1970) and Ivanov (1975, 1984), and is sometimes called the Garnett set or the Garnett–Ivanov set.

A result of Besicovitch says that any $E \subset \mathbb{R}^2$ of finite 1-measure can be decomposed into a regular and irregular part; the regular part projects to zero length in at most one direction and the irregular part projects to zero length in almost every direction. The regular part can be covered by a countable union of rectifiable curves and the irregular part hits any rectifiable curve in zero length.

Falconer (1985) proved using duality that given a measurable set $E \subset \mathbb{R}^2$, there is a measurable set K so that the projection $\Pi_\theta(K)$ onto the line L_θ through the origin equals $E \cap L_\theta$ (up to length zero) for almost every θ (an explicit construction in higher dimensions is given in Falconer (1986)). We will outline the proof of this and some related results in Exercises 9.20–9.27. The three-dimensional version of Falconer's result says that, in theory, one can build a "digital sundial", whose shadow shows the time of day. As of this writing, such devices are actually available for purchase.

Davies (1971) first proved that a planar Besicovitch set must have dimension 2. The Kakeya conjecture states that the dimension of a Besicovitch–Kakeya set in $\mathbb{R}^d$ should be d; this is currently an intensively investigated problem at the crossroads of harmonic analysis, geometric measure theory, additive combinatorics and discrete geometry. Wolff (1995) proved the Hausdorff dimension is $\geq (d+2)/2$ in d dimensions. This improved an estimate $(d+1)/2$ that was previously known and based on estimates for X-ray and k-plane transforms due to Drury (1983) and Christ (1984). Bourgain (1999) improved Wolff's estimate to $\geq \frac{13}{25}d + \frac{12}{25}$, and Katz and Tao (2002) gave $(2 - \sqrt{2})(d-4) + 3$ (which is better than Wolff's result when $d \geq 5$). In the same paper they prove the Minkowski dimension must be at least $d/\alpha + (\alpha - 1)\alpha$ where $\alpha \approx 1.675$.

The Kakeya conjecture would follow from certain estimates for maximal functions defined in terms of averages over long, narrow tubes. See Exercises 9.41–9.44 for more precise statements. A detailed, but highly readable, description of the connections between Kakeya sets, harmonic analysis, PDE and combinatorics is given in Izabella Łaba's survey (Łaba, 2008). An excellent introduction to the deep connections between harmonic analysis and the Kakeya conjecture is Mattila (2015), which includes a detailed discussion of restriction problems, Fourier multipliers, dimensions estimates in higher dimensions and (n, k) Besicovitch sets. Mattila's book also covers deep applications of the

Fourier transforms to other geometric problems, such as Falconer's distance set conjecture: if $E \subset \mathbb{R}^n$ and $\dim(E) > n/2$ then $D(A) = \{|x - y| : x, y \in A\}$ has positive length.

Fefferman's disk multiplier example started the intense investigation of the connections between Besicovitch sets and Fourier analysis, which continues to the present day. It is interesting to note that Fefferman had done so much great work by 1978 that his Fields medal citation by Carleson (1980) doesn't even mention the disk multiplier problem.

If $\mathscr{F}$ is a finite field with q elements and $z, y \in \mathscr{F}^n$ then $L = \{z + ay : a \in \mathscr{F}\}$ is the line through z in direction y. A Kakeya set $K \subset \mathscr{F}^n$ is a subset that contains some line in direction y for every $y \in \mathscr{F}^n$. Zeev Dvir showed in 2009 that any Kakeya set in $\mathscr{F}^n$ has at least $C_n q^n$ elements answering a question of Tom Wolff (1999). Exercises 9.30–9.37 outline the surprisingly short proof. The finite field case captures many of the features of the continuous case, but Dvir's result is the analog of saying a Besicovitch set in $\mathbb{R}^n$ must have positive n-measure; a distinct difference due to there being "more" lines in the finite field case. A near optimal value of C_n is given in Dvir et al. (2009), and further applications of Dvir's method are given in Dvir and Wigderson (2011), Elekes et al. (2011), Ellenberg et al. (2010), Guth (2010), Quilodrán (2009/10), Saraf and Sudan (2008).

The proof of necessity of the open set condition given in Section 9.6 is a simplification of the original proof in Schief (1994), which made use of Ramsey's Theorem on graph coloring.

9.8 Exercises

• **Exercise 9.1** Show that the $1/4$ Cantor set in the plane is irregular, i.e., its intersection with any rectifiable curve has zero length.

• **Exercise 9.2** Construct a closed set K in $\mathbb{R}^2$ of zero area that contains a line in every direction.

Exercise 9.3 If $g : [0, 1] \to [0, 1]$ is smooth, construct a closed set K in $\mathbb{R}^2$ of zero area that contains an arc of the form $\{(t, ag(t) + b)\}$ for every $a \in [0, 1]$. See Figure 9.8.1 for the cases when $g(t) = t^2/2$ and $g(t) = \frac{1}{2}(1 + \sin(t))$.

Exercise 9.4 For $0 < \alpha < 1$, construct a compact set $E \subset [0, 1]$ of dimension α and with zero α-measure that contains a line segment of slope t for every $t \in E$.

Figure 9.8.1

Exercise 9.5 Give an elementary proof of Lemma 9.5.4 by covering F by open intervals $\{I_i\}$ such that $\sum_i |I_i| \le (1-\delta)^{-1}|F|$, and proving one of them has the desired property.

Exercise 9.6 Show that if K is the union of N unit line segments in the plane, then

$$\text{area}(K(\varepsilon)) \ge C\varepsilon N / \log N.$$

• **Exercise 9.7** Prove that the Favard length of the nth generation of K^2 is $\gtrsim n^{-1}$, where K^2 is the four corner Cantor set discussed in Section 9.5.

• **Exercise 9.8** Given two parallel lines in the plane, and any $\varepsilon > 0$, there is a set E of area $< \varepsilon$ so that a unit segment can be continuously moved from one line to the other without leaving E.

Exercise 9.9 Use Exercise 9.8 and the triangles constructed in Section 9.2 to construct a Kakeya set, i.e., a set of arbitrarily small area in which a unit segment can be continuously turned around $180°$.

Exercise 9.10 Show that a Kakeya set (as in Exercise 9.9) must have positive area.

Exercise 9.11 Show that a Kakeya set for a unit needle inside $\{|z| < r\}$ must have area at least $\pi(1-r)^2$.

Exercise 9.12 Show there are Kakeya sets of arbitrarily small area inside $\mathbb{D}$. This is due to Cunningham (1971).

Exercise 9.13 Show that if a great circle on the unit sphere $\mathbb{S}$ in $\mathbb{R}^3$ is continuously moved so that it returns to its original position but with the opposite orientation, the motion must cover every point of the sphere. Show that if a half-great-circle is continuously reversed it must sweep out area at least 2π

(half the sphere) and this can be attained. This is also due to Cunningham (1974).

Exercise 9.14 Show that an arc of a great circle on $\mathbb{S}$ of length $< \pi$ can be continuously moved so that it returns to its original position but with the opposite orientation, inside a set of arbitrarily small area. If the arc has length in $[\pi, 2\pi)$, then area at least 2π is needed and area $2\pi + \varepsilon$ suffices (for any $\varepsilon > 0$). See Cunningham (1974).

Exercise 9.15 Let $\mathbf{C}$ be the middle thirds Cantor set. Show that $\mathbf{C} - \mathbf{C} = \{x - y : x, y \in \mathbf{C}\}$ contains all numbers in $[0, 1]$.

Exercise 9.16 Let $\mathbf{C}$ be the middle thirds Cantor set. Use Exercise 9.15 to show that $K = (\mathbf{C} \times [0,1]) \cup ([0,1] \times \mathbf{C})$ contains rectangles with every pair of side lengths ≤ 1 (Kinney, 1968).

Exercise 9.17 If the projection of $E \subset \mathbb{R}^2$ in direction θ contains an interval, must its projection also contain an interval for all directions close enough to θ? For any directions other than θ?

• **Exercise 9.18** By a theorem of Lebesgue, a function is Riemann integrable on $\mathbb{R}^d$ if the set where it is discontinuous has d-measure zero. Show that there is a Riemann integrable function on $\mathbb{R}^2$ that fails to be Riemann integrable on some line in every direction. Thus the 2-dimensional integral can't be evaluated by iterating 1-dimensional integrals in any coordinate system. It was this problem that led Besicovitch to invent Besicovitch sets.

Exercise 9.19 This exercise outlines a proof for $\mathbb{R}^2$ of Körner's 2003 result that "most" Besicovitch sets have measure zero. Let $\mathcal{K}_t$ be the collection of compact sets K in the plane defined as follows: $K \in \mathcal{K}_t$ if: (1) K consists of finitely many disjoint, non-trivial, closed line segments of slope t; and (2) the vertical projections of these segments are contained in and cover $[0,1]$ and are disjoint except for endpoints. Let $\overline{\mathcal{K}}$ be the closure of $\mathcal{K}_t$ in the Hausdorff metric. Justify the notation by showing the closure is independent of t.

   For $K \in \overline{\mathcal{K}}$, let $K_t = \{(x, y + tx) : (x, y) \in K\}$ and let $m(K, t)$ be the Lebesgue measure of the horizontal projection of K_t. Prove:

1. If $K \in \overline{\mathcal{K}}$, $E = \{\{(t, y + tx) : (x, y) \in K, t \in [0, 1]\}$ is a compact set that contains a unit line segment of every slope in $[0, 1]$.
2. $m(K, t) = 0$ for all $t \in [0, 1]$ implies E has zero area.

3. $m(K,t)$ is upper semi-continuous in K. This means that for any $\varepsilon > 0$, we have $m(K,t) < m(K',t) + \varepsilon$, if K' is close enough to K in the Hausdorff metric (depending on K and t).

4. If $K \in \mathcal{K}_0$, then $m(K,t) \leq \varepsilon$ for $|t| \leq \varepsilon$.

5. For $s \in [0,1]$, $\{K \in \overline{\mathcal{K}} : \exists t \in [s, s+\varepsilon] \text{ s.t. } m(K,t) \geq 2\varepsilon\}$ is closed and disjoint from $\mathcal{K}_{-s}$, hence nowhere dense in $\overline{\mathcal{K}}$.

6. $\{K \in \overline{\mathcal{K}} : \exists t \in [0,1] \text{ s.t. } m(K,t) > 0\}$ is of first category in $\overline{\mathcal{K}}$.

7. Zero area Besicovitch sets form a residual set in $\overline{\mathcal{K}}$.

• **Exercise 9.20** Suppose I is a line segment in $\mathbb{R}^2$ parallel to the x-axis and suppose $\varepsilon, \rho > 0$ are given. Construct a set E consisting of a finite union of line segments contained in a ρ-neighborhood of I, so that for $\theta \in [-\frac{\pi}{4}, 0]$, $\Pi_\theta(E) \supset \Pi_\theta(I)$ and for $\theta \in [0, \frac{\pi}{4}]$, $\Pi_\theta(E)$ has 1-measure less than ε.

• **Exercise 9.21** Suppose I is a line segment in $\mathbb{R}^2$ parallel to the x-axis and suppose $\varepsilon, \rho, \delta, \alpha > 0$ are given. Construct a set E consisting of a finite union of line segments contained in a ρ-neighborhood of I, so that for $\theta \in J_1 = [-\frac{\pi}{4}, -\delta]$, $\Pi_\theta(E) \supset \Pi_\theta(I)$ and for $\theta \in J_2 = [0, \frac{\pi}{4}]$, $\Pi_\theta(E) \setminus \Pi_\theta(I)$ has 1-measure less than ε.

Exercise 9.22 For each $x \in \mathbb{R}^2$, and line L through the origin, the orthogonal projection of x onto L sweeps out a circle with diameter $[0,x]$ as the line L rotates $180°$. We will call this circle $C(x)$ and for a set $E \subset \mathbb{R}^2$ we let $C(E)$ be the union of all $C(x)$, $x \in E$. See Figure 9.8.2. We take as a basis of open sets in $\mathbb{R}^2 \setminus \{0\}$ quadrilaterals bounded by circles $C(x)$ and lines through the origin, as shown in the figure.

Suppose V is an open subset of $\mathbb{R}^2$ and $z \in C(V)$. Show there is a sequence of basis elements N_k that all contain z and sets $E_k \subset V$, each a union of open balls, so that $N_k \subset C(E_k)$ and $C(E_k) \setminus N_k$ has 2-measure tending to 0.

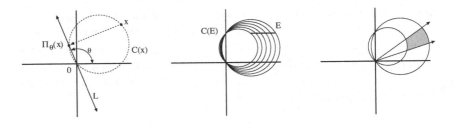

Figure 9.8.2 The definitions for Exercise 9.22.

• **Exercise 9.23** There exists a Borel set $K \subset \mathbb{R}^2$ of zero area so that if $z \notin K$, then there is $x \in K$ so that $C(x) \setminus K = z$.

Exercise 9.24 Show there is a set of open rays in the plane whose union has zero area, but whose endpoints have positive area.

Exercise 9.25 Use Exercise 9.23 and the fact that the map $z \to 1/z$ sends circles through the origin to straight lines to prove there is a set of full measure N in $\mathbb{R}^2$ so that for every $z \in N$ there is a line L so that $L \cap N = z$. This is called a Nikodym set and existence was first proved in Nikodym (1927), with a simplification given in Davies (1952). Also see Theorem 4 of Cunningham (1974).

• **Exercise 9.26** If $A \in \mathbb{R}^2$ is compact, then there is a compact $K \subset \mathbb{R}^2$ so that $A \subset C(K)$ and $C(K) \setminus E$ has zero area. In other words, $\Pi_\theta(K) = E \cap L_\theta$ a.e. for a.e. θ. The 3-dimensional version of this is Falconer's "digital sundial" (Falconer, 1986).

Exercise 9.27 Given a compact $E \subset \mathbb{R}^2$ show that there is a set K that is a union of lines, contains E and has the same area as E.

Exercise 9.28 Construct a compact set of area zero that has zero Favard length, but hits every circle centered at the origin with radius between 1 and 2.

Exercise 9.29 Show there is a compact set of zero area that contains a circle of every radius between 0 and 1. See Figure 9.8.3.

Exercise 9.30 Exercises 9.30–9.37 describe Zeev Dvir's result on finite field Kakeya sets.

Fix d and use induction on n to prove that if $f \in \mathscr{F}[x_1, \ldots, x_n]$ is a non-zero polynomial of degree d then it has at most $d \cdot q^{n-1}$ zeros in $\mathscr{F}^n$. This is due to Schwartz (1980) and Zippel (1979).

Exercise 9.31 Find a bijection between subsets of size k from n elements with repetition allowed, to subsets of k distinct elements from $n + d - 1$ elements. Deduce that the number of monomials of degree d in n variables is $\binom{d+n-1}{n-1}$.

Exercise 9.32 If $K \subset \mathscr{F}^n$ is any subset with fewer than $\binom{d+n-1}{n-1}$ elements then there is a non-zero homogeneous polynomial g of degree d that vanishes on K.

Exercise 9.33 If $K \subset \mathscr{F}^n$, let $K' = \{cx : x \in K, c \in \mathscr{F}\}$. If g, K are as in Exercise 9.32, show g also vanishes on K'.

Figure 9.8.3 A set of zero area containing circles with an interval of radii.

• **Exercise 9.34** We say K is a (δ, γ)-Kakeya set if there is a set $X \subset \mathscr{F}^n$ with at least δq^n elements so that for every $x \in X$ there is a line in direction x that hits K in at least γq points. For such a K show that g also vanishes on X.

Exercise 9.35 Use Exercise 9.30 to show that if K is a (δ, γ)-Kakeya set in $\mathscr{F}^n$ then it has at least $\binom{d+n-1}{n-1}$ elements, $d = \lfloor q \min(\delta, \gamma) \rfloor - 2$.

• **Exercise 9.36** Set $\gamma = \delta = 1$ and deduce that if K is a Kakeya set in $\mathscr{F}^n$ then it has at least $C_n q^{n-1}$ elements.

• **Exercise 9.37** Show that if K is a Kakeya set in $\mathscr{F}^n$ then it has at least $C_n q^n$ elements. Dvir credits the observation that this follows from the previous exercise independently to Noga Alon and Terence Tao.

• **Exercise 9.38** Prove that if $K \subset \mathbb{R}^d$ contains a unit line segment in every direction, then $\overline{\dim}_{\mathscr{M}}(K) \geq (d+1)/2$

• **Exercise 9.39** Khinchin's Inequality: if $a = (a_1, \ldots, a_n)$ is a unit vector in $\mathbb{R}^n$ and $x = (x_1, \ldots, x_n)$ is a random n-vector of ± 1's, prove $\|a \cdot x\|_p \asymp \|a\|_2$.

Exercise 9.40 Use Khinchin's inequality to finish the proof of Fefferman's Disk Multiplier Theorem instead of the vector-valued estimate in Lemma 9.3.3. The idea is to define bounded functions $f = (f_1, \ldots, f_n)$ supported on $R_1, \ldots R_n$, so that applying the disk multiplier gives $g = (g_1, \ldots, g_n)$ so that $|g_j| > c > 0$ on $\tilde{R}_j$. Then Khinchin's inequality implies there is $a = (a_1, \ldots, a_n) \in \{-1, 1\}^n$

such that

$$\|a \cdot g\|_p^p \geq C \left\| \left(\sum 1_{\tilde{R}_j} \right)^{1/2} \right\|_p^p,$$

which implies $a \cdot g$ has large L^p norm compared to $a \cdot f$.

• **Exercise 9.41** Define Bourgain's maximal function on $\mathbb{R}^d$

$$M_B^\delta f(u) = \sup_R \int_R |f(y)| \, dx \, dy,$$

where u is a unit vector in $\mathbb{R}^d$ and the supremum is over all δ-neighborhoods of unit line segments parallel to u. This operator was introduced by Jean Bourgain (1991) who conjectured that it satisfies

$$\|M_B^\delta f(u)\|_{L^p(S^{d-1})} = O(\delta^{(d/p)-1+\varepsilon})\|f\|_{L^p(\mathbb{R}^d)},$$

for all $\varepsilon > 0$ and $1 \leq p \leq d$. Show that if this holds for some $p \in [1, d]$, then any Besicovitch–Kakeya set K in $\mathbb{R}^d$ has Hausdorff dimension at least p.

• **Exercise 9.42** Prove the L^2 bound in $\mathbb{R}^2$ $\|M^\delta f\|_2 \leq (\log \frac{1}{\delta})^{1/2}\|f\|_2$, for Bourgain's maximal function. This gives an alternate proof that a Besicovitch–Kakeya set in $\mathbb{R}^2$ has dimension 2.

Exercise 9.43 Define Córdoba's maximal function on $\mathbb{R}^d$

$$M_C^\delta f(x) = \sup_{x \in R} \int_R |f(y)| \, dx \, dy,$$

where the supremum is over all δ-neighborhoods of unit line segments containing x. Like Bourgain's maximal operator, this is conjectured to satisfy

$$\|M_C^\delta f(x)\|_{L^p(\mathbb{R}^d)} = O(\delta^{(d/p)-1+\varepsilon})\|f\|_{L^p(\mathbb{R}^d)},$$

for all $\varepsilon > 0$ and $1 \leq p \leq d$. Show that if this holds for some $p \in [1, d]$, then any Nikodym set N in $\mathbb{R}^d$ has Hausdorff dimension at least p. (A Nikodym set is defined in Exercise 9.25.)

• **Exercise 9.44** Prove the L^2 bound on $\mathbb{R}^2$, $\|M^\delta f\|_2 \leq (\log \frac{1}{\delta})^{1/2}\|f\|_2$, for Córdoba's maximal function. This proves that a Nikodym set in $\mathbb{R}^2$ has dimension 2.

Exercise 9.45 Show the conjectured L^p bounds for both the Bourgain and Córdoba maximal functions are sharp (except for the multiplicative constant) for all $p \in [2, \infty]$. For $p = 2$ define $f(x) = 1$ if $|x| \leq \delta$, and set $f(x) = \delta/|x|$ if $\delta < |x| \leq 1$ and $f(x) = 0$ otherwise. If $p > 2$, use the indicator function of the union of rectangles constructed in Fefferman's disk multiplier example to show the bound for Córdoba's maximal function is sharp. Use a Besicovitch–Kakeya set to show the bound for Bourgain's is sharp.

• **Exercise 9.46**   Show that a bounded, measurable function m on $\mathbb{R}^d$ defines a L^p bounded Fourier multiplier if and only if there is a constant $C < \infty$ so that

$$\left| \int m(x)\hat{f}(x)\hat{g}(-x)\,dx \right| \leq C\|f\|_p\|g\|_q,$$

where $\frac{1}{p} + \frac{1}{q} = 1$.

Exercise 9.47   If a bounded, measurable function m on $\mathbb{R}^d$ defines an L^p bounded Fourier multiplier, then we denote the operator norm on L^p by $|m|_p$. Show $|m|_q = |m|_p$ where $\frac{1}{p} + \frac{1}{q} = 1$ are conjugate exponents. Thus Fefferman's Theorem for $p > 2$ implies the case $1 < p < 2$.

Exercise 9.48   If a bounded, measurable function m on $\mathbb{R}^d$ defines a L^p bounded Fourier multiplier of norm $|m|_p$ then $\|m\|_\infty \leq |m|_p$. (Use $\|m\|_\infty = |m|_2$, $|m|_p = |m|_q$ (the conjugate exponent) and Riesz interpolation.)

• **Exercise 9.49**   If a continuous function m on $\mathbb{R}^d$ defines an L^p bounded Fourier multiplication operator, show that its restriction to $\mathbb{R}^k \subset \mathbb{R}^d$ defines a L^p bounded multiplier on $\mathbb{R}^k$.

Exercise 9.50   Extend Exercise 9.49 to the indicator function of a disk (which is not continuous). Define a sequence of continuous functions that converge pointwise and boundedly to the indicator of the unit ball and then apply Exercise 9.49 and the Lebesgue Dominated Convergence Theorem. Thus if the ball multiplier were L^p bounded on $\mathbb{R}^d$ for some $1 < p < \infty$, then it would be L^p bounded on $\mathbb{R}^k$ for any $1 \leq k < d$. This result is due to de Leeuw (1965), but the proof we have sketched here is due to Jodeit (1971).

10

The Traveling Salesman Theorem

In this chapter we introduce Peter Jones' β-numbers and use them to estimate the length of a shortest curve Γ containing a given set $E \subset \mathbb{R}^2$ to within a bounded factor:

$$\mathscr{H}^1(\Gamma) \asymp |E| + \sum_{Q \in \mathscr{D}} \beta_E^2(3Q)|Q|,$$

where the sum is over all dyadic squares, $|E| = \operatorname{diam}(E)$, and $\beta_E(3Q)$ measures the deviation of $E \cap 3Q$ from a line segment. This result is Jones' Traveling Salesman Theorem (TST). Finding the absolute shortest path through a finite set is the classical traveling salesman problem (hence the name of the theorem) and is one of the most famous "intractable" problems of combinatorial optimization. Our proof uses Crofton's formula for computing the length of a curve Γ in terms of the number of times a random line hits Γ and interprets the number $\beta_\Gamma^2(Q)$ as approximately the probability that a line hitting Q will hit $\Gamma \cap 3Q$ in at least two points separated by distance approximately $|Q|$. We will also give an application of the TST to estimating the dimension of "wiggly" sets.

10.1 Lines and length

If A, B are quantities that depend on some parameter then $A \lesssim B$ means that A is bounded by a constant times B, where the constant is independent of the parameter. It is equivalent to writing $A = O(B)$ or $A \leq C \cdot B$, with a uniform constant C. If $A \lesssim B$ and $B \lesssim A$ then we write $A \asymp B$, i.e., their ratio is bounded above and below by constants that are independent of the parameter.

Each line $L \subset \mathbb{R}^2 \setminus \{0\}$ is uniquely determined by the point $z \in L$ closest to the origin. If we write $z = re^{i\theta}$ with $(r, \theta) \in \mathbb{R} \times [0, \pi)$, then $d\mu = dr d\theta$ defines

313

a measure on the space of lines (lines through the origin have measure zero). We claim this measure is invariant under Euclidean isometries. First, a rotation by angle ϕ of the plane becomes the map

$$(r, \theta) \to (r, \theta + \phi)$$

on lines and this clearly preserves $dr d\theta$. See Figure 10.1.1. The translation $(x, y) \to (x + t, y)$ in the plane induces the map

$$(r, \theta) \to (r + t \cos \theta, \theta)$$

on lines and this also preserves $dr d\theta$ measure. It is easy to see reflection across the real line preserves μ and hence it is preserved by all Euclidean isometries, as claimed. The dilation $(x, y) \to (ax, ay)$ for $a > 0$ becomes the map

$$(r, \theta) \to (ar, \theta),$$

on lines and hence Euclidean similarities multiply $d\mu$ by the dilation factor a.

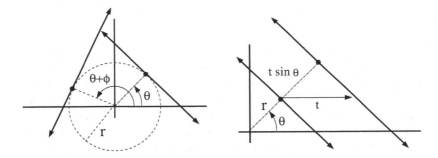

Figure 10.1.1 The effect of isometries on the space of lines. Rotation and translation act as a translation and a skew translation in (r, θ) coordinates.

Finally, consider the affine stretch:

Lemma 10.1.1 *The affine stretch* $(x, y) \to (x, by), b \geq 1,$ *multiplies* μ *measure by at most a factor of* b^2 *above and* b^{-1} *below.*

Proof The vertical affine stretch multiplies the slope of a line by b and hence the induced map on angles is

$$\tau(\theta) = \arctan(b \tan(\theta)).$$

See Figure 10.1.2. Assuming $b > 1$, using the chain rule and applying some basic trigonometric identities, we can compute the derivative of this map as

$$\tau'(\theta) = \frac{b}{1 + (b^2 - 1) \sin^2 \theta}$$

and hence $b^{-1} \leq \tau' \leq b$. The stretch changes r by multiplying it by

$$\lambda = b \frac{\sin \theta}{\sin \tau(\theta)} = \frac{\cos \theta}{\cos \tau(\theta)} = \sqrt{1 + (b^2 - 1) \sin^2 \theta},$$

which is clearly between 1 and b. Thus the affine stretch in (r, θ) coordinates has a derivative matrix that is upper triangular and whose diagonal elements are in $[b^{-1}, b]$ and $[1, b]$ respectively. Therefore the determinant is bounded between b^{-1} and b^2 at every point. □

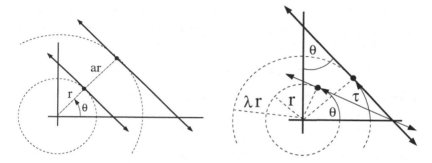

Figure 10.1.2 Dilation in the plane multiplies r and leaves θ alone. A vertical affine stretch $(x, y) \to (x, by)$ sends $(r, \theta) \to (\lambda r, \tau)$ as described in the text. The measure of the space of lines is multiplied by a non-constant function bounded between b^{-1} and b^2.

It is easy to see that both the upper and lower bounds in Lemma 10.1.1 are approximately sharp by fixing some $0 < \beta < 1$ and considering the set of lines that hit both of the vertical line segments $[0, i\beta]$ and $[1, 1 + i\beta]$. The measure of this set is (see Exercise 10.4)

$$2(\sqrt{1 + \beta^2} - 1) \asymp \beta^2.$$

If we apply the vertical stretch $(x, y) \to (x, y/\beta)$ the line segments become unit length and the measure of lines hitting them is $2(\sqrt{2} - 1) \asymp 1$ (again by Exercise 10.4). Thus the measure is multiplied by a factor $\simeq \beta^{-2}$. This is the upper bound in Lemma 10.1.1. On the other hand, if we apply $(x, y) \to (\beta x, y)$ the two segments are mapped to opposite sides of a $\beta \times \beta$ square and the set of lines hitting both of these has measure $2\beta(\sqrt{2} - 1) \asymp \beta$. Thus the inverse map multiplies the x-coordinate by β^{-1} and multiplies the measure by $\simeq \beta$. This is the lower bound in the lemma.

The following simple result is one of the key estimates we will need:

Lemma 10.1.2 *Let $S = [z,w]$ be a unit length segment in the plane and assume $|z| \leq 100$. The μ measure of lines that hit both S and $I = [0,1]$ is $\gtrsim \mathrm{Im}(z)^2$.*

Proof If $|\mathrm{Im}(z)| \geq 1$ the measure of lines hitting both $[0,1]$ and S is non-zero. The subset of such segments among those we are considering is compact, and the measure of the set of lines considered is a continuous function of z, w, so the measure is bounded away from zero in this case, independent of $\mathrm{Im}(z)$.

Otherwise, we may assume $|\mathrm{Im}(z)| < 1$ (but $\mathrm{Im}(z) \neq 0$, otherwise there is nothing to do). Apply the map $(x,y) \to (x,y/\mathrm{Im}(z))$. This sends $[0,1]$ to itself and S to a segment S' that has length at least 1 (the map is an expansion) and one endpoint distance 1 from the real line. By the previous case, the measure of lines hitting both $[0,1]$ and S' is uniformly bounded away from zero, so the lemma follows from the upper bound in Lemma 10.1.1. ◻

The other key fact we need is a well-known formula of Crofton.

Lemma 10.1.3 *If Γ is a planar straight line graph,*

$$\ell(\Gamma) = \frac{1}{2} \int n(\Gamma, L) \, d\mu(L),$$

where $n(\Gamma, \mu) = \#(L \cap \Gamma)$ is the number of points in $\Gamma \cap L$ and $\ell(\Gamma)$ is the length of Γ (the sum of the lengths of its edges).

Proof Both sides are additive for disjoint pieces of Γ, so it suffices to prove it for line segments. By the invariance of μ, the integral is a function of the length of the segment. By the additivity of the integral, this function is additive, it is clearly non-decreasing, hence it is a constant multiple of the identity function. To compute the multiple, note that for $\Gamma = [0,1]$, we have $\ell(\Gamma) = 1$ and

$$\int n(\Gamma, L) \, d\mu(L) = \int_0^\pi |\cos\theta| \, d\theta = 2,$$

so the integral is twice the length of Γ. ◻

A simple computation shows the formula is also correct for circles. It extends to all rectifiable curves by a limiting argument (Exercise 10.6), but to avoid technicalities we will only prove the following version (recall that $\mathcal{H}^1$ denotes 1-dimensional Hausdorff measure):

Lemma 10.1.4 *For any path connected set E,*

$$\mathcal{H}^1(E) \asymp \int n(E, L) \, d\mu(L).$$

Proof Suppose E is covered by disks $D_j = D(x_j, r_j)$ so that $x_j \in E$. For each disk, choose a path $\gamma_j \subset E \cap D_j$ that connects x_j to a point $y_j \in \partial D_j$ (we assume that no individual D_j contains E). Any line that hits the segment $[x_j, y_j]$ must also hit γ_j and hence for any line

$$n(E,L) \geq n\left(\bigcup_j \gamma_j, L\right) \geq n\left(\bigcup_j [x_j, y_j], L\right).$$

Integrating over L gives

$$\int n(E,L)\,d\mu(L) \geq 2\sum_j |y_j - x_j| = 2\sum_j r_j \gtrsim \mathcal{H}^1(E).$$

To prove the other direction let $n_\varepsilon(E,L)$ be the maximal size of an ε-net in $E \cap L$ (a set of points in $E \cap L$ that are at least distance ε apart). As $\varepsilon \searrow 0$, we have $n_\varepsilon(E,L) \nearrow n(E,L)$, so the Monotone Convergence Theorem implies

$$\int n_\varepsilon(E,L)\,d\mu(L) \nearrow \int n(E,L)\,d\mu(L).$$

But if we cover E by disks $\{D_j\}$ with radii $r_j < |E|/2$ and $r_j < \varepsilon/4$, then $n_\varepsilon(E,L)$ is less than $n(\bigcup \partial D_j, L)$ (any two points of the ε-net must be in disjoint disks and hence L crosses two disk boundaries between them). Thus if we cover by disks whose radii satisfy $\sum r_j \leq 2\mathcal{H}^1(E)$, then

$$\int n_\varepsilon(E,L)\,d\mu(L) \leq \int n\left(\bigcup \partial D_k, L\right)d\mu(L) \leq 4\pi \sum r_j \leq 8\pi \mathcal{H}^1(E). \quad \square$$

For any set E let $L(E)$ be the set of lines hitting E, i.e.,

$$L(E) = \{L : L \cap E \neq \emptyset\}.$$

Recall that $|E|$ denotes the diameter of E.

Lemma 10.1.5 *For a compact, path connected set Γ, $|\Gamma| \asymp \mu(L(\Gamma))$.*

Proof Choose $z, w \in \Gamma$ so $|z - w| = |\Gamma|$. Then any line that hits $[z, w]$ also hits Γ so

$$\mu(L(\Gamma)) \geq \mu(L([z, w])) = 2|z - w| = 2|\Gamma|.$$

For the other direction, the disk $D = D(z, |z - w|)$ contains Γ, so

$$\mu(L(\Gamma)) \leq \mu(L(D)) = 2\pi|\Gamma|. \qquad \square$$

In fact, for connected sets, $\mu(L(\Gamma))$ is the perimeter of the convex hull of Γ (Exercise 10.2).

10.2 The β-numbers

In this section we define Peter Jones' β-numbers and prove they can be used to approximate the length of a path connected set (assuming some facts to be proven in the following sections). The result is also true for more general connected sets, but any connected set that is not path connected must have infinite length, so we are most interested in the path connected case, and restricting to this simplifies some arguments.

We start by reviewing the definition of dyadic squares. For $n \in \mathbb{Z}$, we let $\mathscr{D}_n$ denote the grid of nth generation closed dyadic squares

$$Q = [j2^{-n}, (j+1)2^{-n}] \times [k2^{-n}, (k+1)2^{-n}], \quad j, k \in \mathbb{Z}$$

and $\mathscr{D}$ is the union of $\mathscr{D}_n$ over all integers n. The side length of such a square is denoted $\ell(Q) = 2^{-n}$ and (as usual) its diameter is denoted $|Q| = \sqrt{2}\ell(Q)$. It is convenient to expand the collection of dyadic squares by translating every square by the eight complex numbers $\pm 1/3, \pm i/3, \pm(1+i)/3$ and $\pm(1-i)/3$. Let $\mathscr{D}^*$ denote the union of $\mathscr{D}$ and all squares obtained in this way. For a given dyadic square Q let $\mathscr{D}_0^*(Q)$ denote its translates by $\frac{1}{3}\ell(Q)(a+ib)$ where $a, b \in \{-1, 0, 1\}$. It is easy to show $\mathscr{D}_0^*(Q) \subset \mathscr{D}^*$ (see Exercise 10.18).

For $\lambda > 0$, λQ denotes the square concentric with Q such that $|\lambda Q| = \lambda |Q|$. It would be very convenient if λQ were always contained in an ancestor of Q of comparable size, but this is not the case (e.g., $2 \cdot [0,1]^2$ is not contained in any dyadic square). This is why we introduce the translated squares $\mathscr{D}^*$; λQ is contained in an element of $\mathscr{D}^*$ of comparable size. Checking this is left to the reader. See Exercises 10.17 and 10.19.

Fix a set $E \subset \mathbb{R}^2$ and for each square Q define

$$\beta_E(Q) = |Q|^{-1} \inf_{L \in L(Q)} \sup_{z \in E \cap Q} \operatorname{dist}(z, L).$$

This number measures, in a scale invariant way, how far E deviates from being on a line segment. See Figure 10.2.1.

Define the set of lines

$$S(Q, \Gamma) = \{L : L \cap \Gamma \cap \tfrac{5}{3}Q \neq \emptyset \text{ and } L \cap \Gamma \cap 3Q \setminus 2Q \neq \emptyset\}.$$

Note that $S(Q, \Gamma)$ is a subset of the set of lines that hit $3Q \cap \Gamma$ in at least two points that are separated by $\geq \frac{1}{6}\ell(Q)$. This has measure zero if $3Q \cap \Gamma$ lies on a line segment so, like β, it measures how far $3Q \cap \Gamma$ deviates from lying on a line. In fact, we shall prove that it measures this deviation in almost exactly the same way that β^2 does:

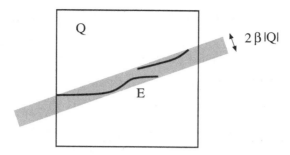

Figure 10.2.1 The definition of $\beta_E(Q)$.

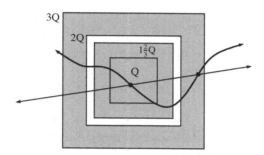

Figure 10.2.2 The definition of $S(Q, \Gamma)$.

Theorem 10.2.1 *Suppose Γ is path connected and $|Q| \leq |\Gamma|$. Then*

$$\mu(S(Q, \Gamma)) \lesssim \beta_\Gamma^2(3Q)|Q|.$$

There is a $Q^* \in \mathscr{D}_0^*(Q)$ so that

$$\beta_\Gamma^2\left(2\frac{11}{12}Q\right)|Q| \lesssim \mu(S(Q^*, \Gamma)).$$

Moreover,

$$\sum_{\mathscr{D}^*} \mu(S(Q^*, \Gamma)) \asymp \sum_{\mathscr{D}} \beta_\Gamma^2(3Q)|Q|.$$

This is proved in Section 10.4. Fix a line L and let $S(L, \Gamma)$ be the collection of squares $Q \in \mathscr{D}^*$ so that $L \in S(Q, \Gamma)$. For any such Q, we can choose points $z \in L \cap \Gamma \cap 1\frac{2}{3}Q$ and $w \in L \cap \Gamma \cap 3Q \setminus 2Q$ that are as close together as possible. This associates to each Q a pair of points in $L \cap \Gamma \cap 3Q$. Since $z, w \in 3Q$ and $|z - w| \asymp |Q|$ only a bounded number of squares can be associated to any pair of points. Thus the number $N(L, \Gamma)$ of squares in $S(L, \Gamma)$ is $O(n(L, \Gamma)^2)$. In fact, a much better two-sided linear bound holds:

Theorem 10.2.2   *With notation as above, $n(L, \Gamma) \asymp 1 + N(L, \Gamma)$.*

These results will be proven in the next two sections. Using them, we can prove the following result of Peter Jones:

Theorem 10.2.3   *If Γ is path connected, then*

$$\mathcal{H}^1(\Gamma) \asymp |\Gamma| + \sum_{Q \in \mathcal{D}} \beta_\Gamma^2(3Q)|Q|. \qquad (10.2.1)$$

Proof   Lemmas 10.1.4 and 10.1.5 plus Theorems 10.2.1 and 10.2.2 give

$$|\Gamma| + \sum_{Q \in \mathcal{D}} \beta_\Gamma^2(3Q)|Q| \asymp |\Gamma| + \sum_{Q^* \in \mathcal{D}^*} \mu(S(Q^*, \Gamma))$$

$$\asymp \int_{L(\Gamma)} d\mu(L) + \int_{L(\Gamma)} \sum_{Q^* \in \mathcal{D}^*} 1_{S(Q^*, \Gamma)} \, d\mu(L)$$

$$\asymp \int_{L(\Gamma)} [1 + N(L, \Gamma)] \, d\mu$$

$$\asymp \int n(L, \Gamma) \, d\mu$$

$$\asymp \mathcal{H}^1(\Gamma). \qquad \square$$

Corollary 10.2.4   *If Γ is path connected, then the number N of dyadic squares of side length 2^{-n} that hit Γ satisfies*

$$N \gtrsim 2^n \left[|\Gamma| + \sum_{Q \in \mathcal{D}_k, k \leq n} \beta_\Gamma^2(3Q)|Q| \right].$$

Proof   The union of the boundaries of the squares in $\mathcal{D}_n$ that hit Γ form a path connected set γ of length $\asymp N2^{-n}$. Moreover, $\Gamma \subset W$, where W is the union of these closed squares. Thus for dyadic squares $Q \in \mathcal{D}_k, k \leq n$ we have $\beta_\gamma(Q) = \beta_W(Q) \geq \beta_\Gamma(Q)$. Thus

$$2^{-n}N \gtrsim |\gamma| + \sum_{Q \in \mathcal{D}} \beta_\gamma^2(3Q)|Q|$$

$$\gtrsim |\gamma| + \sum_{Q \in \mathcal{D}_k, k \leq n} \beta_\gamma^2(3Q)|Q|$$

$$\geq |\Gamma| + \sum_{Q \in \mathcal{D}_k, k \leq n} \beta_\Gamma^2(3Q)|Q|. \qquad \square$$

Theorem 10.2.3 has several important applications including the study of the Cauchy integral and harmonic measure, but these require too much technical background to present here. Instead, we shall use Corollary 10.2.4 to prove that a path connected set that "wiggles" at all points and all scales must have dimension > 1. Although this seems obvious, the only proof known to the authors uses Theorem 10.2.3.

Theorem 10.2.5   *There is a $c > 0$ so the following holds. Suppose Γ is a closed, path connected set in the plane and $\beta_\Gamma(3Q) \geq \beta_0 > 0$ for every square Q with $Q \cap \Gamma \neq \emptyset$ and $|Q| \leq \frac{1}{3}|\Gamma|$. Then $\dim(\Gamma) \geq 1 + c\beta_0^2$.*

Proof   Suppose Q is any dyadic square hitting Γ with $|Q| \leq \frac{1}{3}|\Gamma|$. To simplify notation, we rescale so $Q = [0,1]^2$ and $|\Gamma| > 3\sqrt{2}$. Fix some integer $k > 0$ and fill $2Q \setminus Q$ with dyadic squares of size 2^{-k}. Group them into 2^{k-1} concentric square "annuli" of thickness 2^{-k}. Since Γ connects Q to $2Q^c$ it must hit each of these annuli and hence at least 2^{k-1} of the sub-squares. See Figure 10.2.3. Thus

$$\sum_{Q \in \mathscr{D}_k} \beta_\Gamma^2(3Q)|Q| \geq 2^{k-1}\beta_0^2 2^{-k} = \frac{1}{2}\beta_0^2,$$

and hence for $n \geq 1$

$$\sum_{k=1}^{n} \sum_{Q \in \mathscr{D}_k} \beta_\Gamma^2(3Q)|Q| \geq \frac{1}{2}n\beta_0^2.$$

By Corollary 10.2.4 the number of dyadic squares in $\mathscr{D}_n$ hitting $\Gamma \cap 2Q$ is $N \gtrsim n\beta_0^2 2^n$. By throwing away at most $\frac{8}{9}$ of the squares, we can assume the remaining ones have disjoint doubles and these doubles lie inside $2Q$. By choosing n large enough (but still $n \lesssim \beta_0^{-2}$), we can assume $N \geq 2^{n+1}$.

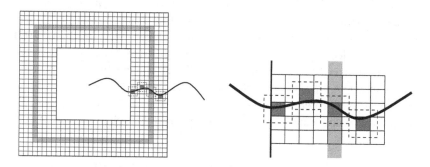

Figure 10.2.3 As Γ crosses $2Q \setminus Q$, it intersects 2^{k-1} "rings" of smaller dyadic squares and so hits at least one from each ring.

Now fix n as above and a square Q_0 that hits Γ and satisfies $|Q_0| \leq \frac{1}{3}|\Gamma|$. Now build generations of sub-squares of Q_0, so that each jth generation square Q has side length $2^{-jn}\ell(Q_0)$, so that all these squares have disjoint doubles $2Q$, so that each Q hits Γ, and each doubled square $2Q$ contains 2^{n+1} squares of generation $j+1$. We define a measure μ by setting $\mu(2Q) = 2^{-j(n+1)}$ for each of the $2^{j(n+1)}$ distinct jth generation squares Q. Thus

$$\mu(2Q) = 2^{-j(n+1)} = (2^{-jn})^{1+\frac{1}{n}} \leq |2Q|^{1+c\beta_0^2},$$

since $n \lesssim \beta_0^{-2}$. From this it is easy to see that $\mu(Q) \lesssim |Q|^{1+c\beta_0^2}$ for any $Q \in \mathscr{D}$ and this proves the theorem. $\square$

A set E satisfying the hypothesis of Theorem 10.2.5 is called **uniformly wiggly**. Such sets occur in many parts of analysis and dynamics, especially in situations where a set has some form of self-similarity. Some examples are discussed in Exercises 10.30–10.34. Note that the proof of Theorem 10.2.5 only requires the $\gtrsim$ direction of (10.2.1) in Theorem 10.2.3 (although this is the more difficult direction).

10.3 Counting with dyadic squares

In this section we prove Theorem 10.2.2; this describes an alternative way of counting how many times a line L hits a path connected set Γ using dyadic squares. The proof follows the n-dimensional argument of Okikiolu (1992), although the notation has been changed somewhat.

Recall from above that $\mathscr{D}_n$ denotes the dyadic squares of side length 2^{-n} and $\mathscr{D} = \bigcup_n \mathscr{D}_n$. Each $Q \in \mathscr{D}_n$ is contained in a unique dyadic square $Q^\uparrow \in \mathscr{D}_{n-1}$ called its "parent". More generally, $Q^{\uparrow k}$ denotes the unique $(n-k)$th generation dyadic square containing Q (Q's kth ancestor). In the other direction, $\mathscr{D}(Q)$ denotes the collection of all dyadic sub-squares of Q (Q's descendants), $\mathscr{D}_k(Q)$ is the collection of $(n+k)$th generation dyadic sub-squares (the kth generation descendants) and $\mathscr{D}_1(Q)$ are the children of Q.

The proof of Theorem 10.2.2 is broken into a number of simple lemmas. We need to give an upper and lower bound, and we will do each of these, first in $\mathbb{R}$ and then in $\mathbb{R}^2$.

Lemma 10.3.1 *If $0 \leq x < y \leq 1$ then there is at least one dyadic interval in $I \in \mathscr{D}$ so that one of these points is in I and the other is in $3I \setminus 2I$. Moreover, we have $\frac{1}{2}|x-y| \leq |I| \leq 2|x-y|$.*

Proof Let J be the smallest dyadic interval containing both x and y (there is one since $x, y \in [0, 1]$ and the points are distinct). If m is the midpoint of J then $x < m < y$ (otherwise both points would be in the left or right half of J, a smaller dyadic interval). Let J_1 be the smallest dyadic interval containing x whose right endpoint is m and let J_2 be the smallest dyadic interval containing

y whose left endpoint is m. First suppose $|J_1| \leq |J_2|$. Note that

$$\frac{1}{2}|x-y| \leq \frac{1}{2}(|J_1| + |J_2|) \leq |J_2| \leq 2|m-y| \leq 2|x-y|.$$

Let I be the dyadic interval so that $|I| = |J_2|$ and $J_1 \subset I$. Then $x \in J_1 \subset I$ and $y \in J_2 \subset 3I$ but $y \notin 2I$ since y must be in the right half of J_2 by the minimality of J_2. If $|J_1| > |J_2|$ then the roles of x and y may be reversed. $\qquad\square$

Lemma 10.3.2 *If $0 \leq x < y \leq 1$ then the number of closed dyadic intervals I with $y \in 2I$ and $x \in 3I \backslash 2I$ is uniformly bounded.*

Proof If $x, y \in 3I$, then $|I| \geq (y-x)/3$, so there are only a bounded number of dyadic intervals that satisfy the lemma and have length $\leq 4|y-x|$. We will show there is at most one interval I satisfying the lemma with length $> 4|x-y|$, by assuming there are two such and deriving a contradiction.

If $|I| \geq 4|x-y|$, let I' be the adjacent dyadic interval of the same length to the left of I. The center of I' is exactly the right endpoint of $2I$. Since x is not in $2I$ we deduce that the center of I' is between x and y. Now suppose there is a second pair of such intervals J, J'. Without loss of generality, assume $|J| \geq 2|I| \geq 8|x-y|$. Then since $I' \cap J' \neq \emptyset$ (they both contain x) we have $I' \subset J'$. Thus I' lies in one half of J' or the other. In particular the centers of I' and J' cannot coincide and in fact they must differ by at least $\frac{1}{2}|I'| \geq 2|x-y|$. However, both centers are between x and y so differ by at most $|x-y|$. This is a contradiction, so there is no second interval J. $\qquad\square$

Next we give the 2-dimensional versions of the last two results.

Lemma 10.3.3 *If $z, w \in [0,1]^2$ are distinct, then there is $Q \in \mathscr{D}^*$ so that one of these points is in $1\frac{2}{3}Q$, the other is in $3Q \backslash 2Q$.*

Proof The line L through z and w makes an angle of $\leq 45°$ with either the vertical or horizontal axis. Assume it is the horizontal axis (otherwise we simply reverse the roles of the two axes in the following argument). Let x and y be the vertical projections of z and w onto the horizontal axis. By assumption, $|z-w|/\sqrt{2} \leq |x-y| \leq |z-w|$. By Lemma 10.3.1 there is at least one dyadic interval in $I \in \mathscr{D}$ with one of x, y in I, the other in $3I \backslash 2I$ and

$$\frac{1}{2\sqrt{2}}|z-w| \leq \frac{1}{2}|x-y| \leq |I| \leq 2|x-y| \leq 2|z-w|.$$

By relabeling the points, if necessary, we may assume $x \in I$.

Let x' and y' be the horizontal projections of z, w onto the vertical axis and suppose x' is below y' (otherwise reflect the following proof over a horizontal line). Choose a closed dyadic interval J' on the vertical axis containing x' with $|J'| = |I|$. Our assumption on L implies

$$|x' - y'| \leq |x - y| \leq 3|J'|.$$

If x' is in the upper third of J', then let $J = J' + i|J'|$ (vertical translation by $|J'|$). See Figure 10.3.1. If x' is in the middle third of J', let $J = J' + i\frac{2}{3}|J'|$ and if x' is in the lower third of J', let $J = J' + i\frac{1}{3}|J'|$. In every case, it is easy to check that $x' \in 1\frac{2}{3}J$ and $y' \in 3J$. In every case set $Q = I \times J \in \mathscr{D}^*$ and

$$\frac{1}{2\sqrt{2}}|z - w| \leq \frac{1}{2}|x - y| \leq \ell(Q) \leq 2|x - y| \leq 2|z - w|.$$

Note that $z \in I \times 1\frac{2}{3}J \subset 1\frac{2}{3}Q$ and $w \in 3Q$, but $w \notin 2Q$ since $y \notin 2I$. This proves the lemma. □

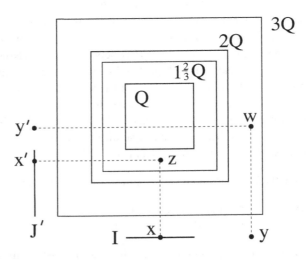

Figure 10.3.1 The proof of Lemma 10.3.3. This illustrates the case when x is in the upper third of J'. The other cases are similar.

Lemma 10.3.4 *If $z, w \in \mathbb{R}^2$ are distinct, then there are only a uniformly bounded number of squares $Q \in \mathscr{D}^*$ such that $z \in 2Q$ and $w \in 3Q \setminus 2Q$.*

Proof If Q has the desired property, then projecting onto one of the two coordinate axes, we must get a dyadic interval $I = P(Q)$ so that $x = P(z) \in 2I$ and $y = P(w) \in 3I \setminus 2I$. By Lemma 10.3.2 there are only a finite number of possible lengths for I, and thus for Q. Also, $z \in 2Q$ for only boundedly many Q of a given size, which proves the lemma. □

Proof of Theorem 10.2.2 Suppose $L \subset \mathbb{R}^2$ is a line and as before, let $n(L, \Gamma)$ be the number of points in $L \cap \Gamma$ and let $N(L, \Gamma)$ be the number of squares in $\mathscr{D}^*$ so that $L \in S(Q, \Gamma)$. Either number may be infinite.

Suppose $z, w \in L \cap \Gamma$ are distinct. By Lemma 10.3.3, there is at least one Q so that $L \in S(Q, \{z, w\}) \subset S(Q, \Gamma)$. The boundary of $2Q$ must hit the segment $(w, z]$ and can only hit L twice, so there are at least half as many Qs as there are disjoint segments on L with endpoints in Γ. Thus

$$n(L, \Gamma) - 1 \le 2N(L, \Gamma),$$

and this holds even if the left side is infinite.

Next we prove the opposite direction. If $n(L, \Gamma) = \infty$, we already know $N(L, \Gamma) = \infty$, so we may assume $L \cap \Gamma$ is a finite set, say with n elements. Given any $Q \in S(L, \Gamma)$, by definition we can choose $z \in L \cap \Gamma \cap 1\frac{2}{3}Q$ and $w \in (L \cap \Gamma \cap 3Q) \setminus 2Q$. Therefore by finiteness, there is a point $z' \in L \cap \Gamma \cap 2Q$ that is closest to w and a point $w' \in (L \cap \Gamma \cap 3Q) \setminus 2Q$ that is closest to z. Note that z', w' must be adjacent on L in the sense that they are distinct and no other point of $L \cap \Gamma$ separates them on L. Since $L \cap \Gamma$ has n points, there are at most $n - 1$ such adjacent pairs z', w', and by Lemma 10.3.4, only a bounded number M of Qs can be associated to this pair in the way described. Thus $N(L, \Gamma) \le M \cdot (n(L, \Gamma) - 1)$. This and the inequality derived earlier imply

$$\frac{1}{M}(N(L, \Gamma) + 1) \le n(L, \Gamma) \le 2(N(L, \Gamma) + 1),$$

as desired. □

10.4 β and μ are equivalent

Next we prove Theorem 10.2.1, starting with the easier direction:

Lemma 10.4.1 *Suppose $|Q| \le |\Gamma|$. Then*

$$\mu(S(Q, \Gamma)) \lesssim \beta_\Gamma^2(3Q)|Q|.$$

Proof Let $\beta = \beta_\Gamma(3Q)$. Since $\mu(L(Q)) \asymp |Q|$, the inequality is trivially true if $\beta \ge 1/100$ so assume $\beta < 1/100$. Suppose W is a $2\beta|Q|$-wide closed strip containing $\Gamma \cap 3Q$. Any line in $S(Q, \Gamma)$ must hit both short sides of some rectangle $R \subset W \cap (2Q \setminus 1\frac{2}{3}Q)$ that has long sides of length $\asymp |Q|$ on the sides of W and short sides of length $\asymp \beta|Q|$. By Exercise 10.4, the measure of this set is $\lesssim \beta^2|Q|$. □

For the other direction, we start with a simple fact about $\mathscr{D}_0^*(Q)$.

Lemma 10.4.2 *If $Q^* \in \mathscr{D}_0^*(Q)$ then $A = (3Q \setminus 2Q) \cap (3Q^* \setminus 2Q^*)$ is a topological annulus whose inner and outer boundaries are at least distance $\frac{1}{6}\ell(Q)$ apart. Furthermore, there is a curve $\gamma \subset \overline{A}$ that separates the boundaries of $2\frac{1}{12}Q$ and $2\frac{11}{12}Q$ and is distance $\geq \frac{1}{12}\ell(Q)$ from each of these boundaries.*

The proof is a picture; see Figure 10.4.1.

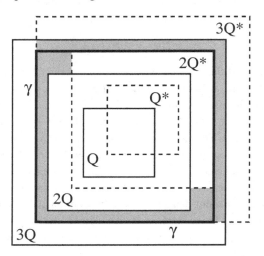

Figure 10.4.1 A typical case of Lemma 10.4.2. The shaded region is $A = (3Q \setminus 2Q) \cap (3Q^* \setminus 2Q^*)$. The bold square is the curve γ in Lemma 10.4.2.

Lemma 10.4.3 *Suppose that Γ is path connected, $Q \in \mathscr{D}$, $\Gamma \cap Q \neq \emptyset$ and $|Q| \leq \frac{1}{3}|\Gamma|$. Then there is a $Q^* \in \mathscr{D}_0^*(Q)$ so that*

$$\mu(S(Q^*, \Gamma)) \gtrsim \beta_\Gamma^2 \left(2\frac{11}{12}Q\right) |Q|.$$

Proof Without loss of generality we may assume Q has side length 1. Note that Γ hits both Q and $3Q^c$. Let $\beta = \beta_\Gamma(2\frac{11}{12}Q)$. Choose $z \in \Gamma \cap \partial Q$ and define the disk $D_1 = D(z, \frac{1}{12})$. Choose $w \in \Gamma \cap \partial D_1$ so that z and w are connected by a component of $\Gamma \cap D_1$. Let $S_1 = [z, w]$ be the segment connecting them. Then $S_1 \subset 1\frac{2}{3}Q$ and any line that hits S_1 must also hit $\Gamma \cap 1\frac{2}{3}Q$. Let L be the line that contains S_1 and let $W_0 \subset W_1$ be the strips of width $\beta/1000$ and $\beta/2$ respectively, both with axis L. See Figure 10.4.2.

Case 1: Suppose there is a point

$$v \in \Gamma \cap \left(2\frac{11}{12}Q \setminus 2\frac{1}{12}Q\right) \setminus W_0.$$

Since Γ is path connected and has diameter $\geq 3|Q|$, v can be connected to a point $u \in \Gamma$ with $|u - v| = \frac{1}{24}$ by a subarc of Γ that stays inside the disk $D_2 = D(v, \frac{1}{24})$. Let $S_2 = [u, v]$ be the segment connecting these two points and note that any line that hits S_2 also hits $\Gamma \cap D_2 \subset 3Q \setminus 2Q$. Since S_2 has an endpoint outside W_0, the measure of the set of lines that hits both S_1 and S_2 is $\gtrsim \beta^2$ by Lemma 10.1.2.

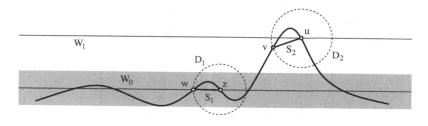

Figure 10.4.2 In Case 1, Γ is close to a line inside $2\frac{1}{12}Q$, but not outside this square. Therefore we can find inscribed arcs that are well separated and each is not close to the line containing the other. This means the measure of lines hitting both segments is bounded away from zero in terms of β.

Case 2: Suppose that

$$\Gamma \cap \left(2\frac{11}{12}Q \setminus 2\frac{1}{12}Q\right) \setminus W_0 = \emptyset.$$

Thus there must be a point $p \in \Gamma \cap (2\frac{1}{12}Q \setminus W_1)$. Choose q with $|p - q| \leq \frac{1}{12}$ that is connected to p by a subarc of Γ inside the disk $D_3 = D(p, \frac{1}{12}) \subset 2\frac{1}{3}Q$ and let $S_3 = [p, q]$. As before, any line that hits S_3 must also hit $\Gamma \cap D_3$. Choose an element $Q^* \in \mathcal{D}_0^*(Q)$ (possibly Q itself) so that $D_3 \subset 1\frac{2}{3}Q^*$ (it is easy to check there is at least one). Let γ be the curve from Lemma 10.4.2. Γ must hit γ at some point u, and $u \in 2\frac{11}{12}Q \setminus 2\frac{1}{12}Q$ and is distance $\geq \frac{1}{12}$ from the boundary of $2\frac{11}{12}Q \setminus 2\frac{1}{12}Q$. Thus if D_4 is a disk of radius $\frac{1}{12}$ centered at u, then $D_4 \subset 2\frac{11}{12}Q \setminus 2\frac{1}{12}Q$. As before, we can find a radius S_4 of D_4 so that any line that hits S_4 also hits $\Gamma \cap D_4$. Moreover, p must be outside the strip of width $\beta/1000$ whose axis is the line containing S_4 (this strip intersected with the cube is inside W_1 because our assumption implies S_4 is close to W_0 and close to parallel to L). Hence the measure of the set of lines that hit both S_4 and S_3 is $\gtrsim \beta^2$ by Lemma 10.1.2.   □

The two lemmas show that for a path connected set Γ and a square Q there

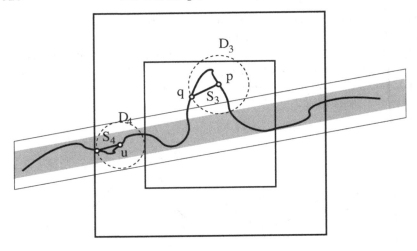

Figure 10.4.3 In Case 2, we can still find two inscribed segments, S_3 and S_4, that are well separated and not near the same line, implying the set of lines hitting both has large measure (at least $\simeq \beta^2$).

is a $Q^* \in \mathscr{D}_0^*(Q)$ so that

$$\beta_\Gamma \left(2\frac{11}{12} Q \right)^2 |Q| \lesssim \mu(S(Q^*, L))$$

$$\lesssim \beta_\Gamma(3Q^*)^2 |Q| \lesssim \beta_\Gamma(3Q^\uparrow)^2 |Q|, \quad (10.4.1)$$

since $3Q^* \subset 3Q^\uparrow$. The asymmetry between the left and right sides of (10.4.1) can be removed by summing over all dyadic squares as follows. Relabeling $Q^\uparrow$ by Q,

$$\sum_{Q^* \in \mathscr{D}^*} \mu(S(Q^*, \Gamma)) \lesssim \sum_{Q \in \mathscr{D}} \beta_\Gamma^2(3Q) |Q|.$$

For $Q \in \mathscr{D}$, let $Q^{\uparrow *} \in \mathscr{D}_0^*(Q^\uparrow)$ be the square corresponding to $Q^\uparrow$ by Lemma 10.4.3. Note that $3Q \subset 2Q^\uparrow \subset 2\frac{11}{12}Q^\uparrow$, so we have

$$\beta_\Gamma(3Q)^2 |Q| \lesssim \beta_\Gamma \left(2\frac{11}{12} Q^\uparrow \right)^2 |Q^\uparrow| \lesssim \mu(S(Q^{\uparrow *}, L)),$$

and hence (relabeling $Q^{\uparrow *}$ by Q^*),

$$\sum_{Q \in \mathscr{D}} \beta_\Gamma^2(3Q) |Q| \lesssim \sum_{Q^* \in \mathscr{D}^*} \mu(S(Q^*, \Gamma)). \quad (10.4.2)$$

This completes the proofs of Theorems 10.2.1 and 10.2.3.

10.5 β-sums estimate minimal paths

For a path connected set Γ, the β-sum estimates $\mathscr{H}^1(\Gamma)$ up to a bounded factor. For a general set E it estimates the length of a shortest connected set containing E. This result is known as Jones' Traveling Salesman Theorem, the analysts' Traveling Salesman Theorem or sometimes simply as the TST.

Theorem 10.5.1 *If $E \subset \mathbb{R}^2$ then there is a path connected set Γ containing E so that*

$$\mathscr{H}^1(\Gamma) \asymp |E| + \sum_{Q \in \mathscr{D}} \beta_E^2(3Q)|Q|,$$

and this set has length comparable to the infimum of the lengths of all path connected sets containing E.

Proof We only need to prove $\lesssim$ since the other direction is immediate from Theorem 10.2.3 and the fact that $\beta_E(Q) \le \beta_\Gamma(Q)$ if $E \subset \Gamma$. Our proof is adapted from an argument given by Garnett and Marshall (2005).

We describe an inductive construction of nested, path connected, closed sets $\Gamma_0 \supset \Gamma_1 \supset \Gamma_2 \supset \cdots$ that all contain E and whose limit will be the desired curve Γ. Each Γ_n is a union of closed convex sets (this collection is denoted $\mathscr{R}_n$) and line segments $\bigcup_{k \le n} \mathscr{S}_k$. Each $R \in \mathscr{R}_n$ is the convex hull of $R \cap E$ and Γ_{n+1} is obtained from Γ_n by replacing each $R \in \mathscr{R}_n$ by a union of two convex subsets that have disjoint interiors and possibly a line segment.

If R is a closed convex region, we let

$$\beta(R) = \sup_L \sup_{z \in R} \operatorname{dist}(z, L)/|R|,$$

where the first supremum is over all chords of R of length $|R|$. A diameter of R will be a choice of L where the supremum is attained.

To start the induction, let $\mathscr{R}_0$ consist of one set; the convex hull of E. In general, suppose $R \in \mathscr{R}_n$ and let I be a diameter. There are two cases to consider. Let K be the middle third of I and let P be the orthogonal projection on I. Then either $P(E \cap R) \cap K$ contains a point or it does not.

Case 1: Assume there is a point $z \in R \cap E$ with $P(z) \in K$. Then $P(z)$ divides I into two closed segments I_1 and I_2. Define $E_1 = P^{-1}(I_1) \cap E \cap R$ and similarly let $E_2 = P^{-1}(I_2) \cap E \cap R$. Let R_1 and R_2 be the convex hulls of these sets. Clearly these are both subsets of R, they both contain z on their boundaries and $\partial R \cap E \subset \partial R_1 \cup \partial R_2$, so that replacing R by $R_1 \cup R_2$ keeps Γ_{n+1} connected. See Figure 10.5.1. We let S be the degenerate segment consisting of the point z.

Case 2: Assume $P(E \cap R) \cap K = \emptyset$. Let I_1, I_2 be the components of $I \setminus K$ and define E_1, E_2 to be (as above) the parts of $E \cap R$ that project onto I_1, I_2 respectively and let R_1, R_2 be the convex hulls of these sets. We also add the shortest possible segment S that connects E_1 and E_2. Again, replacing R by the union of these three pieces keeps Γ_{n+1} connected.

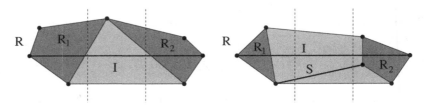

Figure 10.5.1 Cases 1 and 2 of the construction. The convex set R is replaced by two convex subsets and possibly a line segment connecting them.

In each case, we call R_1, R_2 the "children" of R and S the segment associated to R. The collection $\mathscr{R}_{n+1}$ is defined by replacing each element of $\mathscr{R}_n$ by its two children and $\mathscr{S}_{n+1}$ consists of all the segments associated to Case 2 sets in $\mathscr{R}_n$. We define Γ_n as the union of all sets in $\mathscr{R}_n$ and $\bigcup_{k \leq n} \mathscr{S}_k$. Then $\{\Gamma_n\}$ is a nested sequence of compact sets and $\Gamma = \cap \Gamma_n$ is the desired set. To prove the theorem, it suffices to show Γ has length bounded by the β-sum for E, and this follows from

$$\sup_n \left[\sum_{R \in \mathscr{R}_n} |R| + \sum_{k=1}^{n} \sum_{S \in \mathscr{S}_k} |S| \right] \lesssim |E| + \sum_{Q \in \mathscr{D}} \beta_E^2(3Q)|Q|, \qquad (10.5.1)$$

and the fact that the diameters of sets in $\mathscr{R}_n$ tend to zero with n (this follows from Lemma 10.5.3 below).

Lemma 10.5.2 *If R is replaced by R_1, R_2, S in the construction, then*

$$|R_1| + |R_2| + \frac{1}{2}|S| \leq |R| + O(\beta^2(R)|R|).$$

Proof In Case 1, $|I_1|, |I_2|$ are both $\geq |R|/3$ and $|S| = 0$. Since R_1 is contained in a $|I_1| \times 2\beta(R)|R|$ rectangle it has diameter bounded by

$$\sqrt{|I_1|^2 + 4\beta^2(R)|R|^2} \leq |I_1| + O(\beta^2(R)|R|).$$

Adding this to the analogous estimate for R_2 gives the desired estimate. Case 2 is easier, for

$$|R_1| + |R_2| + \frac{1}{2}|S| \leq \frac{5}{2}|R| \leq |R| + 2250\beta^2(R)|R|,$$

if $\beta(R) \geq 1/30$ and if $\beta(R) < 1/30$, then

$$|R_1| + |R_2| + \frac{1}{2}|S|$$

$$\leq |I_1| + |I_2| + 4\beta(R)|R| + \frac{1}{2}(|R| - |I_1| - |I_2| + 2\beta(R)|R|)$$

$$\leq \frac{1}{2}|R| + \frac{1}{2}|I_1| + \frac{1}{2}|I_2| + 5\beta(R)|R|$$

$$\leq |R|. \qquad \Box$$

Note that

$$\frac{1}{2}\beta(R)|R|^2 \leq \operatorname{area}(R) \leq 2\beta(R)|R|^2,$$

since there is a triangle of height $\beta|R|$ and base $|R|$ contained in R and R is contained in a $2\beta(R)|R| \times |R|$ rectangle. See Figure 10.5.2. The part of the triangle that projects onto I_1 has base length at least $|R|/3$ and height $\geq \beta(R)|R|/3$ and hence area $\geq \beta(R)|R|^2/18 \geq \operatorname{area}(R)/36$. Since R_2 always omits the part of R that projects onto I_1, we have

$$\operatorname{area}(R_2) \leq \frac{35}{36}\operatorname{area}(R),$$

and the same holds for R_1. See Figure 10.5.2.

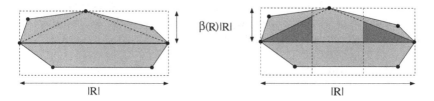

Figure 10.5.2 On the left is the proof that the area of a convex region is comparable to $\beta(R)|R|^2$. The right shows that the areas of R_1 and R_2 are smaller than the area of R by a uniform factor.

Lemma 10.5.3 *There is a $M < \infty$ so that if $R \in \mathscr{R}_n$, $R' \in \mathscr{R}_{n+M}$ and $R' \subset R$, then $|R'| \leq \frac{1}{2}|R|$.*

Proof Let $\lambda = 35/36$. By our previous remarks, after M steps the area goes down by λ^M and hence if $R' \subset R$ and $R' \in \mathscr{R}_{n+k}$, then

$$\frac{1}{2}\beta(R')|R'|^2 \leq \operatorname{area}(R') \leq \lambda^k \operatorname{area}(R) \leq \lambda^k 2\beta(R)|R|^2.$$

If $|R'| \geq |R|/2$, then solving for $\beta(R')$ gives that

$$\beta(R') \leq 16\lambda^k \beta(R) \leq 16\lambda^k$$

is as small as we wish. But if $\beta(R')$ is small enough, then it is easy to see that each of its "children" has diameter $\leq \frac{3}{4}|R'|$. So in either case, after $k+1$ generations, the diameters must drop by a factor of at least $3/4$. Thus after $M = 3(k+1)$, the diameters must drop by $(3/4)^3 < 1/2$. $\quad\square$

For $Q \in \mathscr{D}$ let $\mathscr{R}(Q,n)$ be the collections of convex sets $R \in \mathscr{R}_n$ such that $\ell(Q)/2 < |R| \leq \ell(Q)$ and $R \cap Q \neq \emptyset$. Let $\mathscr{R}(Q) = \bigcup_{n \geq 0} \mathscr{R}(Q,n)$. Elements of $\mathscr{R}(Q,n)$ have disjoint interiors, but elements of $\mathscr{R}(Q)$ need not. However, if two elements overlap, then one is contained in the other and the smaller is obtained from the larger by applying the construction a finite number of times. Because of Lemma 10.5.3, the number of times is uniformly bounded by M. Thus elements of $\mathscr{R}(Q)$ have bounded overlap, i.e., any point is contained in at most $O(1)$ elements of $\mathscr{R}(Q)$.

Lemma 10.5.4 *With notation as above,*

$$\sum_{R \in \mathscr{R}(Q)} \beta^2(R)|R| \lesssim \beta_E^2(3Q)|Q|.$$

Proof Let W be the strip in the definition of $\beta_E(3Q)$. Every R that hits Q and has diameter $\leq \ell(Q)$ is contained in $3Q$. Since W is convex, every such R is also contained in W. Since no point is in more than M of the Rs, and since the boundary of a convex set has zero area,

$$\sum_{\mathscr{R}(Q)} \beta(R)|Q|^2 \lesssim \sum \text{area}(R) \leq M \cdot \text{area}(3Q \cap W) \leq 18M\beta_E(3Q)|Q|^2,$$

and hence

$$\sum_{R \in \mathscr{R}(Q)} \beta(R) \lesssim \beta_E(3Q).$$

For positive numbers, $\sum a_n^2 \leq (\sum a_n)^2$, and this proves the lemma. $\quad\square$

Lemma 10.5.2 implies

$$\sum_{R \in \mathscr{R}_{k+1}} |R| + \frac{1}{2} \sum_{S \in \mathscr{S}_{k+1}} |S| - \sum_{\mathscr{R}_k} |R| \lesssim \sum_{\mathscr{R}_k} \beta^2(R)|R|.$$

Now sum this for $k = 1, \ldots, n$, use telescoping series and the fact that every R

is in some $\mathscr{R}(Q)$,

$$\sum_{\mathscr{R}_{n+1}} |R| + \frac{1}{2} \sum_{k=1}^{n} \sum_{S \in \mathscr{S}_{k+1}} |S| \lesssim |E| + \sum_{k=1}^{n} \sum_{R \in \mathscr{R}_k} \beta^2(R)|R|$$

$$\lesssim |E| + \sum_{Q \in \mathscr{D}} \sum_{R \in \mathscr{R}(Q)} \beta^2(R)|R|$$

$$\lesssim |E| + \sum_{Q \in \mathscr{D}} \beta_E^2(Q)|Q|.$$

If we multiply both sides by 2, we get (10.5.1). This completes the proof of Theorem 10.5.1.   □

Corollary 10.5.5 *If Γ is path connected, then the number N of dyadic squares of side length 2^{-n} that hit Γ satisfies*

$$N \asymp 2^n \left[|\Gamma| + \sum_{Q \in \mathscr{D}_k, k \le n} \beta_\Gamma^2(3Q)|Q| \right].$$

Proof The $\gtrsim$ direction was proven in Corollary 10.2.4. For the other direction let $E \subset \Gamma$ be a 2^{-n+2}-net, i.e., a finite set so that any two points of E are at least distance 2^{-n+2} apart but every point of Γ is within 2^{-n+3} of E. If E has M points then $M \asymp N$. Any path containing E has length $\gtrsim 2^{-n}M$ since the path has length $\ge 2^{-n}$ in each of the disjoint disks $D(z, 2^{-n})$, $z \in E$. Thus if γ is a path connected set containing E such that $\mathscr{H}^1(\gamma)$ is within a factor of 2 of the infimum over all path connected sets containing E, then

$$N \lesssim 2^n \mathscr{H}^1(\gamma) \lesssim 2^n \left(|E| + \sum_{Q \in \mathscr{D}} \beta_E^2(3Q)|Q| \right)$$

$$\lesssim 2^n \left(|\Gamma| + \sum_{Q \in \mathscr{D}_k, k \le n} \beta_\Gamma^2(3Q)|Q| \right),$$

since $\beta_E(Q) \le \beta_\Gamma(Q)$ for any Q and $\beta_E(3Q) = 0$ if $|3Q| \le 2^{-n+2}$.   □

Theorem 10.5.6 *If Γ is path connected and bounded, and if $\beta_\Gamma(3Q) \le \beta$ for every $Q \in \mathscr{D}$, then the upper Minkowski dimension of Γ is $1 + O(\beta^2)$.*

Proof We may suppose that $\Gamma \subset [0, 1]^2$ and that β is small. Let $\mathscr{C}_n \subset \mathscr{D}_n$ be the squares that hit Γ and let $N_n = \#(\mathscr{C}_n)$. Let

$$b_n = |\Gamma| + \sum_{Q \in \mathscr{D}_k, k \le n} \beta_\Gamma^2(3Q)|Q|.$$

By Corollary 10.5.5 there is an $M < \infty$ so that $N_n \le M2^n b_n$. Thus

$$b_n \le b_{n-1} + \beta^2 N_n 2^{-n} \le b_{n-1} + \beta^2 M b_n,$$

and so,

$$b_n \leq \frac{1}{1 - \beta^2 M} b_{n-1} \leq (1 + 2M\beta^2) b_{n-1} \leq (1 + 2M\beta^2)^n |\Gamma|,$$

if β is small enough. Thus

$$N_n \lesssim 2^n (1 + 2M\beta^2)^n \leq (2^n)^{1 + \log_2(1 + 2M\beta^2)} = (2^n)^{1 + O(\beta^2)},$$

which proves the theorem. □

10.6 Notes

The classical traveling salesman problem asks for the shortest path that visits every point of a finite set in the plane or an edge-weighted network. It is a famous example of an NP-hard problem (meaning that if it could be solved in polynomial time, so could a host of other apparently difficult problems). For subsets of $\mathbb{R}^d$, for any fixed dimension d, it is possible to build a $(1 + \varepsilon)$-approximation to an optimal path for any $\varepsilon > 0$ in polynomial time due to celebrated independent results of Arora (1998) and Mitchell (1999), for which they were both awarded the 2010 Gödel prize. Such a method is called a polynomial time approximation scheme (PTAS). For subsets of a metric space, $\frac{3}{2}$-approximation is the best known (see Exercise 10.9) and for a general weighted graph any C-approximation is NP-hard (see Exercise 10.11). For more background and details see Har-Peled (2011) and Mitchell (2004).

Jones originally proved his traveling salesman theorem to study the L^p boundedness of the Cauchy integral on graphs of Lipschitz functions and it has found many applications to this and other problems. For example, it forms part of the deep work of David, Mattila, Melnikov, Semmes, Tolsa, Verdera and others on analytic capacity and Vitushkin's conjecture, see e.g., Jones (1991), David (1998), Tolsa (2003), Mattila et al. (1996), Dudziak (2010). Bishop and Jones used it to study the geometric properties of 2-dimensional harmonic measure (Bishop and Jones, 1990, 1994a).

Jones' original proof of Theorem 10.2.3 (Jones, 1990) uses quite a bit of machinery from complex analysis. First, he shows that the bound (10.2.1) is correct if Γ is a Lipschitz graph. Then he shows that a simply connected region Ω that is bounded by a rectifiable curve can be decomposed into a union of domains $\Omega = \bigcup \Omega_j$, each of which are Lipschitz domains and so that $\sum \ell(\partial \Omega_j) \lesssim \ell(\partial \Omega)$. This step uses a powerful technique of Carleson called the "corona construction" (e.g., Garnett, 1981) applied to the Riemann mapping from the disk onto Ω. Jones then shows that the β-sum for the original domain can be bounded using the β-sums and areas of the Lipschitz subdomains. Finally, if Γ is any rectifiable planar curve, then Γ divides the plane into at most countably many simply connected regions $\{W_j\}$ and he estimates the β-sum for Γ in terms of the β-sums for these regions. This proves the desired result, but this approach is limited to two dimensions (because it uses the Riemann mapping) and is rather indirect compared to the proof of sufficiency. Okikiolu (1992) replaces Jones complex analytic approach by a much more geometric argument (our proof was based on hers) that extends to $\mathbb{R}^n$, and Schul (2005, 2007b, 2007c) has shown TST holds in Hilbert space (in particular, the constants on $\mathbb{R}^n$ do not blow up with n). His theorem requires a reformulation where dyadic cubes are replaced by 2^{-n}-nets centered on the set; the moral is that in high dimensions one must concentrate on the set and ignore the empty space around it.

If $E \subset \mathbb{R}^2$ is a compact set of diameter 1 and if for every $x \in E$

$$\sum_{x \in Q} \beta_E(3Q)^2 \leq M,$$

then K lies in a rectifiable curve of length at most Ce^{CM}. If Γ is a curve, then except for a set of $\mathscr{H}^1$ measure 0, x is a tangent point of Γ if and only if

$$\sum_{x \in Q} \beta_E(Q)^2 < \infty.$$

Both these results are from (Bishop and Jones, 1994b). The characterization of tangent points in terms of a β-sum solves a simpler version of a problem of Carleson that is still open (the "ε^2-conjecture"): are tangent points of a closed curve Γ almost everywhere characterized by

$$\int_0^1 \varepsilon^2(t,x) \frac{dt}{t} < \infty,$$

where $\varepsilon(x,t) = \max(|\pi - \theta_1|, |\pi - \theta_2|)$, and θ_1, θ_2 are the angle measures of the longest arcs of $\{z : |z - x| = t\} \setminus \Gamma$ on either side of Γ?

Fang (1990) showed that for every $\varepsilon > 0$ there is a $\delta > 0$ so that if F is a set with $\beta_F(Q) \leq \delta$ for every Q then F lies on a curve Γ with $\beta_\Gamma(Q) \leq \varepsilon$ for every Q. Thus "flat sets" lie on "flat curves".

Jones' TST has also been considered on spaces other than Euclidean space, e.g., the Heisenberg group. See Ferrari et al. (2007) and Juillet (2010). A version for metric spaces (using Menger curvature in place of the βs) has been given by Hahlomaa (2005, 2008, 2007) and Schul (2007b, 2007a). TST has also been generalized from sets to measures by Lerman (2003): he considers the question of finding conditions on a positive measure in $\mathbb{R}^d$ that imply a certain fraction of its mass will live on a curve of given length. See also Hahlomaa (2008).

The uniformly wiggly condition arises in various contexts that involve some form of self-similarity (see Exercises 10.30–10.34), but many interesting sets satisfy only a "β is large at most points and most scales" condition. This is common in Julia sets, Kleinian limit sets and many random examples, but so far dimension estimates need to be handled on a case-by-case basis, using extra information about each situation. For example, Jones' Theorem was exploited in Bishop et al. (1997) to show that the frontier of planar Brownian motion (i.e., the boundaries of the complementary components) has dimension > 1 (and amusingly makes use of a tiling of the plane by fractal Gosper islands instead of the usual dyadic squares). This result has since been superseded by the work of Lawler, Schramm and Werner; see the papers Lawler (1996), Lawler et al. (2001a, 2001b, 2001c, 2002). Among the results obtained is that the Brownian frontier has dimension $4/3$, verifying a conjecture of Mandelbrot. Also see related work of Aizenman and Burchard (1999).

A version of Theorem 10.2.5 for "wiggly metric spaces" is given by Azzam (2015).

Theorem 10.2.5 can be used to prove Bowen's dichotomy: a connected limit set of a co-compact Kleinian group either is a circle (or line) or has dimension > 1 (indeed, it is uniformly wiggly). In many cases of interest Ω is a planar domain and G is a group of Möbius transformations acting discontinuously on Ω, $\Lambda = \partial\Omega$ is the limit set (the accumulation set of the orbit of any point) and $R = \Omega/G$ is a Riemann surface. Sullivan (1984) showed Bowen's result was true if $R = \Omega/G$ is a Riemann surface of finite area, although the uniform wiggliness may fail. Bishop (2001) showed dim > 1 still holds if $R = \Omega/G$ is recurrent for Brownian motion (equivalently, R has no Green's function). However, this is sharp: Astala and Zinsmeister showed that for any Riemann surface R that is transient for Brownian motion we can write $R = \Omega/G$ for some simply connected domain with rectifiable (but non-circular) boundary. Related examples are given in Bishop (2002). In Bishop and Jones (1997) it is proven that the limit set of a finitely generated Kleinian group either is totally disconnected, is a circle or has dimension > 1. There are similar generalizations of Bowen's dichotomy to non-hyperbolic, connected Julia sets. See Zdunik (1990), Urbański (1991), Hamilton (1995).

Theorem 10.5.6 is due to Mattila and Vuorinen (1990). It implies that if f is a K-quasiconformal mapping of the plane to itself, then $\dim(f(\mathbb{R})) = 1 + O(k^2)$

where $k = (K-1)/(K+1)$. A mapping is K-quasiconformal if for all $x \in \mathbb{R}^2$,

$$\limsup_{r \to 0} \frac{\max_{|x-y|=r} |f(x) - f(y)|}{\min_{|x-y|=r} |f(x) - f(y)|} \le K.$$

More recently, Smirnov (2010) has proved that $\dim(f(\mathbb{R})) \le 1 + k^2$, verifying a conjecture of Astala. See Prause and Smirnov (2011). Ivrii (2016) has recently proven that the $1 + k^2$ is not sharp, using estimates of Hedenmalm (2015) on the Beurling transform.

10.7 Exercises

Exercise 10.1   Suppose γ is a curve connecting points $a, b \in \mathbb{R}^2$ and S is the segment $[a, b]$. Prove that any line that hits S also hits γ (this fact was used several times throughout the chapter).

Exercise 10.2   If K is convex, prove that $\mu(L(K))$ is the boundary length of K. If K is connected, then it is the boundary length of the convex hull of K.

Exercise 10.3   Show that if $K_1 \subset K_2$ are convex sets with perimeters L_1, L_2, then the probability that a line that hits K_2 also hits K_1 is L_1/L_2.

• **Exercise 10.4**   Suppose $\{a, b, c, d\}$ are the vertices (in counterclockwise order) of a convex quadrilateral Q. Show that the measure of the set S of lines that hit both $[a, b]$ and $[c, d]$ is

$$\mu(S) = |a - c| + |b - d| - |b - c| - |a - d|.$$

• **Exercise 10.5**   What is the probability that a unit line segment dropped at random onto the plane will hit $\Gamma = \{(x, y) : y \in \mathbb{Z}\}$? This is the Buffon Needle problem.

Exercise 10.6   Suppose Γ is a rectifiable Jordan arc (a 1-to-1 image of an interval $[a, b]$ under an absolutely continuous map γ with $|\gamma'| = 1$ a.e.). Show that $\ell(\Gamma) = |b - a| = \frac{1}{2} \int n(\Gamma, L) \, d\mu(L)$.

Exercise 10.7   If K is convex then $A = \text{area}(K) = \int |L \cap K| \, dL$ (recall $|E|$ is the diameter of E). Thus the average length of chord of K is $\pi A/L$, where L is the perimeter of K.

Exercise 10.8   The minimal spanning tree (MST) of a planar point set V is a tree with these points as vertices that has minimum total edge length. Show that the MST of an n point set can be constructed in time $O(n^2 \log n)$ using

Kruskal's algorithm (Kruskal, 1956): add edges in order of increasing length, but skipping edges that would create a cycle. Deduce that the solution to TSP on V can be computed to within a factor of 2 in polynomial time.

Exercise 10.9   There must be an even number $2N$ of vertices O where the minimal spanning tree (MST) has odd degree, so we can choose a minimal perfect matching (N non-adjacent edges of the complete graph on O that minimize total length). Prove that adding these edges to the MST creates a graph G with all even degrees whose total edge length is less than $3/2$ the TSP solution for V. Taking an Eulerian circuit for G and shortening it by skipping previously visited vertices gives the desired path. This is the Christofides algorithm.

Exercise 10.10   Show the Christofides algorithm works in polynomial time. The main point is to show the minimal perfect matching M can be constructed in polynomial time. An $O(n^4)$ method is due to Edmonds (1965). See also Gabow (1990).

Exercise 10.11   Show that if there is a polynomial time algorithm for computing a C-approximation to TSP for a general weighted graph for any $C < \infty$, then finding a Hamiltonian path is also possible in polynomial time. The latter is a well-known NP-hard problem, so even approximately solving TSP on a general weighted graph is NP-hard. There are polynomial time $(1 + \varepsilon)$-approximation methods for TSP in $\mathbb{R}^2$, due to Arora (1998) and Mitchell (1999).

Exercise 10.12   Give an explicit estimate for the constant in Lemma 10.1.2.

Exercise 10.13   Given a set E and a disk $D = D(x,t)$ define

$$\beta_E(x,t) = \frac{1}{t} \inf_L \sup_{z \in E \cap D} \operatorname{dist}(z,L),$$

where the infimum is taken over all straight lines L hitting D. Show that

$$\int_0^\infty \int_{\mathbb{R}^2} \beta_E(x,t) \frac{dxdt}{t}$$

is comparable to the sum over dyadic squares in (10.2.1).

Exercise 10.14   If E is a set in the plane, show that there is a connected set Γ containing E that attains the minimum 1-dimensional Hausdorff measure over all connected sets containing E.

Exercise 10.15   Show that the shortest Γ in Exercise 10.14 need not be unique.

Exercise 10.16   Consider the snowflake curves in Figure 10.7.1 formed by a variation on the usual construction, where at the nth stage we add to each edge of length r a centered isosceles triangle of base $r/3$ and height $ra_n/6$. Show that the resulting curve has finite length if and only if $\sum_n a_n^2 < \infty$.

Figure 10.7.1 A rectifiable and non-rectifiable snowflake. These were formed with the sequences $a_n = n^{-1}$ and $a_n = n^{-1/2}$. See Exercise 10.16.

Exercise 10.17 Prove that for any disk D in the plane there are squares $Q_1, Q_2 \in \mathscr{D}^*$ with $Q_1 \subset D \subset Q_2$ and such that $|Q_2|/|Q_1|$ is uniformly bounded.

Exercise 10.18 Prove that $\mathscr{D}_0^*(Q) \subset \mathscr{D}^*$. In particular, $\mathscr{D}_0^*(Q)$ consists of the elements of $\mathscr{D}^*$ that are the same size as Q and are contained in $1\frac{2}{3}Q$.

Exercise 10.19 Show that for any $Q \in \mathscr{D}$ and $\lambda > 0$, there is a $Q' \in \mathscr{D}^*$ so that $\lambda Q \subset Q'$ and $|Q'| \asymp \lambda|Q|$. The number of different Q's that are associated to a single Q' in this way is uniformly bounded.

Exercise 10.20 If Γ is a connected rectifiable set, show there is an arc-length preserving map γ to some circle to Γ that is 2 to 1 almost everywhere.

Exercise 10.21 A set Γ is called Ahlfors regular (or Ahlfors–David regular or David regular) if there is a $C < \infty$ such that $C^{-1}r \le \mathscr{H}^1(\Gamma \cap B(x,r)) \le Cr$ for every $x \in \Gamma$ and $r \in (0, |\Gamma|]$. Show that a curve Γ is Ahlfors regular if and only if there is a $C < \infty$ such that

$$\sum_{Q' \subset Q} \beta_\Gamma(Q')^2 |Q'| \le C|Q|,$$

for all dyadic squares Q. David (1984) showed that these are exactly the class of curves for which the Cauchy integral is a bounded operator on $L^2(\Gamma)$. They also arise in the Hayman–Wu problem (Bishop and Jones, 1990).

Exercise 10.22 A curve Γ is called locally flat if $\beta_\Gamma(Q) \to 0$ as $\ell(Q) \to 0$. Construct an example of a non-rectifiable, locally flat curve.

Exercise 10.23 If Γ is a locally flat curve, show $\dim(\Gamma) = 1$.

Exercise 10.24 A Jordan curve Γ satisfies Ahlfors' three-point condition if there is a $C < \infty$ such that for any two points $x, y \in \Gamma$ and $z \in \Gamma(x, y)$ we have $|x - z| + |z - y| \le (1 + C)|x - y|$. Show that a locally flat curve must satisfy the three-point condition.

Exercise 10.25   Curves satisfying the three-point condition in Exercise 10.24 are called quasicircles. Show that a quasicircle in $\mathbb{R}^2$ is porous and must have upper Minkowski dimension strictly less than 2.

Exercise 10.26   A rectifiable Jordan curve Γ is called chord-arc (or Lavrentiev) if there is a $C < \infty$ such that for any two points $x, y \in \Gamma$, the shorter arc on Γ between x and y has length $\le C|x - y|$. Show that Γ is chord-arc if and only if it is both a quasicircle and Ahlfors regular.

Exercise 10.27   Construct a rectifiable quasicircle that is not a chord-arc curve.

Exercise 10.28   Show that the estimate $\varepsilon = C\beta_0^2$ is sharp in Theorem 10.2.5 (except for the choice of C) by building snowflakes that attain this bound.

Exercise 10.29   Is Theorem 10.2.5 still true if we drop the assumption that E is connected?

Exercise 10.30   If Γ is a self-similar connected set in $\mathbb{R}^2$, show that it is either a line segment or uniformly wiggly.

• **Exercise 10.31**   Suppose Ω is a bounded, simply connected plane domain and G is a group of Möbius transformations each of which maps Ω 1–1 onto itself and so that the quotient space Ω/G is compact. Show that $\partial\Omega$ is either a circle or uniformly wiggly.

• **Exercise 10.32**   The Julia set $\mathcal{J}$ of a polynomial P is a compact invariant set for P and is called hyperbolic if all critical points of P and their iterates remain outside an open neighborhood U of $\mathcal{J}$. Prove that a connected hyperbolic Julia set is either an analytic curve or uniformly wiggly.

• **Exercise 10.33**   A holomorphic function on the unit disk is called Bloch if

$$\|g\|_{\mathscr{B}} = |g(0)| + \sup_{z \in \mathbb{D}} |g'(z)|(1 - |z|^2) < \infty.$$

This is a Möbius invariant norm. g is called "maximal Bloch" if

$$\sup_{0 < r < 1} \inf_{\tau} \|g \circ \tau(rz)\|_{\mathscr{B}} > 0,$$

where the infimum is over Möbius self-maps of the unit disk. Show that if $f : \mathbb{D} \to \Omega$ is conformal and $g = \log f'$ is maximal Bloch, then $\partial\Omega$ is uniformly wiggly.

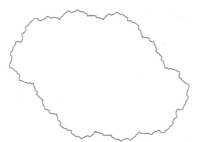

Figure 10.7.2 f-image of circle where $f'(z) = \exp(.4\sum_{n=1}^{10} e^{in}z^{2^n})$. This is a uniformly wiggly curve coming from a maximal Bloch function. See Exercise 10.34.

• **Exercise 10.34** Show $g(z) = \sum_{n=1}^{\infty} z^{b^n}$ is maximal Bloch if b is large enough. It is a theorem of Nehari that if $\log f' = \varepsilon g$, then f is conformal if ε is small enough, so this provides an example of a domain with uniformly wiggly boundary via Exercise 10.33. See Figure 10.7.2.

• **Exercise 10.35** Construct f so $\Omega = \partial f(\mathbb{D})$ is uniformly wiggly, but f is not maximal Bloch.

Exercise 10.36 If f is a continuous function on $\mathbb{R}$ whose graph Γ is uniformly wiggly, show that f is nowhere differentiable. Is there such a function?

Exercise 10.37 If f is a function on $[0,1]$, define its deviation from linearity on an interval $I \subset [0,1]$ as

$$\alpha(f,I) = \inf_{a,b\in\mathbb{R}} \sup_{x\in I} \frac{|f(x) - (ax+b)|}{|I|}.$$

Let $\mathscr{C}_\varepsilon \subset \mathscr{D}$ be the collection of dyadic intervals such that $\alpha(f,I) \geq \varepsilon$. If f is 1-Lipschitz, show that

$$\sum_{I\in\mathscr{C}_\varepsilon} |I| \lesssim \varepsilon^{-2}.$$

Thus the graph of f is close to a line at "most" points and scales. This is a quantitative version of Rademacher's Differentiation Theorem for Lipschitz functions and has been extended in several directions by Cheeger and others. See Cheeger and Kleiner (2009) and its references.

Exercise 10.38 Three distinct points $x,y,z \in \mathbb{R}^2$ determine a circle (or line) in the plane and we let $c(x,y,z)$ be the reciprocal of the radius of this circle

$(c = 0$ if the points are collinear). Show that

$$c(x,y,z) = \frac{2\operatorname{dist}(x,L_{yz})}{|x-y| \cdot |x-z|},$$

where L_{yz} is the line through y and z.

- **Exercise 10.39**   If $Q \in \mathscr{D}^*$, $x \in 1\frac{2}{3}Q$ and $y \in 3Q \setminus 2Q$, show that

$$\iint_{\gamma \cap 3Q} c^2(x,y,z)d\mu(z) \lesssim \beta_\gamma^2(3Q)|Q|.$$

- **Exercise 10.40**   The Menger curvature of a measure μ on $\mathbb{R}^2$ is defined as

$$c^2(\mu) = \iiint c^2(x,y,z)d\mu(x)d\mu(y)d\mu(z).$$

If μ is arclength measure on an Ahlfors regular curve Γ, show $c^2(\mu) \lesssim \mathscr{H}^1(\Gamma)$. This inequality was used by Melinkov and Verdera to give a short, geometric proof of the boundedness of the Cauchy integral on Lipschitz graphs (Melnikov and Verdera, 1995). See Tolsa (2014) for much more about Menger curvature, rectifiable measures, singular integrals and analytic capacity.

Exercise 10.41   Show the inequality in Exercise 10.39 does not reverse, i.e., if $\gamma = [-1,1] \cup [0,i\beta]$, and $Q = [-1,1]^2$, then

$$\iint_{\gamma \cap 3Q} c^2(x,y,z)d\mu(z) \lesssim \beta_\gamma^3(3Q)|Q|.$$

(However, the inequality in Exercise 10.40 does reverse, e.g., Theorem 38 of Pajot (2002).)

Appendix A

Banach's Fixed-Point Theorem

Here we record the well-known fact that a strict contraction on a complete metric space has a unique fixed point, followed by proofs of completeness for two of the most interesting metric spaces: the space of compact subsets of a complete space and the space of measures on a compact space.

A.1 Banach's Fixed-Point Theorem

Theorem A.1.1 *Let K be a complete metric space. Suppose that $T : K \to K$ satisfies $d(Tx, Ty) \le \lambda d(x, y)$ for all $x, y \in K$, with $0 < \lambda < 1$ fixed. Then T has a unique fixed point $z \in K$. Moreover, for any $x \in K$, we have*

$$d(T^n x, z) \le \frac{d(x, Tx)\lambda^n}{1 - \lambda}.$$

Proof If $Tx = x$ and $Ty = y$, then

$$d(x, y) = d(Tx, Ty) \le \lambda d(x, y).$$

Thus, $d(x, y) = 0$, so $x = y$, so a fixed point is unique (if it exists).

As for existence, given any $x \in K$, we define $x_n = Tx_{n-1}$ for each $n \ge 1$, setting $x_0 = x$. Set $a = d(x_0, x_1)$, and note that $d(x_n, x_{n+1}) \le \lambda^n a$. If $k > n$, then by triangle inequality,

$$d(x_n, x_k) \le d(x_n, x_{n+1}) + \cdots + d(x_{k-1}, x_k)$$
$$\le a(\lambda^n + \cdots + \lambda^{k-1}) \le \frac{a\lambda^n}{1-\lambda}.$$

This implies that $\{x_n : n \in \mathbb{N}\}$ is a Cauchy sequence. The metric space K is complete, whence $x_n \to z$ as $n \to \infty$. Note that

$$d(z, Tz) \le d(z, x_n) + d(x_n, x_{n+1}) + d(x_{n+1}, Tz) \le (1 + \lambda)d(z, x_n) + \lambda^n a \to 0$$

as $n \to \infty$. Hence, $d(Tz,z) = 0$, and $Tz = z$.

Thus, letting $k \to \infty$ in (A.1) yields

$$d(T^n x, z) = d(x_n, z) \le \frac{a\lambda^n}{1 - \lambda}. \qquad \square$$

It is not sufficient, however, for distances to decrease in order for there to be a fixed point, as the following example shows.

Example A.1.2 Consider the map $T : \mathbb{R} \to \mathbb{R}$ given by

$$T(x) = x + \frac{1}{1 + \exp(x)}.$$

Note that, if $x < y$, then

$$T(x) - x = \frac{1}{1 + \exp(x)} > \frac{1}{1 + \exp(y)} = T(y) - y,$$

implying that $T(y) - T(x) < y - x$. Note also that

$$T'(x) = 1 - \frac{\exp(x)}{\left(1 + \exp(x)\right)^2} > 0,$$

so that $T(y) - T(x) > 0$. Thus, T decreases distances, but it has no fixed points. This is not a counterexample to Banach's Fixed-Point Theorem, however, because there does not exist any $\lambda \in (0, 1)$ for which $|T(x) - T(y)| < \lambda |x - y|$ for all $x, y \in \mathbb{R}$.

This requirement can sometimes be relaxed, in particular for compact metric spaces.

Theorem A.1.3 (Compact Fixed-Point Theorem) *If K is a compact metric space and $T : K \to K$ satisfies $d(T(x), T(y)) < d(x, y)$ for all $x \ne y \in K$, then T has a fixed point $z \in K$. Moreover, for any $x \in K$, we have $T^n(x) \to z$.*

Proof Let $f : K \to \mathbb{R}$ be given by $f(x) = d(x, Tx)$. It is easy to check that f is continuous; since K is compact, there exists $z \in K$ such that

$$f(z) = \min_{x \in K} f(x). \qquad (A.1)$$

If $Tz \ne z$, then $f(T(z)) = d(Tz, T^2 z) < d(z, Tz) = f(z)$, and we have a contradiction to the minimizing property (A.1) of z. This implies that $Tz = z$.

Finally, we observe that iteration converges from any starting point x. Let $x_n = T^n x$, and suppose that x_n does not converge to z. Then for some $\varepsilon > 0$, the set $S = \{n \mid d(x_n, z) \ge \varepsilon\}$ is infinite. Let $\{n_k\} \subset S$ be an increasing sequence such that $y_k = x_{n_k} \to y \ne z$. Now

$$d(Ty_k, z) \to d(Ty, z) < d(y, z). \qquad (A.2)$$

But $T^{n_{k+1}-n_k-1}(Ty_k) = y_{k+1}$, so

$$d(Ty_k, z) \geq d(y_{k+1}, z) \to d(y, z)$$

contradicting (A.2). $\qquad\qquad\qquad\qquad\qquad\qquad\qquad\qquad\qquad\qquad\qquad\qquad\square$

A.2 Blaschke Selection Theorem

Let (X, d) be a metric space and let $\mathbf{Cpt}(X)$ denote the collection of non-empty compact subsets of X. Endow $\mathbf{Cpt}(X)$ with the Hausdorff metric, whose definition we recollect below.

Definition A.2.1 For a compact subset K of a metric space (X, d) define K^ε to be the set of points at distance less than ε from K, that is $K^\varepsilon = \{x \in X : d(x, K) < \varepsilon\}$. If K_1 and K_2 are two compact subsets of (X, d) define the Hausdorff distance between K_1 and K_2 as

$$d_H(K_1, K_2) = \inf\{\varepsilon > 0 : K_1 \subset K_2^\varepsilon, K_2 \subset K_1^\varepsilon\}.$$

Theorem A.2.2 (Blaschke Selection Theorem) *If (X, d) is a compact metric space, then $(\mathbf{Cpt}(X), d_H)$ is also compact.*

In fact, the following two lemmas show something slightly stronger.

Lemma A.2.3 *If (X, d) is a complete metric space, then $(\mathbf{Cpt}(X), d_H)$ is also complete.*

Lemma A.2.4 *If (X, d) is totally bounded, then $(\mathbf{Cpt}(X), d_H)$ is also totally bounded.*

It is a basic result in metric topology (see, for example, Folland (1999)) that a metric space is compact if and only if it is complete and totally bounded, so the above two lemmas imply Theorem A.2.2. This result will be used throughout this section.

Let $B(x, \varepsilon)$ denote the ball in the metric space (X, d) with the center x and radius ε.

Proof of Lemma A.2.3 Let $\{K_n\}$ be a Cauchy sequence in $\mathbf{Cpt}(X)$ and define $K = \bigcap_n \overline{\bigcup_{j>n} K_j}$. We claim that $K \in \mathbf{Cpt}(X)$ and that the sequence (K_n) converges to K in metric d_H.

To prove compactness of K, we show that K is complete and totally bounded. Since X is complete, the closed subset K is also complete. To prove that K is totally bounded, fix $\varepsilon > 0$. Consider N such that $d_H(K_n, K_N) < \varepsilon$, for all

$n \geq N$. Cover the compact set K_N with L balls of radius ε and denote the centers of these balls by $\{x_i\}$, $i = 1, \ldots, L$. Because $K_n \subset K_N^\varepsilon$, for all $n \geq N$ and $K_N \subset \bigcup_{i=1}^L B(x_i, \varepsilon)$, we see that $K_n \subset \bigcup_{i=1}^L B(x_i, 3\varepsilon)$ (actually the sets $\overline{B(x_i, 2\varepsilon)}$ would suffice to cover K_n, but since we are considering open balls, we use the larger sets $B(x_i, 3\varepsilon)$). Since K is also equal to $\bigcap_{n \geq N} \overline{\bigcup_{j > n} K_j}$, we get that $K \subset \bigcup_{i=1}^L B(x_i, 4\varepsilon)$. This proves that K is totally bounded. Hence, K is compact, that is, an element of $\mathbf{Cpt}(X)$.

Next, we prove that the sequence (K_n) converges to K in the Hausdorff metric d_H. As in the previous paragraph, given an $\varepsilon > 0$ we choose an integer N so that $d_H(K_n, K_N) < \varepsilon$ for all $n \geq N$. Hence we have the inclusion $\overline{\bigcup_{n > N} K_n} \subset K_N^{2\varepsilon}$, which implies $K \subset K_N^{2\varepsilon}$. To prove the opposite inclusion, for all $i \geq 0$ let N_i be a sequence of integers such that $d_H(K_{N_i}, K_n) < \varepsilon/2^i$ for all $n \geq N_i$ (we take $N_0 = N$). Fix an arbitrary $y_0 \in K_N$ and for all $i \geq 1$ choose points $y_i \in K_{N_i}$ such that $d(y_i, y_{i-1}) < \varepsilon/2^i$. Then (y_i) is a Cauchy sequence and, since (X, d) is complete, converges to some $y_\infty \in X$. By construction $y_m \in \bigcup_{j \geq N_i} K_j$ for all $m \geq i$, which implies that $y_\infty \in \overline{\bigcup_{j \geq N_i} K_j}$. Since this holds for any i, we obtain $y_\infty \in K$. Clearly we have that $d(y_0, y_\infty) < 2\varepsilon$, which implies that $K_N \subset K^{2\varepsilon}$, hence $d_H(K_N, K) \leq 2\varepsilon$. Therefore, $\lim_n d_H(K_n, K) = 0$. This establishes the fact that $(\mathbf{Cpt}(X), d_H)$ is complete. $\qquad\square$

Proof of Lemma A.2.4   We begin by fixing an $\varepsilon > 0$. Since (X, d) is totally bounded, there exists an ε-net $\{x_j\}_{j=1}^L$ for X, that is $X \subset \bigcup_{j=1}^L B(x_j, \varepsilon)$. Let S be the collection of all non-empty subsets of $\{x_j\}_{j=1}^L$. Clearly, S is a finite subset of $\mathbf{Cpt}(X)$. Given any $K \in \mathbf{Cpt}(X)$, consider the set $A = \{x_j : d(x_j, K) < \varepsilon\}$. Observe that $K \subset \bigcup \{B(x_j, \varepsilon) : x_j \in A\}$, which implies $K \subset A^\varepsilon$. Moreover, since $d(x, K) < \varepsilon$, for any $x \in A$, we have $A \subset K^\varepsilon$. Hence, $d_H(K, A) \leq \varepsilon$. This proves that S is a finite ε-net for $\mathbf{Cpt}(X)$. Since ε was arbitrary, we conclude that $(\mathbf{Cpt}(X), d_H)$ is totally bounded. $\qquad\square$

A.3 Dual Lipschitz metric

In this section we study the properties of the space of probability measures on a compact metric space. As before let (X, d) be a metric space. By $C(X)$ denote the space of real-valued continuous functions on X and let $\mathrm{Lip}(X)$ stand for the set of functions in $C(X)$ which are Lipschitz continuous. For $g \in \mathrm{Lip}(X)$, let $\mathrm{Lip}(g)$ stand for the Lipschitz constant of g, that is

$$\mathrm{Lip}(g) = \sup_{x \neq y} \frac{|g(x) - g(y)|}{d(x, y)}.$$

If $g \in \text{Lip}(X)$ is a 1-Lipschitz function, that is $\text{Lip}(g) \leq 1$, we will simply call it a Lip-1 function. The aim of this section is to prove the compactness of the space defined as follows.

Definition A.3.1 Let $\mathbf{P}(X)$ denote the set of probability measures on a compact metric space (X, d). The metric L on the space $\mathbf{P}(X)$ given by

$$L(v, v') = \sup_{\text{Lip}(g) \leq 1} \left| \int g \, dv - \int g \, dv' \right|$$

is called *the dual Lipschitz metric*.

Observe that

$$\left| \int g \, dv - \int g \, dv' \right| = \left| \int (g + c) \, dv - \int (g + c) \, dv' \right|,$$

for all constants c, Lip-1 functions g and measures $v, v' \in \mathbf{P}(X)$. Thus in the definition of L it suffices to take supremum over Lip-1 functions g which vanish at some fixed point $x_0 \in X$. Since X is compact, for any such function g we have $\|g\|_\infty \leq \text{diam}(X)$. We will use this observation in the proofs that follow.

First of all, we need to demonstrate that L as defined above is indeed a metric. This easy proof is deferred to Lemma A.3.5. Given the fact that $(\mathbf{P}(X), L)$ is a metric space, we proceed to prove that it is compact.

Theorem A.3.2 *For a compact metric space (X, d), the space of probability measures on X with the dual Lipschitz metric $(\mathbf{P}(X), L)$ is compact.*

We will give two proofs of this theorem. The first one has the virtue of being elementary, whereas the second one involves a nice characterization of the topology induced by the metric L.

We think of a measure as a functional on an appropriate function space (the space $C(X)$ or $\text{Lip}(X)$), defined in the canonical way by integration:

$$\mu(f) = \int f \, d\mu.$$

We will use the fact that Lipschitz functions are dense in the set of continuous functions on the compact metric space X in the usual sup norm; the proof of this is deferred to Lemma A.3.4.

First proof of Theorem A.3.2 In order to prove that $(\mathbf{P}(X), L)$ is compact, we show that it is complete and totally bounded.

To show completeness, let (μ_n) be a Cauchy sequence in $(\mathbf{P}(X), L)$. By the definition of L, the real numbers $\mu_n(g)$ form a Cauchy sequence for each Lip-1 function g and by scaling, for any Lipschitz function g. Since Lipschitz functions are dense in $C(X)$ (see Lemma A.3.4), this is true for any function

$f \in C(X)$. As a result, and since $\mathbb{R}$ is complete, for each $f \in C(X)$ the sequence $(\mu_n(f))$ converges to a limit $\lambda(f)$. Since $f \mapsto \mu_n(f)$ are positive linear functionals on $C(X)$, the same is true for $f \mapsto \lambda(f)$.

Also note that

$$|\lambda(f)| = \lim_n \left| \int f \, d\mu_n \right| \leq \lim_n \int \|f\|_\infty \, d\mu_n \leq \|f\|_\infty.$$

This implies that λ is a bounded linear functional on $C(X)$. By the Riesz Representation Theorem, the functional λ is given by integration against a nonnegative Borel measure μ; that is

$$\lambda(f) = \int f \, d\mu,$$

for all $f \in C(X)$. By taking f to be the constant function 1, we deduce that μ is in fact a probability measure. We claim that $\lim_n L(\mu_n, \mu) = 0$. To see this, recall that μ_n is a Cauchy sequence in L, fix $\varepsilon > 0$ and let N be such that $|\mu_n(g) - \mu_m(g)| < \varepsilon$ for all Lip-1 functions g, and all $m, n \geq N$. Letting $m \to \infty$ implies that $|\mu_n(g) - \mu(g)| \leq \varepsilon$, for all Lip-1 functions g and for all $n \geq N$. Thus $L(\mu_n, \mu) \leq \varepsilon$ for all $n \geq N$. Since $\varepsilon > 0$ was arbitrary, we have that $\lim_n L(\mu_n, \mu) = 0$.

Next we want to show that $(\mathbf{P}(X), L)$ is totally bounded. Fix $\varepsilon > 0$ and take an ε-net $\{x_j\}_{j=1}^N$ for the compact space (X, d). Write X as the disjoint union $X = \bigcup_{j=1}^N D_j$, where $D_j = B(x_j, \varepsilon) \setminus \bigcup_{i<j} B(x_i, \varepsilon)$. Given any $\mu \in \mathbf{P}(X)$ construct its discrete approximation

$$\tilde{\mu} = \sum_{j=1}^N \mu(D_j) \delta_{x_j},$$

where δ_{x_j} is the delta mass at the point x_j. For any Lip-1 function g, we have $|\int g \, d\mu - \int g \, d\tilde{\mu}| \leq \sum_{j=1}^L \mu(D_j) \varepsilon = \varepsilon$. Therefore, $L(\mu, \tilde{\mu}) \leq \varepsilon$. We will now approximate the set of all possible such $\tilde{\mu}$ (as μ ranges over $\mathbf{P}(X)$) by a finite set of discrete probability measures. For this, first fix an integer K. Given any $\mu \in \mathbf{P}(X)$, set $p_j = \frac{1}{K} \lfloor K\mu(D_j) \rfloor$, for all $1 \leq j \leq N-1$ and $p_N = 1 - \sum_{j=1}^{N-1} p_j$. Define the measure $\gamma(\mu) = \sum_{j=1}^N p_j \delta_{x_j}$. For any Lip-1 function g vanishing at some fixed $x_0 \in X$ we have

$$\left| \int g \, d\tilde{\mu} - \int g \, d\gamma(\mu) \right| \leq \sum_{j=1}^N \|g\|_\infty |\mu(D_j) - p_j|$$

$$\leq \text{diam}(X) \left(\sum_{j=1}^{N-1} \frac{1}{K} + \frac{N-1}{K} \right).$$

By the remark after Definition A.3.1, we obtain

$$L(\tilde{\mu}, \gamma(\mu)) \leq \operatorname{diam}(X) \frac{2(N-1)}{K}.$$

Choosing $K = 2N \operatorname{diam}(X)/\varepsilon$, we get $L(\tilde{\mu}, \gamma(\mu)) < \varepsilon$. Thus the set $\{\gamma(\mu), \mu \in \mathbf{P}(X)\}$ is a 2ε-net in $(\mathbf{P}(X), L)$. Note that the set

$$\{(p_1, \ldots, p_N) : Kp_j \in \mathbb{Z}^+, \sum_{j=1}^{N} p_j = 1\}$$

is finite, which implies that the set $\{\gamma(\mu), \mu \in \mathbf{P}(X)\}$ is finite. Therefore, the space $(\mathbf{P}(X), L)$ is totally bounded. This completes the proof of the theorem.

$\square$

Next, we embark upon the second approach to Theorem A.3.2. First define $\mathcal{M}(X)$ as the space of finite (signed) Borel measures on X. The basic idea is to show that $\mathbf{P}(X)$ is a closed subset of the unit ball of $\mathcal{M}(X)$ in the weak* topology. The proof is then completed by showing that L metrizes the weak* topology on $\mathbf{P}(X)$, induced from $\mathcal{M}(X)$. First we recall some basic notions required for this approach.

We consider the Banach space $(C(X), \|.\|_{\infty})$ and recall the standard fact (see, for example, Rudin (1987)) that its dual is $\mathcal{M}(X)$ with the norm

$$\|\mu\| = \sup_{f \in C(X), \|f\|_{\infty} \leq 1} \left| \int f \, d\mu \right|. \tag{A.1}$$

As before, we view the elements of $\mathcal{M}(X)$ as linear functionals on $C(X)$ acting via integration: $\mu(f) = \int f \, d\mu$. For a fixed f and varying μ, the same recipe defines a (continuous) linear functional on $\mathcal{M}(X)$, namely $\mu \mapsto \mu(f)$. The weak* topology on $\mathcal{M}(X)$ is given by the convergence of all such linear functionals, that is, it is the topology in which a sequence (μ_n) converges to μ if and only if $\int f \, d\mu_n \to \int f \, d\mu$, for all $f \in C(X)$.

Lemma A.3.3 *Let (X, d) be a compact metric space. Let $\mathcal{M}(X)$ be the set of finite signed Borel measures on the space X. Then the dual Lipschitz metric L on $\mathbf{P}(X) \subset \mathcal{M}(X)$ metrizes the weak* topology on $\mathbf{P}(X)$ (induced from the weak* topology on $\mathcal{M}(X)$).*

Proof The fact that L is indeed a metric will be demonstrated in the subsequent Lemma A.3.5. Given that L is a metric on $\mathbf{P}(X)$, we show that a sequence of probability measures (μ_n) converges to $\mu \in \mathbf{P}(X)$ in the weak* topology if and only if $\lim_n L(\mu_n, \mu) = 0$.

Let us suppose that $\lim_n L(\mu_n, \mu) = 0$. By the definition of the metric L, for each Lip-1 function g we have $\int g \, d\mu_n \to \int g \, d\mu$. By scaling, this is true for any Lipschitz function g on X, and since by Lemma A.3.4 the Lipschitz functions are dense in $(C(X), \|.\|_\infty)$, we obtain that $\lim_n \int f d\mu_n = \int f d\mu$, for each $f \in C(X)$. Therefore $\mu_n \to \mu$ in the weak* sense.

For the reverse direction, let $\mu_n \to \mu$ in the weak* sense. We want to show that $\lim_n L(\mu_n, \mu) = 0$. Fix an arbitrary point $x_0 \in X$ and by Lip$_1$ denote the space of Lip-1 functions which vanish at x_0, endowed with the sup norm. By Lemma A.3.6 this space is compact. Given $\varepsilon > 0$ consider an ε-net $\{g_k\}_{k=1}^N$ for Lip$_1$. Let n_0 be an integer such that for any $n \geq n_0$ we have

$$\left| \int g_k d\mu_n - \int g_k d\mu \right| \leq \varepsilon$$

for all $1 \leq k \leq N$. Then, for any $g \in \text{Lip}_1$, choosing g_i to be an element of the ε-net such that $\|g - g_i\|_\infty \leq \varepsilon$, we obtain for any $n \geq n_0$,

$$\left| \int g d\mu_n - \int g d\mu \right| \leq \left| \int (g - g_i) d\mu_n \right|$$
$$+ \left| \int (g - g_i) d\mu \right| + \left| \int g_i d\mu_n - \int g_i d\mu \right| \leq 3\varepsilon.$$

Thus by the remark after Definition A.3.1, we have that $L(\mu_n, \mu) \leq 3\varepsilon$ for any $n \geq n_0$. Since $\varepsilon > 0$ was arbitrary, this implies that the sequence (μ_n) converges to μ in the dual Lipschitz metric L. $\qquad\square$

Next we deduce Theorem A.3.2 from Lemma A.3.3.

Second proof of Theorem A.3.2   The set of probability measures $\mathbf{P}(X)$ is precisely the set of elements of $\mathcal{M}(X)$ which satisfy

(i) $\mu(f) \geq 0$ for all $f \in C(X)$ such that $f \geq 0$,
(ii) $\mu(1) = 1$.

Both of these conditions are weak* closed. Moreover,

$$\|\mu\| = \sup_{\|f\|_\infty \leq 1} \left| \int f d\mu \right| = 1,$$

for all $\mu \in \mathbf{P}(X)$. Thus $\mathbf{P}(X)$ is a weak* closed subset of the unit ball of $(\mathcal{M}(X), |\cdot|)$. By the Banach–Alaoglu Theorem, the unit ball in $(\mathcal{M}(X), |\cdot|)$ is compact in the weak* topology. Therefore the set $\mathbf{P}(X)$ is compact in the weak* topology. By Lemma A.3.3, the topology on $\mathbf{P}(X)$ induced by the metric L is the same as the weak* topology inherited from $\mathcal{M}(X)$. This completes the proof that $(\mathbf{P}(X), L)$ is a compact metric space. $\qquad\square$

We now complete the proof of the auxiliary lemmas which were invoked in the above proof several times. Some of them are useful observations in their own right.

Lemma A.3.4 *For any compact metric space (X,d), the space $\mathrm{Lip}(X)$ of Lipschitz functions is dense in $C(X)$.*

Proof We first prove that $\mathrm{Lip}(X)$ is an algebra over $\mathbb{R}$. Indeed, if $f, g \in \mathrm{Lip}(X)$ and $\alpha \in \mathbb{R}$, then $f + g, \alpha g$ are in $\mathrm{Lip}(X)$. For the product fg we write

$$(fg)(x) - (fg)(y) = f(x)g(x) - f(y)g(y)$$
$$= f(x)(g(x) - g(y)) + g(y)(f(x) - f(y)).$$

Since $\|f\|_\infty$ and $\|g\|_\infty$ are finite, this representation allows us to conclude that fg is also a Lipschitz function. Hence $\mathrm{Lip}(X)$ is an algebra which contains all constant functions on X. Hence, by the Stone–Weierstrass Theorem, the algebra $\mathrm{Lip}(X)$ is dense in $C(X)$. ☐

Lemma A.3.5 *The space $(\mathbf{P}(X), L)$ as defined in Definition A.3.1 is a metric space.*

Proof We note that reflexivity and the triangle inequality for L are completely straightforward. It only remains to prove that if $L(\mu_1, \mu_2) = 0$ then $\mu_1 = \mu_2$. The condition $L(\mu_1, \mu_2) = 0$ implies that $\int g \, d\mu_1 = \int g \, d\mu_2$ for all $g \in \mathrm{Lip}(X)$. By Lemma A.3.4, the set $\mathrm{Lip}(X)$ is dense in $C(X)$, so $\int f \, d\mu_1 = \int f \, d\mu_2$ for all $f \in C(X)$. By Riesz Representation Theorem, this gives $\mu_1 = \mu_2$, as desired. ☐

Lemma A.3.6 *For any compact metric space (X,d) and $x_0 \in X$, the set of Lip-1 functions which vanish at x_0 is compact in $(C(X), \|.\|_\infty)$.*

Proof We will show that the set of Lip-1 functions which vanish at $x_0 \in X$ form a bounded, equicontinuous and closed subset of $(C(X), \|.\|_\infty)$. The desired result will then follow from the Arzela–Ascoli Theorem. Equicontinuity is a direct consequence of the fact that these functions are Lip-1. Closedness in $\|.\|_\infty$ is also clear, because given a sequence (g_n) of functions in $C(X)$ converging to g and satisfying both conditions $|g_n(x) - g_n(y)| \le d(x,y)$ and $g_n(x_0) = 0$, the limit function g also satisfies both conditions. Because g is Lip-1 and satisfies $g(x_0) = 0$ and X is a compact metric space, we have $\|g\|_\infty \le \mathrm{diam}(X)$, which proves the boundedness. This completes the proof of the lemma. ☐

We note that on $\mathbf{P}(X)$, the dual Lipschitz metric

$$L(v, v') = \sup_{\mathrm{Lip}(g) \le 1} \left| \int g \, dv - \int g \, dv' \right|,$$

is the same as the Wasserstein metric given by

$$W(v, v') = \inf_\mu \int \int d(x,y) \, d\mu(x,y),$$

where μ ranges over all possible couplings of ν and ν'. We refer the interested reader to pages 420–421 of Dudley (2002).

Appendix B

Frostman's Lemma for analytic sets

In Chapter 3 we proved Frostman's Lemma for compact sets, but stated that it held for all Borel sets. Here we prove that claim. Not only does this generalize Frostman's Lemma, it allows for easy generalizations of many other results whose proofs used Frostman's Lemma. We shall actually define an even larger class of sets, the analytic sets, and prove two facts. First, the analytic sets form a σ-algebra containing the open sets and hence every Borel set is analytic. Second, every analytic set of positive α-measure contains a compact subset of positive α-measure.

B.1 Borel sets are analytic

A **Polish space** is a topological space that can be equipped with a metric that makes it complete and separable.

Definition B.1.1 If Y is Polish, then a subset $E \subset Y$ is called **analytic** if there exists a Polish space X and a continuous map $f : X \to Y$ such that $E = f(X)$.

Analytic sets are also called Souslin sets in honor of Mikhail Yakovlevich Souslin. The analytic subsets of Y are often denoted by $A(Y)$ or $\Sigma_1^1(Y)$. In any uncountable Polish space there exist analytic sets which are not Borel sets, see e.g., Proposition 13.2.5 in Dudley (2002) or Theorem 14.2 in Kechris (1995). By definition, if $g : Y \to Z$ is a continuous mapping between Polish spaces and $E \subset Y$ is analytic, then $g(E)$ is also analytic. In other words, continuous images of analytic sets are themselves analytic, whereas it is known that continuous images of Borel sets may fail to be Borel sets. This fact is the main reason why it can be useful to work with analytic sets instead of Borel sets. The next couple of lemmas prepare for the proof that every Borel set in a Polish space is analytic.

We first show that analytic sets have a nice representation in terms of sequences. Let $\mathbb{N}^\infty$ be the space of all infinite sequences of non-negative integers equipped with the metric given by $d((a_n),(b_n)) = e^{-m}$, where $m = \max\{n \geq 0: a_k = b_k \text{ for all } 1 \leq k \leq n\}$ (this space is also sometimes denoted $\mathbb{N}^{\mathbb{N}}$).

Lemma B.1.2   *For every Polish space X there exists a continuous mapping $f: \mathbb{N}^\infty \to X$ such that $X = f(\mathbb{N}^\infty)$. Moreover, for all $(b_n) \in \mathbb{N}^\infty$, the sequence of diameters of the sets $f(\{(a_n): a_n = b_n \text{ for } 1 \leq n \leq m\})$ converges to zero, as $m \uparrow \infty$.*

Proof   Given a Polish space X we construct a continuous and surjective mapping $f: \mathbb{N}^\infty \to X$. Fix a metric ρ making X complete and separable. By separability, we can cover X by a countable collection of closed balls $X(j), j \in \mathbb{N}$ of diameter one. We continue the construction inductively. Given closed sets $X(a_1, \ldots, a_k)$, for $(a_1, \ldots, a_k) \in \mathbb{N}^k$, we write $X(a_1, \ldots, a_k)$ as the union of countably many non-empty closed sets $X(a_1, \ldots, a_k, j), j \in \mathbb{N}$ of diameter at most 2^{-k}; we can do this by covering $X(a_1, \ldots, a_k)$ by countably many closed balls of diameter $\leq 2^{-k}$ with centers in $X(a_1, \ldots, a_k)$ and then intersecting these balls with $X(a_1, \ldots, a_k)$. Given $(a_n) \in \mathbb{N}^\infty$ the set $\bigcap_{k=1}^\infty X(a_1, \ldots, a_k)$ has diameter zero, hence contains at most one point. By construction all the sets are non-empty and nested, so if we choose a point $x_k \in X(a_1, \ldots, a_k)$ it is easy to see this forms a Cauchy sequence and by completeness it converges to some point x. Since each $X(a_1, \ldots, a_k)$ is closed it must contain x and hence $\bigcap_{k=1}^\infty X(a_1, \ldots, a_k)$ contains x. Define $f((a_n)) = x$.

By construction, if $(b_n) \in \mathbb{N}^\infty$, the set $f(\{(a_n): a_n = b_n \text{ for } 1 \leq n \leq m\})$ has diameter at most $2^{-m+3} \to 0$, which implies continuity of f. Finally, by the covering property of the sets, every point $x \in X$ is contained in a sequence of sets $B(a_1, \ldots, a_k), k \in \mathbb{N}$, for some infinite sequence (a_k) which implies, using the nested property, that $f((a_n)) = x$ and hence $f(\mathbb{N}^\infty) = X$, as required.   □

Lemma B.1.3   *If $E \subset X$ is analytic, then there exists a continuous mapping $f: \mathbb{N}^\infty \to X$ such that $E = f(\mathbb{N}^\infty)$. Moreover, for any sequence $(k_n) \in \mathbb{N}^\infty$,*

$$\bigcap_{m=1}^\infty \overline{f(\{(a_n): a_n \leq k_n \text{ for } 1 \leq n \leq m\})} = f(\{(a_n): a_n \leq k_n \text{ for all } n \geq 1\}).$$

Proof   If $E \subset X$ is analytic, there exists a Polish space Y and $g_1: Y \to X$ continuous with $g_1(Y) = E$. From Lemma B.1.2 we have a continuous mapping $g_2: \mathbb{N}^\infty \to Y$ such that $Y = g_2(\mathbb{N}^\infty)$. Letting $f = g_1 \circ g_2: \mathbb{N}^\infty \to X$ gives $f(\mathbb{N}^\infty) = E$.

Fix a sequence of positive integers $k_1, k_2, \ldots$ and note that the inclusion $\supset$ in the displayed equality holds trivially. If x is a point in the set on the left-hand side, then there exist a_n^m with $a_n^m \leq k_n$ for $1 \leq n \leq m$ such that $\rho(x, f((a_n^m : n \geq 1))) < \frac{1}{m}$. We successively pick integers $a_1, a_2, \ldots$ with the property that, for every n, the integer a_n occurs infinitely often in the collection $\{a_n^m : a_1^m = a_1, \ldots, a_{n-1}^m = a_{n-1}\}$. Then there exists $m_j \uparrow \infty$ such that $a_1^{m_j} = a_1, \ldots, a_j^{m_j} = a_j$ and hence $(a_n^{m_j} : n \geq 1)$ converges, as $j \to \infty$, to (a_n). Since f is continuous, $\rho(x, f((a_n^{m_j} : n \geq 1)))$ converges to $\rho(x, f((a_n)))$ and by construction also to zero, whence $x = f((a_n))$. This implies the x is contained in the set on the right-hand side in the displayed equation. □

Our next task is to show that every Borel set is analytic. We will use several times the simple fact that the finite or countable product of Polish spaces is Polish, which we leave for the reader to verify.

Lemma B.1.4 *Open and closed subsets of a Polish space are Polish, hence analytic.*

Proof Let X be Polish and ρ a metric making X a complete and separable metric space. If $C \subset X$ is closed then the metric ρ makes C complete and separable, hence C is a Polish space. If $O \subset X$ is open, let $Y = \{(x, y) : y = 1/\rho(x, O^c)\} \subset X \times \mathbb{R}$. Then Y is a closed subset of the Polish space $X \times \mathbb{R}$ and hence itself a Polish space. As Y and O are homeomorphic, O is Polish. □

Lemma B.1.5 *Let $E_1, E_2, \ldots \subset X$ be analytic sets. Then*

1. $\bigcup_{i=1}^{\infty} E_i$ *is an analytic set;*

2. $\bigcap_{i=1}^{\infty} E_i$ *is an analytic set.*

Proof For every analytic set $E_i \subset X$ there exists a continuous $f_i \colon \mathbb{N}^\infty \to X$ such that $f_i(\mathbb{N}^\infty) = E_i$. Then $f \colon \mathbb{N}^\infty \to X$ given by $f((a_n)) = f_{a_1}((a_{n+1}))$ is continuous and satisfies $f(\mathbb{N}^\infty) = \bigcup_{i=1}^{\infty} E_i$, as required to show (a).

Now look at continuous mappings $f_i \colon X_i \to X$ with $f_i(X_i) = E_i$. Define a continuous mapping $g \colon \prod_{i=1}^{\infty} X_i \to X^\infty$ by $g(x_1, x_2, \ldots) = (f_1(x_1), f_2(x_2), \ldots)$. The diagonal $\Delta \subset X^\infty$ is closed, and so is $g^{-1}(\Delta)$, by continuity of g. In particular, $Y = g^{-1}(\Delta)$ is a Polish space and it is easy to see that $f_1(Y) = \bigcap_{i=1}^{\infty} E_i$, proving (b). □

Lemma B.1.6 *If X is Polish, then every Borel set $E \subset X$ is analytic.*

Proof The collection $\{S \subset X \colon S \text{ and } S^c \text{ are analytic}\}$ of sets contains the open sets by Lemma B.1.4 and is closed under countable unions by Lemma B.1.5. As it is obviously closed under taking the complement it must contain the

Borel sets, which, by definition, is the smallest collection of sets with these properties.   □

It is a theorem of Souslin that if both A and A^c are analytic, then A is Borel. Thus the collection of sets considered in the previous proof are exactly the Borel sets.

B.2 Choquet capacitability

The main step in the extension of Frostman's Lemma to Borel sets is a technical device called the *Choquet Capacitability Theorem*, which we now introduce.

Definition B.2.1   Let X be a Polish space. A set function Ψ defined on all subsets of X is called a **Choquet capacity** if

(a) $\Psi(E_1) \leq \Psi(E_2)$ whenever $E_1 \subset E_2$;

(b) $\Psi(E) = \inf\limits_{O \supset E \text{ open}} \Psi(O)$ for all $E \subset X$;

(c) for all increasing sequences $\{E_n \colon n \in \mathbb{N}\}$ of sets in X,

$$\Psi\Big(\bigcup_{n=1}^{\infty} E_n\Big) = \lim_{n\to\infty} \Psi(E_n).$$

Given Ψ we can define a set function Ψ_* on all sets $E \subset X$ by

$$\Psi_*(E) = \sup_{F \subset E \text{ compact}} \Psi(F).$$

A set E is called **capacitable** if $\Psi(E) = \Psi_*(E)$.

Theorem B.2.2 (Choquet Capacitability Theorem)   *If Ψ is a Choquet capacity on a compact metric space X, then all analytic subsets of X are capacitable.*

Proof   Let $E = f(\mathbb{N}^{\infty}) \subset X$ be analytic. We define sets

$$S^k = f\big(\{(a_n) \colon a_1 \leq k\}\big).$$

The sequence of sets $S^k, k \geq 1$, is increasing and their union is E. Hence, by (c), given $\varepsilon > 0$ we can find $k_1 \in \mathbb{N}$ such that $S_1 := S^{k_1}$ satisfies

$$\Psi(S_1) \geq \Psi(E) - \frac{\varepsilon}{2}.$$

Having found $S_1, \dots, S_{m-1}$ and $k_1, \dots, k_{m-1}$ we continue the sequence by defining

$$S_m^k = f\big(\{(a_n) \colon a_i \leq k_i \text{ for } i \leq m-1, \, a_m \leq k\}\big),$$

and as the sequence of sets S_m^k, $k \geq 1$, is increasing and their union is S_{m-1} we find $k_m \in \mathbb{N}$ such that $S_m := S_m^{k_m}$ satisfies $\Psi(S_m) \geq \Psi(S_{m-1}) - \varepsilon 2^{-m}$. We conclude that

$$\Psi(S_m) \geq \Psi(E) - \varepsilon \qquad \text{for all } m \in \mathbb{N}.$$

Denoting by $\overline{S_m}$ the closure of S_m we now define a compact set

$$S = \bigcap_{m=1}^{\infty} \overline{S_m}.$$

(It is easy to check that S is complete and totally bounded, hence it is compact.) By Lemma B.1.3 S is a subset of E. Now take an arbitrary open set $O \supset S$. Then there exists m such that $O \supset \overline{S_m}$ and hence $\Psi(O) \geq \Psi(S_m) \geq \Psi(E) - \varepsilon$. Using property (b) infer that $\Psi(S) \geq \Psi(E) - \varepsilon$ and therefore, as $\varepsilon > 0$ was arbitrary, that $\Psi_*(E) \geq \Psi(E)$. As $\Psi_*(E) \leq \Psi(E)$ holds trivially, we get that E is capacitable. $\qquad\square$

We now look at the boundary ∂T of a tree. Recall from Lemma 3.1.4 that ∂T is a compact metric space. For any edge e we denote by $T(e) \subset \partial T$ the set of rays passing through e. Then $T(e)$ is a closed ball of diameter $2^{-|e|}$ (where $|e|$ is tree distance to the root vertex of the endpoint of e further from the root) and also an open ball of diameter r, for $2^{-|e|} < r \leq 2^{-|e|+1}$. Moreover, all closed balls in ∂T are of this form. If $E \subset \partial T$ then a set Π of edges is called a **cut-set** of E if every ray in E contains at least one edge of Π or, equivalently, if the collection $T(e)$, $e \in \Pi$, is a covering of E. Recall from Definition 4.5 that the α-Hausdorff content of a set $E \subset \partial T$ is defined as

$$\mathscr{H}_\infty^\alpha(E) = \inf \left\{ \sum_{i=1}^{\infty} |E_i|^\alpha : E_1, E_2, \ldots \text{ is a covering of } E \right\}$$

$$= \inf \left\{ \sum_{e \in \Pi} 2^{-\alpha|e|} : \Pi \text{ is a cut-set of } E \right\},$$

where the last equality follows from the fact that every closed set in ∂T is contained in a closed ball of the same diameter.

Lemma B.2.3 *The set function Ψ on ∂T given by $\Psi(E) = \mathscr{H}_\infty^\alpha(E)$ is a Choquet capacity.*

Proof Property (a) holds trivially. For (b) note that given E and $\varepsilon > 0$ there exists a cut-set Π such that the collection of sets $T(e)$, $e \in \Pi$, is a covering of E with $\Psi(E) \geq \sum_{e \in \Pi} 2^{-\alpha|e|} - \varepsilon$. As $O = \bigcup_{e \in \Pi} T(e)$ is an open set containing E and $\Psi(O) \leq \sum_{e \in \Pi} 2^{-\alpha|e|}$ we infer that $\Psi(O) \leq \Psi(E) + \varepsilon$, from which (b) follows.

We now prove (c). Suppose $E_1 \subset E_2 \subset \ldots$ and let $E = \bigcup_{n=1}^{\infty} E_n$. Fix $\varepsilon > 0$

and choose cut-sets Π_n of E_n such that

$$\sum_{e\in\Pi_n} 2^{-\alpha|e|} \leq \Psi(E_n) + \frac{\varepsilon}{2^{n+1}}. \tag{B.1}$$

For each positive integer m we will prove that

$$\sum_{e\in\Pi_1\cup\cdots\cup\Pi_m} 2^{-\alpha|e|} \leq \Psi(E_m) + \varepsilon. \tag{B.2}$$

Taking the limit as $m \to \infty$ gives

$$\Psi(E) \leq \sum_{e\in\Pi_1\cup\cdots} 2^{-\alpha|e|} \leq \lim_{m\to\infty} \Psi(E_m) + \varepsilon.$$

Taking $\varepsilon \to 0$ gives $\Psi(E) \leq \lim_{m\to\infty} \Psi(E_m)$. Since the opposite inequality is obvious, we see that (B.2) implies (c).

For every ray $\xi \in E$ we let $e(\xi)$ be the edge of smallest order in $\xi \cap \bigcup_n \Pi_n$. Note that $\Pi = \{e(\xi): \xi \in E\}$ is a cut-set of E and no pair of edges $e_1, e_2 \in \Pi$ lie on the same ray. Fix a positive integer m and let $Q_m^1 \subset E_m$ be the set of rays in E_m that pass through some edge in $\Pi\cap\Pi_1$. Let Π_m^1 be the set of edges in Π_m that intersect a ray in Q_m^1. Then Π_m^1 is a cut-set for Q_m^1 and hence for $Q_m^1 \cap E_1$, and hence $\Pi_m^1 \cup (\Pi_1 \setminus \Pi)$ is a cut-set for E_1. From our choice of Π_1 in (B.1) and the fact that $\Psi(E_1)$ is a lower bound for any cut-set sum for E_1 we get

$$\sum_{e\in\Pi_1} 2^{-\alpha|e|} \leq \Psi(E_1) + \frac{\varepsilon}{4} \leq \sum_{e\in\Pi_m^1\cup(\Pi_1\setminus\Pi)} 2^{-\alpha|e|} + \frac{\varepsilon}{4}.$$

Now subtract the contribution from edges in $\Pi_1 \setminus \Pi$ on both sides, to get

$$\sum_{e\in\Pi\cap\Pi_1} 2^{-\alpha|e|} \leq \sum_{e\in\Pi_m^1} 2^{-\alpha|e|} + \frac{\varepsilon}{4}.$$

Now iterate this construction. Suppose $1 \leq n < m$ and $\Pi_m^1, \ldots, \Pi_m^n$ are given. Set $\Pi_{n+1}^* = \Pi_{n+1} \setminus (\Pi_1 \cup \cdots \cup \Pi_n)$ and let Q_m^{n+1} be the set of rays in E_m that pass through some edge in $\Pi\cap\Pi_{n+1}^*$. Let Π_m^{n+1} be the set of edges in Π_m that intersect a ray in Q_m^{n+1}. Then, as above, Π_m^{n+1} is a cut-set for $Q_m^{n+1} \cap E_{n+1}$, and hence $\Pi_m^{n+1} \cup (\Pi_{n+1} \setminus (\Pi\cap\Pi_{n+1}^*))$ is a cut-set for E_{n+1}. Using (B.1) and subtracting equal terms as before gives

$$\sum_{e\in\Pi\cap\Pi_{n+1}^*} 2^{-\alpha|e|} \leq \sum_{e\in\Pi_m^{n+1}} 2^{-\alpha|e|} + \frac{\varepsilon}{2^{n+2}}.$$

Adding up the m inequalities thus obtained and using (B.1) we get

$$\sum_{e\in\Pi\cap(\Pi_1\cup\cdots\cup\Pi_m)} 2^{-\alpha|e|} \leq \sum_{n=1}^{m}\left(\sum_{e\in\Pi_m^n} 2^{-\alpha|e|} + \frac{\varepsilon}{2^{n+1}}\right) \leq \sum_{e\in\Pi_m} 2^{-\alpha|e|} + \frac{\varepsilon}{2}.$$

This and (B.1) imply (B.2). This completes the proof. $\square$

The following is immediate from Lemma B.2.3.

Corollary B.2.4 *If an analytic set $E \subset \partial T$ has $\mathscr{H}^\alpha(E) > 0$, then there exists a compact set $A \subset E$ with $\mathscr{H}^\alpha(A) > 0$.*

Proof Recall that by Proposition 1.2.6 Hausdorff content and Hausdorff measure vanish simultaneously. Hence, if E is analytic and $\mathscr{H}^\alpha(E) > 0$, then $\Psi(E) > 0$. Lemma B.2.3 implies that there exists a compact set $A \subset E$ with $\Psi(A) > 0$, and therefore $\mathscr{H}^\alpha(A) > 0$.   □

All that remains to be done now is to transfer this result from the boundary of a suitable tree to Euclidean space.

Theorem B.2.5 *Let $E \subset \mathbb{R}^d$ be a Borel set and assume $\mathscr{H}^\alpha(E) > 0$. Then there exists a closed set $A \subset E$ with $\mathscr{H}^\alpha(A) > 0$.*

Proof Find a hypercube $Q \subset \mathbb{R}^d$ of unit sidelength with $\mathscr{H}^\alpha(E \cap Q) > 0$, and a continuous mapping $\Phi \colon \partial T \to Q$ from the boundary of the 2^d-ary tree T to Q, mapping closed balls $T(e)$ onto compact dyadic subcubes of sidelength $2^{-|e|}$. The α-Hausdorff measure of images and inverse images under Φ changes by no more than a constant factor. Indeed, for every $B \subset \partial T$, we have that $|\Phi(B)| \leq \sqrt{d}|B|$. Conversely, every set $B \subset Q$ of diameter 2^{-k} lies in the interior of the union of no more than 3^d compact dyadic cubes of sidelength 2^{-k}, whence the edges corresponding to these dyadic cubes form a cut-set of $\Phi^{-1}(B)$. Therefore, $\mathscr{H}^\alpha(E \cap Q) > 0$ implies $\mathscr{H}^\alpha(\Phi^{-1}(E \cap Q)) > 0$. As $\Phi^{-1}(E \cap Q)$ is a Borel set and hence analytic we can use Corollary B.2.4 to find a compact subset A with $\mathscr{H}^\alpha(A) > 0$. Now $\Phi(A) \subset E \cap Q$ is a compact subset of E and $\mathscr{H}^\alpha(\Phi(A)) > 0$, as required.   □

Given a Borel set $E \subset \mathbb{R}^d$ with $\mathscr{H}^\alpha(E) > 0$ we can now pick a closed set $A \subset E$ with $\mathscr{H}^\alpha(A) > 0$, apply Frostman's Lemma to A and obtain a probability measure on A (and, by extension, on E) such that $\mu(D) \leq C|D|^\alpha$ for all Borel sets $D \subset \mathbb{R}^d$. The proof of Theorem B.2.5 also holds for Borel sets E with Hausdorff measure $\mathscr{H}^\varphi(E) > 0$ taken with respect to a gauge function φ.

We have adapted the proof of Theorem B.2.5 from Carleson (1967). For a brief account of Polish spaces and analytic sets see Arveson (1976). For a more comprehensive treatment of analytic sets and the general area of descriptive set theory, see the book of Kechris (1995). There are several variants of Choquet's Capacitability Theorem. See, for example, Bass (1995) for an alternative treatment and for other applications of the result, e.g., the infimum of the hitting times of Brownian motion in a Borel set defines a stopping time.

Appendix C

Hints and solutions to selected exercises

Solution 1.12: ≈ 1.63772 and ≈ 1.294874

Hint 1.25: Show A^n has the form

$$\begin{pmatrix} F_{n+1} & F_n \\ F_n & F_{n-1} \end{pmatrix},$$

where F_n is the nth Fibonacci number. Alternatively, compute the eigenvectors and eigenvalues of A.

Solution 1.58: Consider 5-adic intervals and consider an iterative construction that maps the first, third and fifth subintervals isometrically, shrinks the second and expands the fourth.

Hint 1.59: Define a positive sequence of functions on $[0,1]$ by $m_0 = 1$, and $m_{n+1} = m_n + (2x_n - 1)\mathbf{1}_{m_n \geq 0}$. The limit defines a positive measure μ on $[0,1]$ that satisfies $\mu(I) = O(|I|\log|I|^{-1})$ for all intervals. Show that the support of μ has positive φ-measure for the gauge $\varphi(t) = t \log \frac{1}{t}$. Deduce that the random walk $\{s_n(x)\}$ tends to infinity on a set of $x's$ of dimension 1.

In fact, stronger estimates are true, but harder to prove. There is a $C > 0$ so that

$$\int_0^1 \exp\left(\frac{C_1}{n}m_n^2\right) dx < C_2.$$

This is a very special case of a result for square integrable martingales, e.g., Chang et al. (1985). Assuming this, one can show that

$$\mathscr{L}_1(\{x : m_n(x) > \lambda n\}) \leq C_2 \exp(-\lambda^2/C_1).$$

360

This implies $m_n(x) = O(\sqrt{n \log n})$ on a set of full μ measure and that the support of μ has positive φ-measure for the gauge

$$\varphi(t) = t\sqrt{\log \frac{1}{t} \log \log \frac{1}{t}}.$$

The final result in this direction is that the support of μ has finite and positive φ-measure for the gauge $\varphi(t) = t\sqrt{\log \frac{1}{t} \log \log \log \frac{1}{t}}$. See Makarov (1989).

Solution 1.60: Let A_k denote the event that k is minimal such that $|S_k| \geq h$. Then by Pythagoras, $\mathbb{E}S_n^2 1_{A_k} \geq \mathbb{E}S_k^2 1_{A_k} \geq h^2 \mathbb{P}(A_k)$.

Hint 2.11: Since the sets are self-similar, the α-measure of any subarc is finite and positive. Consider the arc from an endpoint x to an interior point y; and map y to the point on the second arc whose corresponding arc has the same measure. Show this is bi-Lipschitz.

Hint 2.14: Let $\{B(x_i, \frac{\delta}{4})\}_{i=1}^m$ be a maximal collection of disjoint balls of radius $\frac{\delta}{4}$ contained in $B(0, 1)$. Attach to each $K \leq N$ the collection of balls $B(x_i, \frac{\delta}{4})$ which intersect E_K, and verify this mapping is one to one. Also see Lemma A.2.4.

Hint 2.19: Let $\mathscr{T}_n$ denote all subtrees of the binary tree that contain the root and all of whose leaves are distance n from the root. Enumerate $\{T_k\} = \bigcup_n \mathscr{T}_n$. Then form a subtree by setting $\Gamma_1 = t_1$ and, in general, attaching T_n to leaf of Γ_{n-1} of minimal distance to the root. The compact set in $[0, 1]$ corresponding to this tree has the desired property.

Hint 2.25: Given $\varepsilon > 0$, use the pigeonhole principle to find $n < m$ so that $0 < (n - m)\alpha < \varepsilon$ mod 1. Thus multiples of $(n - m)\alpha$ are ε-dense mod 1.

Hint 2.27: Consider the sets $E_n = \mathbb{R} \setminus \frac{\mathbb{Z} + (-\varepsilon, \varepsilon)}{h_n}$ that are unions of closed intervals of length $(1 - 2\varepsilon)/h_n$ separated by open intervals of length $2\varepsilon/h_n$. If $h_{n+1}/h_n \geq q > 2$ then it is easy to check that each component interval of E_n contains a component interval of E_{n+1}, if ε is small enough (depending on q). By induction, there a nested sequence of closed components, these have non-empty intersection. Any point $\alpha \in E = \cap_n E_n$ has the desired property. The case $1 < q < 2$ is more difficult and was first proved by Khintchine (1926). For applications and improved bounds, see Katznelson (2001) and Peres and Schlag (2010).

Hint 2.28: By taking every dth term, any lacunary sequence with constant q can be split into d sequences all with constant q^d. If we choose d so $q^d > 2$, then by the "easy case" of Exercise 2.27, there is a d-vector $\alpha = (\alpha_1, \dots, \alpha_d)$ so that α_k has the desired property with respect to the kth sequence. Thus for each n, we have $\mathrm{dist}(h_n \alpha_k, \mathbb{Z}) > \varepsilon$ for at least one k, which proves the result.

Hint 2.32: Remove open intervals of length 4^{-n} from $[0,1]$ to form a Cantor set K of positive length and $\alpha(K) = 0$.

Hint 3.6: Take E_i so that the corresponding tree has 1 child at levels $(4k + 2i) \leq n < (4k + 2i + i)$ and 2 children otherwise.

Hint 3.8: Take $A = \mathbb{R} \times Y$, and $B = X \times \mathbb{R}$ where $X, Y \subset [0,1]$ are Borel sets of dimension zero so that $\dim(X \times Y) = 1$ as in Example 3.2.3. Observe that $A \cap B = X \times Y$, and $\dim(A \cap (B + x)) = \dim(X \times Y)$ for any x. This example is due to Krystal Taylor.

Hint 3.9: Let $B = \bigcup_n B_n$ where B_n corresponds to $\varepsilon = 1/n$ and the B_n are chosen to have a compact union.

Hint 3.11: To prove the result, suppose $E = \cap E_n$ and $F = \cap F_n$ and prove that if E_n and F_n have component intervals that intersect, then so do E_{n+1} and F_{n+1}. Suppose not. Then there are components I_1 of E_n and I_2 of F_n so that none of the components of $I_1 \setminus J(I_1)$ and $I_2 \setminus J(I_2)$ intersect. Thus there is a component I_3 of $I_1 \setminus J(I_1)$ that is contained in $J(I_2)$ and a component I_4 of $I_2 \setminus J(I_2)$ that is contained in $J(I_2)$. Since I_3 is closed and $J(I_2)$ is open the former is strictly shorter than the latter. Thus

$$|I_3| < |J(I_2)| \leq |I_4| / \tau_F < |J(I_1)| / \tau_F \leq |I_3| / (\tau_F \tau_E),$$

which implies $\tau_E \tau_F < 1$.

Hint 3.12: Use Exercise 3.11 and the fact that the line $y = ax + c$ hits the product set $E \times E$ if and only if the Cantor set $E - c$ hits the Cantor set aE (both of which have the same thickness as E). Also use the fact that the thickness of $C_{1/3}$ is clearly 1.

Hint 3.17: Given a measure on the boundary of the tree, form a new measure by swapping two subtrees with the same parent. The average of these two measures has smaller or equal energy and gives the same mass to both subtrees. Doing this for all pairs gives the evenly distributed measure in the limit.

Hint 3.24: The variance of Z_n can be given in terms of $f_n'(1)$ and $f_n''(1)$, which can be computed by induction and the chain rule.

Hint 3.37: Suppose $\{z_k\}_1^{n+1}$ is optimal for $d_{n+1}(E)$. Dropping one point gives a product bounded by d_n. If we multiply the resulting $n+1$ inequalities, we get p_n^{n+1} on the left and all possible distinct products among the $\{z_k\}$ on the right, each repeated $n-1$ times. Thus

$$p_{n+1}^{n-1} \leq p_n^{n+1} \quad \Rightarrow \quad d_{n+1}^{n(n+1)(n-1)/2} \leq d_n^{n(n-1)(n+1)/2} \quad \Rightarrow \quad d_{n+1} \leq d_n.$$

Hint 4.15: Define a measure on K by $\mu(I) = \mathbb{P}(I \cap K \cap A \neq \emptyset)$. Show the random affine set in the previous problem has upper Minkowski dimension $\leq 2 - b$. At generation k partition each of the approximately $(nmp)^k$ surviving rectangles into $(n/m)^k$ squares of side length n^{-k}. This gives approximately $(pn^2)^k = n^{(2-b)k}$ squares of size n^{-k}.

Hint 4.16: Consider horizontal slices. Each such slice is a random Cantor set obtained by partitioning an interval into n subintervals and retaining each with probability p. If $np > 1$, these sets have Hausdorff dimension $1 - b$. Now use the Marstrand Slicing Theorem.

Hint 5.6: If x is a local strict maximum of f then there is a horizontal line segment of positive length r in $\mathbb{R}^2$ that is centered at $(x, f(x))$ and only hits the graph of f at this point. If x and y are two such points with the same r, then it is easy to check $|x - y| > r/2$ (use the Intermediate Value Theorem). Thus there are only countably many points with $r > 1/n$ for any positive integer n and hence at most countably many such points.

Hint 5.9: Prove this for step functions and then approximate.

Hint 5.10: Use the natural parameterization of a snowflake curve to build examples.

Hint 5.11: First show $E = f[0, 1]$ has positive area. Then show there is a square Q so that if we divide it into n^2 equal sized disjoint subsquares, then every subsquare is hit by E. Label them in the order they are hit and show adjacent squares have labels with bounded differences. Prove there is no such labeling.

Hint 5.13: In this case f' exists and f is the integral of f'. Apply the Riemann–Lebesgue Lemma to f'.

Hint 5.14: There are at least three ways to do this: (1) directly from the definition, (2) use the previous exercise or (3) prove that a Lipschitz function must be differentiable somewhere.

Hint 5.16: Use the trigonometric identity

$$\sin\left(k+\frac{1}{2}\right)x - \sin\left(k-\frac{1}{2}\right)x = 2\sin\frac{x}{2}\cos kx,$$

to "telescope" the sum

$$2\sin(x/2)\left[\frac{1}{2}+\cos(x)+\cdots+\cos(nx)\right] = \sin(n+\frac{1}{2})x - \sin(x/2).$$

Hint 5.17: Use the previous exercise to write

$$F_n(x) = \frac{1}{2n\sin(x/2)}\left[\sin x/2 + \cdots + \sin\left(n-\frac{1}{2}\right)x\right]$$

$$= \frac{1}{2n\sin(x/2)}[\text{Im}(e^{ix/2}) + \cdots + \text{Im}(e^{i(n-\frac{1}{2})x})]$$

$$= \frac{1}{2n\sin(x/2)}\text{Im}[e^{ix/2}(1+\cdots+(e^{i(n-1)x})]$$

$$= \frac{1}{2n\sin(x/2)}\text{Im}\left[e^{ix/2}\left(\frac{1-e^{inx}}{1-e^{ix}}\right)\right].$$

Now use the fact that $\sin(x) = (e^{ix} - e^{-ix})/2i$ to get

$$F_n(x) = \frac{1}{2n\sin(x/2)}\text{Im}\left(e^{inx/2}\frac{\sin(nx/2)}{\sin(x/2)}\right)$$

$$= \frac{1}{2n\sin(x/2)}\left(\frac{\sin(nx/2)^2}{\sin(x/2)}\right)$$

$$= \frac{1}{2n}\left(\frac{\sin(nx/2)}{\sin(x/2)}\right)^2.$$

Hint 5.19: Use Plancherel's Formula and the summation formulas for k and k^2.

Hint 5.20: If F_n is the Fejér kernel then $f * F_n$ is a trigonometric polynomial. Use the estimates of the previous exercise to show $f * F_n \to f$ uniformly as $n \to \infty$.

Hint 5.25: Show the derivative of f is zero at local extreme points. Then for intervals (a,b) consider $f(x) - L(x)$ where L is affine and $L(a) = f(a)$ and $L(b) = f(b)$.

Hint 5.29: Find a smooth function g so that $|f(x) - g(x)| \le \varepsilon |x|/2$. Then apply the Weierstrass Approximation Theorem to g'.

Hint 5.34: Choose coefficients b_n so that

$$b_k b_{k+1} = 0, \quad \sum_{k=-\infty}^{\infty} b_k (n_{k+1} - n_k)^\alpha < \infty, \quad \limsup_{k \to \infty} b_k n_k^\alpha = \infty,$$

and prove

$$f(x) = \sum_{k=-\infty}^{\infty} b_k (e^{in_k x} - e^{in_{k+1} x})$$

has the desired properties. Use Exercise 5.12 and the estimate

$$|e^{inx} - e^{imx}| \le C_\alpha |m - n|^\alpha |x|^\alpha.$$

Hint 5.48: Take $\sum_{n=1}^{\infty} 2^{-n} f(x - r_n)$ where f is the Cantor singular function and $\{r_n\}$ is dense in $\mathbb{R}$.

Solution 6.3: Suppose Ω is a complementary component. $\partial \Omega$ is locally connected, so it has a continuous parameterization. Consider any arc γ of this parameterization. If it contains a double point, show this must also be a double point for the Brownian path. If the arc has no double point, it is a Jordan arc. Assume it contains no double point of Brownian motion. Then there must be an arc of the Brownian path that parameterizes γ, for if it ever leaves γ at an interior point of γ, it must revisit that point later, thus creates a double point on γ. Thus the Brownian path has a subarc with no double points, contradicting Exercise 6.2.

Solution 6.8: Let $A_{k,n}$ be the event that $B((k+1)2^{-n}) - B(k2^{-n}) > c\sqrt{n}2^{-n/2}$. Then Lemma 6.1.6 implies

$$\mathbb{P}(A_{k,n}) = \mathbb{P}(B(1) > c\sqrt{n}) \ge \frac{c\sqrt{n}}{c^2 n + 1} e^{-c^2 n/2}.$$

If $0 < c < \sqrt{2\log(2)}$ then $c^2/2 < \log(2)$ and $2^n \mathbb{P}(A_{k,n}) \to \infty$. Therefore,

$$\mathbb{P}(\bigcap_{k=1}^{2^n} A_{k,n}^c) = [1 - \mathbb{P}(A_{k,n})]^{2^n} \le e^{-2^n \mathbb{P}(A_{k,n})} \to 0 \quad \text{as } n \to \infty.$$

The last inequality comes from the fact that $1 - x \le e^{-x}$ for all x. By considering $h = 2^{-n}$, one can see that

$$P\left(\forall h \ B(t+h) - B(t) \le c\sqrt{h \log_2 h^{-1}}\right) = 0 \quad \text{if } c < \sqrt{2\log 2}.$$

Solution 6.10: Suppose that there is a $t_0 \in [0,1]$ such that

$$\sup_{h\in[0,1]} \frac{B(t_0+h)-B(t_0)}{h^\alpha} \leq 1, \quad \text{and}$$

$$\inf_{h\in[0,1]} \frac{B(t_0+h)-B(t_0)}{h^\alpha} \geq -1.$$

If $t_0 \in \left(\frac{k-1}{2^n}, \frac{k}{2^n}\right)$ for $n > 2$, then the triangle inequality gives

$$\left| B\left(\frac{k+j}{2^n}\right) - B\left(\frac{k+j-1}{2^n}\right) \right| \leq 2\left(\frac{j+1}{2^n}\right)^\alpha.$$

Fix $l \geq 1/(\alpha - \frac{1}{2})$ and let $\Omega_{n,k}$ be the event

$$\left(\left| B\left(\frac{k+j}{2^n}\right) - B\left(\frac{k+j-1}{2^n}\right) \right| \leq 2\left[\frac{j+1}{2^n}\right]^\alpha \text{ for } j = 1,2,\ldots,l \right).$$

Then

$$\mathbb{P}(\Omega_{n,k}) \leq \left[\mathbb{P}\left(|B(1)| \leq 2^{n/2} \cdot 2 \cdot \left(\frac{l+1}{2^n}\right)^\alpha \right) \right]^l \leq \left[2^{n/2} \cdot 2 \cdot \left(\frac{l+1}{2^n}\right)^\alpha \right]^l$$

since the normal density is less than $1/2$. Hence

$$\mathbb{P}\left(\bigcup_{k=1}^{2^n} \Omega_{n,k} \right) \leq 2^n \cdot \left[2^{n/2} \cdot 2 \left(\frac{l+1}{2^n}\right)^\alpha \right]^l = C\left[2^{(1-l(\alpha-1/2))} \right]^n,$$

which sums. Thus

$$\mathbb{P}\left(\limsup_{n\to\infty} \bigcup_{k=1}^{2^n} \Omega_{n,k} \right) = 0.$$

Solution 6.12: Define the event $\Omega_{n,k}$ as in the proof of Theorem 6.3.6, and let Ω_∞ be the set of continuous functions on $[0,1]$ for which the event $\Omega_{n,k}$ holds infinitely often. We claim that $\Omega_\infty + f$ has Wiener measure zero for any $f \in C[0,1]$ and thus Ω_∞ is negligible. Since every continuous function that is differentiable at even one point is in Ω_∞, this claim proves the nowhere differentiable functions are prevalent. To prove the claim, note that $B_t + f \in \Omega_\infty$ means that infinitely often the event $E(f,k,n)$ defined by

$$\left| \left(B(\frac{k+j}{2^n}) - B(\frac{k+j-1}{2^n})\right) + \left(f(\frac{k+j}{2^n}) - f(\frac{k+j-1}{2^n})\right) \right| \leq M(2j+1)2^{-n}$$

holding for $j = 1,2,3$. But $\Delta f = f(\frac{k+j}{2^n}) - f(\frac{k+j-1}{2^n})$ is just a real number, the probability of this event is the probability that a Brownian increment over an interval of length 2^{-n} lands in a fixed interval of length $2M(2j+1)2^{-n}$. As in the proof of Theorem 6.3.6, this probability is $O(2^{-n/2})$, since the normal density is $O(2^{n/2})$ (in fact, the normal density takes its maximum at the origin, so if $\Delta f \neq 0$, the probability of $E(f,k)$ is even smaller). The proof finishes with an application of the Borel–Cantelli lemma, just as before.

Solution 6.13: $\{\tau_A \leq t\} = \bigcap_{n \geq 1} \bigcup_{s \in [0,t] \cap \mathbb{Q}} \{\text{dist}(B(s), A) \leq \frac{1}{n}\} \in \mathscr{F}_0(t)$.

Solution 6.14: Conditional on $B(a) = x > 0$ we have

$$\mathbb{P}(\exists t \in (a, a+\varepsilon) : B(t) = 0 | B(a) = x) = \mathbb{P}(\min_{a \leq t \leq a+\varepsilon} B(t) < 0 | B(a) = x).$$

But the right-hand side is equal to

$$\mathbb{P}(\max_{0 < t < \varepsilon} B(t) > x) = 2\mathbb{P}(B(\varepsilon) > x),$$

using the reflection principle.

By considering also the case where x is negative we get

$$\mathbb{P}(\exists t \in (a, a+\varepsilon) : B(t) = 0) = 4 \int_0^\infty \int_x^\infty \frac{e^{-\frac{y^2}{2\varepsilon} - \frac{x^2}{2a}}}{2\pi\sqrt{a\varepsilon}} \, dy \, dx.$$

Computing this last integral explicitly, we get

$$\mathbb{P}(\exists t \in (a, a+\varepsilon) : B(t) = 0) = \frac{2}{\pi} \arctan\sqrt{\frac{\varepsilon}{a}}.$$

Solution 6.15: $0.471573, 0.355983, 0.283858, 0.235402$.

Solution 7.2: One can mimic the proof of Kaufman's dimension doubling theorem in the plane. Balka and Peres (2016) investigate which subsets of $[0, 1]$ have the property that uniform dimension doubling holds for one-dimensional Brownian motion.

Solution 7.3: Consider (B, W) as a 2-dimensional Brownian motions and Z as the preimage of the real line. Then use Kaufman's theorem in the 2-dimensional case.

Solution 7.5: The expected exit time is

$$\mathbb{E}_x \int \mathbf{1}_{B(t) \in \Omega}(x) dt$$

and the second moment is

$$\mathbb{E}_x \int_0^\infty \int_0^\infty \mathbf{1}_{B(t) \in \Omega}(x) \mathbf{1}_{B(s) \in \Omega}(x) \, ds \, dt.$$

The integral is written as the sum of two integrals over $s \leq t$ and $t \leq s$ respectively. Then

$$
\mathbb{E}_x \int_\infty \int_{s \leq t} \mathbf{1}_{B(t) \in \Omega}(x) \mathbf{1}_{B(s) \in \Omega}(x) \, ds \, dt
$$

$$
= \mathbb{E}_x \int_0^\infty \int_s^t \mathbf{1}_{B(t) \in \Omega}(x) \mathbf{1}_{B(s) \in \Omega}(x) \, ds \, dt
$$

$$
= \mathbb{E}_x \int_0^\infty \int_s^\infty \mathbf{1}_{B(t-s)+B(s) \in \Omega}(x) \mathbf{1}_{B(s) \in \Omega}(x) \, dt \, ds
$$

$$
= \mathbb{E}_x \int_0^\infty \int_0^\infty \mathbf{1}_{B(u)+B(s) \in \Omega}(x) \mathbf{1}_{B(s) \in \Omega}(x) \, du \, ds
$$

$$
= \mathbb{E}_x \int_0^\infty \mathbf{1}_{B(u)+B(s) \in \Omega}(x) \, du \int_0^\infty \mathbf{1}_{B(s) \in \Omega}(x) \, ds
$$

$$
= \alpha^2.
$$

The $s \geq t$ also contributes at most α^2, so the second moment is at most $2\alpha^2$.

Hint 7.6: Suppose $f = u + iv$. Considering $u(z) = \mathrm{Re}(z)$ and $\mathrm{Im}(z)$ shows that u and v are harmonic functions, so f is smooth. Considering $\mathrm{Re}(z^2)$ and $\mathrm{Im}(z^2)$ shows that uv and $u^2 - v^2$ are harmonic. Taking the Laplacians of these shows the vectors (u_x, u_y) and (v_x, v_y) are perpendicular and the same length, and these facts imply the Cauchy–Riemann Equations hold for either $u + iv$ or $u - iv$.

Solution 7.11: By Lemma 7.9.2, the exit time is finite if and only if the power series for the conformal map satisfies $\sum_{n=0}^\infty |a_n|^2 < \infty$. This is equivalent to the boundary values of the conformal map being in L^2 with respect to Lebesgue measure on the circle. The conformal map from the disk to the sector is of the form $\left(\frac{1-z}{1+z}\right)^\alpha$; a Möbius transformation to the right half-plane followed by a power, and this is in L^2 if and only if $\alpha < 1$.

Solution 7.12: Green's function $G(x,y)$ is the density of the expected time a Brownian path started at y spends at x. The conformal map sends a Brownian path in V to a Brownian path in u changing the time by a factor of $|f'(x)|^2$ while the path is near x. This factor is also the Jacobian of f, so area changes by the same factor. Thus the density is invariant under f, as claimed.

Solution 7.13: Let $p(x,y,t) = (1/2\pi t)\exp(-|x-y|^2/2t)$ be the transition probabilities for Brownian motion in the plane. What is the expected time between 0 and T that a path started at y spends at 0 before leaving $\mathbb{D}$? The total time it spends at 0, minus the time it spends there after its first exit from $\mathbb{D}$ is

$$\int_0^T p(s,x,y)ds - \int_0^{T-t} p(s,x,z)dsd\mu(z,t),$$

where μ is the hitting distribution of a Brownian motion started at y hitting $z \in \partial\mathbb{D}$ at time $t \leq T$. Let

$$\mu(T) = \mu(\partial\mathbb{D} \times [0,T]) = \mathbb{P}(B_y \text{ hits } \partial\mathbb{D} \text{ before } T).$$

Then the difference of integrals above can be written as

$$(1 - \mu(T)) \int_0^T p(s,x,y)ds$$
$$+ \int \int_0^T p(s,x,y) - p(s,x,z)dsd\mu(z,t)$$
$$+ \int_{T-t}^T p(s,x,z)dsd\mu(z,t).$$

The first term is the probability that a Brownian motion started at y has not yet hit the unit circle, times the expected time it spends at 0 before T. The first factor dies exponentially fast, while the second is trivially bounded by T, so this term tends to zero. The third term is the expected time that a Brownian motion started at y spends at 0 between $T - t$ and T. This is bounded above by $O((T-t)/\sqrt{T})$, so for a fixed t this tends to zero as $T \to \infty$, so the integral against the measure μ also tends to zero.

The main term is the middle term and to complete the proof we claim that for $y \in \mathbb{D}, z \in \partial\mathbb{D}$,

$$\int_0^T p(s,0,y) - p(s,0,z)ds \to \frac{1}{\pi} \log \frac{1}{|y|}.$$

To prove this, note that the integral on the left tends to

$$\int_0^\infty p(s,0,y) - p(s,0,z)ds = \frac{1}{2\pi} \int_0^\infty (\exp(-|x-y|^2/2t) - \exp(-1/2t)) \frac{dt}{t}$$
$$= \frac{1}{2\pi} \int_0^\infty \int_{|x-y|^2/2t}^{1/2t} \exp(-s)ds \frac{dt}{t}$$
$$= \frac{1}{2\pi} \int_0^\infty \exp(-s) \int_{|x-y|^2/2s}^{1/2s} \frac{1}{t}dtds$$
$$= \frac{1}{\pi} \log \frac{1}{|y|}.$$

Thus $G_{\mathbb{D}}(0,y) = \frac{1}{\pi} \log \frac{1}{|y|}$. Using the conformal self-map of the disk given by

$f(z) = (z-x)/(1-\bar{x}z)$ we see that

$$G_{\mathbb{D}}(x,y) = G_{\mathbb{D}}(f(x), f(y)) = \frac{1}{\pi} \log \left| \frac{1-\bar{x}y}{x-y} \right|.$$

Hint 7.14: Show that reflection over the unit sphere is conformal (the map is given by $\mapsto x/|x|^2$). But the image of Brownian motion under this map converges to 0 almost surely if $d \geq 3$.

The proof of Theorem 7.9.1 uses the fact that, in the plane, the composition of a harmonic function with a conformal mapping is harmonic. This is false in higher dimensions, e.g., consider $|x|$ and $|x|^{-1}$ in $\mathbb{R}^3$, which are related by a conformal sphere inversion.

Hint 7.15: Let $f : \mathbb{D} \to \Omega$ be conformal and define

$$v(z) = \frac{1}{1-|z|^2} - \log|f'(z)|$$

on $\mathbb{D}$ and $V(z) = v(f^{-1}(z))$ on Ω. Assume the distortion estimate (Koebe's $\frac{1}{4}$-Theorem)

$$|f'(z)|(1-|z|^2) \simeq \mathrm{dist}(f(z), \partial\Omega)$$

and use it to prove that

$$V(z) = \frac{1}{\mathrm{dist}(z, \partial\Omega)} + O(1),$$

and then show

$$0 < c \leq \mathbb{E}V(z_{n+1}) - V(z_n) \leq C < \infty,$$

by considering what happens to v on $\mathbb{D}$. This shows the expected distance to the boundary decays exponentially. Use the Law of Large Numbers to show that almost every walk converges exponentially quickly to $\partial\Omega$.

Hint 7.17: Assume to the contrary that there is a set $A \subset [0,1]$ such that $\dim A > \frac{1}{2}$ and f is α-Hölder continuous on A. As f is still α-Hölder continuous on the closure of A, we may assume that A itself is closed. Let Z be the zero set of W, then Exercise 7.16 implies that $\dim(A \cap Z) > 0$ with positive probability. Then the α-Hölder continuity of $f|_A$ and Lemma 5.1.3 imply that, with positive probability,

$$\dim(f, W)(A \cap Z) = \dim(f(A \cap \mathscr{Z}) \times \{0\}) = \dim f(A \cap Z)$$
$$\leq \frac{1}{\alpha} \dim(A \cap Z) < 2\dim(A \cap Z),$$

which contradicts the fact that (f, W) is almost surely dimension doubling. See Balka and Peres (2014).

Hint 7.18: Let $\{W(t) \colon 0 \leq t \leq 1\}$ be a linear Brownian motion which is independent of B. By Kaufman's Dimension Doubling Theorem (B, W) is dimension doubling with probability 1, thus applying Exercise 7.17 for an almost sure path of B finishes the proof. See Balka and Peres (2014).

Hint 8.15: Fix $\beta = \log_b d$. Let μ_n denote the uniform distribution on the set A_n defined in Example 8.1.5. The argument in that example implies that, for any integer y, we have $\int_{A_n} F_\beta(x, y) d\mu_n \leq Cn$ where C is a constant. On the other hand, if $|\ell - n| > 1$ and $y \in A_\ell$, direct inspection gives the better bound $\int_{A_n} F_\beta(x, y) d\mu_n \leq C$, possibly using a bigger constant C. Thus taking $\nu_m = \frac{1}{m} \sum_{n=m+1}^{2m} \mu_n$, we deduce that

$$\int F_\beta(x, y) d\nu_m \leq \frac{1}{m}\left(6Cm + \sum_{n=m+1}^{2m} C\right) \leq 7C.$$

Since m is arbitrary, this implies positive asymptotic capacity of A at the critical exponent β.

Solution 8.34: Set $Y_n = X_n/\mathbb{E}X_n$ and note that $\mathbb{E}Y_n = 1$ and $\mathbb{E}Y_n^2 = M < \infty$. Let $\lambda > 0$ and note

$$1 = \mathbb{E}Y_n \leq \mathbb{E}Y_n \mathbf{1}_{Y_n < \lambda} + \frac{1}{\lambda}\mathbb{E}Y_n^2.$$

Using this and Fatou's lemma,

$$1 - \frac{M}{\lambda} \leq \limsup \mathbb{E}(\mathbf{1}_{Y_n < \lambda} Y_n) \leq \mathbb{E}(\limsup \mathbf{1}_{Y_n < \lambda} Y_n) \leq \mathbb{E}(\limsup Y_n).$$

Take $\lambda \to \infty$ to deduce

$$1 \leq \mathbb{E}(\limsup Y_n) = \mathbb{E}\left(\limsup \frac{X_n}{\mathbb{E}(X_n)}\right),$$

and thus $X_n \geq \mathbb{E}(X_n)$ infinitely often on a set of positive measure. This is (1). (2) is similar. To prove (3), take $0 < \lambda < 1$ and note by Cauchy–Schwarz that

$$(1 - \lambda)^2 \leq (\mathbb{E}Y_n \mathbf{1}_{Y_n \geq \lambda})^2 \leq \mathbb{E}Y_n^2 \mathbb{E}\mathbf{1}_{Y_n \geq \lambda}^2 = \mathbb{E}Y_n^2 \mathbb{E}\mathbf{1}_{Y_n \geq \lambda} = \mathbb{E}Y_n^2 \mathbb{P}(Y_n \geq \lambda),$$

so by Fatou's lemma again

$$\mathbb{P}\left(\limsup \frac{X_n}{\mathbb{E}X_n} \geq \lambda\right) = \mathbb{P}(\limsup Y_n \geq \lambda) \geq \limsup \mathbb{P}(Y_n \geq \lambda)$$

$$\geq (1 - \lambda)^2 \limsup 1/\mathbb{E}Y_n = (1 - \lambda)^2 \limsup (\mathbb{E}X_n)^2/\mathbb{E}X_n^2.$$

Taking $\lambda \to 0$ gives (3). This is from Kochen and Stone (1964).

Hint 8.35: Let X_n be the random variable denoting the number of $\{E_1, \ldots E_n\}$ that occur and apply Exercise 8.34. The quasi-independence implies

$$\mathbb{E}X_n^2 = \sum_{1 \le j,k \le n} \mathbb{P}(E_j \cap E_k) \le \left(\sum_{1 \le k \le n} \mathbb{P}(E_k) \right)^2 = (\mathbb{E}X_n)^2,$$

which is the hypothesis of that result. Then part (2) says $\limsup X_n / \mathbb{E}X_n \ge 1$ with positive probability, and since $\mathbb{E}X_n = \sum_{k=1}^n \mathbb{P}(E_k) \to \infty$, we get $X_n \to \infty$ with positive probability.

Hint 8.36: First assume g (8.3.1) holds for all k, j. by Theorem 8.2.4,

$$\mathbb{P}(A_k)\mathbb{P}(A_j) \gtrsim \mathrm{Cap}_K(\Lambda \cap Y(k)) \cdot \mathrm{Cap}_K(\Lambda \cap Y(j)).$$

Clearly,

$$\mathbb{P}(A_k \cap A_j) \le \mathbb{P}(\exists n,m : X_n \in \Lambda \cap Y(k), X_{n+m} \in \Lambda \cap Y(j))$$
$$+ \mathbb{P}(\exists n,m : X_n \in \Lambda \cap Y(j), X_{n+m} \in \Lambda \cap Y(k)).$$

If $k < j$, the first term is bounded above by

$$\mathbb{P}_\rho(\Lambda \cap Y(k)) \cdot \max_{y \in Y(k)} \mathbb{P}_y(\Lambda \cap Y(j)).$$

The first term is bounded by $\mathrm{Cap}_K(\Lambda \cap Y(k)))$ and the second is bounded by the capacity with respect to Martin kernel with initial state y. By (8.3.1) this kernel is bounded below by a multiple of the Martin kernel with initial state ρ; therefore the capacity with initial state y is bounded above by the capacity with initial state ρ, so

$$\max_{y \in Y(k)} \mathbb{P}_y(\Lambda \cap Y(j)) = O(\mathrm{Cap}_K(\Lambda \cap Y(j))).$$

Thus the first term in the sum above is $O(\mathrm{Cap}_K(\Lambda \cap Y(k))\mathrm{Cap}_K(\Lambda \cap Y(j)))$. The second term is estimated in the same way, but using the second part of (8.3.1). This proves quasi-independence when we have (8.3.1) for all pairs j, k.

In Lamperti's test it suffices to only assume (8.3.1) when k, j are sufficiently far apart, say $|j - k| > M$. To do this, we consider a subset of events A_{n_k}, chosen so that the indices are separated by M, but $\sum \mathbb{P}(A_{n_k}) = \infty$. Then the proof above applies, and the Borel–Cantelli lemma for quasi-independent events implies infinitely many of the A_{n_k} occur (and so infinitely many of the A_n occur).

Hint 9.1: Consider a family of disjoint square annuli which surround each generational square in the standard construction. A curve hitting E must cross many of these annuli and hence have infinite length.

Hint 9.2: Modify the construction in Theorem 9.1.1, so that the $\{a_k\}$ are dense in the whole real line. Alternatively, extend the segments in the proof of Theorem 9.5.2 to lines and show the resulting set still has zero area.

Hint 9.7: Using the Cauchy–Schwarz inequality as in the proof of Theorem 9.1.3, show it suffices to prove $1 \le \int (\sum_Q \mathbf{1}_{\Pi_\theta(Q)}(x))^2 d\theta = O(n)$ where the sum is over the 4^n nth generation squares. Prove this estimate by expanding the sum and grouping pairs of squares (Q, Q') according to $k = g(Q, Q')$, the maximal k so that Q, Q' are contained in the same kth generation square. For each k there are about 4^{2n-k} pairs with $g(Q, Q') = k$ and for each such pair $\int \mathbf{1}_{\Pi_\theta(Q)}(x)\mathbf{1}_{\Pi_\theta(Q')}(x)d\theta = O(4^{-n} \cdot 4^{k-n})$.

Hint 9.8: See Figure C.1.

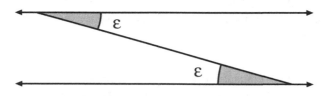

Figure C.1 Hint for Exercise 9.8.

Hint 9.18: Choose f to be zero on the complement of a closed Besicovitch set and discontinuous everywhere when restricted to any segment in the set. For example, set it to be 0 or 1 at (x, y) depending on whether x is rational or not, and adjust the definition for vertical lines.

Hint 9.20: Consider Figure C.2 that shows replacing a segment I by a union of segments E. Note that for $\theta \in [-\frac{\pi}{2}, 0]$ the projection of E covers the projection of I. Iterate the construction using segments with angles between 0 and $\pi/4$.

Figure C.2

Hint 9.21: First do this for $J_1 = [-\frac{\pi}{4}, -\delta]$, and $J_2 = [0, \frac{\pi}{4}]$, using the previous exercise. Then do it for $J_1 = [\frac{\pi}{2} - \alpha, \pi - \delta]$ and $J_2 = [0, \frac{\pi}{2} - \alpha - \delta]$ using the observation that there is an affine transformation of the plane that sends lines with angles $0, \frac{\pi}{4}, -\frac{\pi}{4}$ into lines with angles $0, \frac{\pi}{2} - \alpha - \delta, -\frac{\pi}{2}$ respectively. Finally, apply this case twice: first to the original segment I and then to all the small segments in E but with the construction reflected with respect to the y-axis.

Hint 9.23: For each $j \geq 1$ write $\mathbb{R}^2 \setminus \{0\} = \bigcup_k N_k^j$ as a union of open basic neighborhoods so that each N_k^{j+1} is contained in some N_p^j and choose $\{E_k^j\}$ so that $C(E_k^j)$ covers N_k^j but $\text{area}(C(E_k^j) \setminus N_k^j) \leq 2^{-k-j}$, each E_k^{j+1} is a union of disks and is contained in a component of some E_p^j. Let $K_j = \bigcup_k \overline{(C(E_k^j) \setminus N_k^j)}$ and $K = \liminf C_j = \bigcup_n \bigcap_{j>n} K_j$. Show K has zero area. Show that if $z \neq 0$, then z is in a nested sequence $N_{k_1}^{j_1} \supset N_{k_2}^{j_2} \supset \cdots$ and $z \in C(B_k)$ where $B_1 \supset B_2 \supset \cdots$ are component disks of the E_k^j. Then $x \in \bigcap \overline{B_k}$ satisfies $z \in C(x)$. Show that z is the only point of K^c in $C(x)$.

Hint 9.26: Write A as an intersection of nested open sets $A_1 \supset A_2 \supset \cdots$ so that $\text{area}(A_m \setminus A) < 2^{-m}$. Use Exercise 9.22 to inductively build sets E_m that are unions of balls so that $A_m \subset C(E_m)$, $\text{area}(C(E_m) \setminus A_m) < 2^{-m}$ and $\overline{E_{m+1}} \subset E_m$. Show that $K = \bigcap \overline{E_m}$ satisfies the result.

Hint 9.34: Show that if $y \in X$ then there is $z \in \mathscr{F}^n$ so that the line through z in direction y hits k in at least $d+1$ non-zero points $\{z + a_k y\}$, where $d = \lfloor q \min(\delta, \gamma) \rfloor - 2$. Show the $d+1$ points $w_k = a_k^{-1} z + y$ are in K' and hence $g(b_k) = 0$ for each k. If $z = 0$ then $w_k = y$ and we are done. Otherwise, g has degree d and vanishes at $d+1$ points on a single line. Deduce that g vanishes on the whole line and hence at y.

Hint 9.37: Prove that K has at least $\binom{q+n-1}{n}$ elements by contradiction. If K is smaller than this, show there is a non-zero polynomial g of degree $d \leq q-1$ (not necessarily homogeneous) that vanishes on K. Let h be the homogeneous part of g of degree d and prove h is the zero polynomial; a contradiction. To do this, fix $y \in \mathscr{F}^n$ and restrict g to a line $L = \{z + ay : a \in \mathscr{F}\}$. Since g vanishes on K, $g(z + ay)$ vanishes on $\mathscr{F}$ and is hence the zero polynomial. Thus the coefficient of a^d is zero, but this is the same as $h(y)$. Since y was arbitrary h vanishes everywhere.

Hint 9.38: Suppose not, i.e., $\mathrm{area}(K(\delta)) < \delta^{d-\alpha}$ for small enough δ. Choose $N \sim \delta^{1-\delta}$ unit segments in K with angles differing by at least δ and let $\{T_k\}$ denote δ-neighborhoods of these segments. Show there is a point in at least $\delta^{\alpha-d}$ different tubes. Show that the unions of these tubes have volume $\gtrsim \delta^{\alpha-1}$ and this gives a contradiction. This argument is due to Bourgain (1991).

Solution 9.39: Suppose $2 < p < \infty$. Then $\geq$ is obvious by Hölder's inequality. To prove the other direction, let $S = a \cdot x$ and first prove $\mathbb{P}(S > r) < e^{-r^2/2}$ by proving

$$\mathbb{E}e^{\lambda x_k} = \cosh(\lambda) \leq e^{\lambda^2/2},$$

$$\mathbb{E}e^{\lambda S} = \mathbb{E}e^{\lambda a \cdot x} \leq e^{\lambda^2/2},$$

and applying Chebyshev's inequality. If $p \geq 2$, by using the distribution function formula

$$\exp |S|^p = \int_0^\infty \mathbb{P}(|S| > r)\,dr \leq 2 \int \exp(-r^{2/p}/2)\,dr = C_p.$$

The argument for $1 < p < 2$ is analogous.

Hint 9.41: Suppose K is covered by dyadic boxes and B_j is the union of the boxes of size 2^{-j} and choose $\{\eta_j\} > 0$ so that $\sum_j \eta_j \leq 1/100$, for example, take $\eta_j = O(1/j\log^2 j)$. Then if $\delta = 2^{-j_0-1}$,

$$\sum_{j \geq j_0} M_{\mathrm{B}}^\delta \mathbf{1}_{B_j}(u) \geq \frac{1}{10}$$

for some j_0 and some $D \subset D^{d-1}$ of positive measure. Hence there is a $j > j_0$ so that $\int_D M_{\mathrm{B}}^\delta \mathbf{1}_{B_j}(u) \geq \eta_j \mathrm{vol}(D)$. However, if $\delta = 2^{-j+1}$, then

$$\|M_{\mathrm{B}}^\delta \mathbf{1}_{B_j}\|_p \lesssim 2^{-j((d/p)-1+\varepsilon)} \mathrm{vol}(B_j)^{1/p},$$

from which we can deduce there are more than

$$2^{j(p-\varepsilon)}(j\log^2 j)^{-p}$$

cubes in B_j.

Hint 9.42: Choose a smooth φ so that $\mathbf{1}_{[0,1]} \leq \varphi \leq 2$ and whose Fourier transform $\hat{\varphi} \geq 0$ is supported in $[-10, 10]$. If $w = (x, y)$, let $\psi(w) = \delta^{-1}\varphi(x)\varphi(\delta^{-1}y)$.

If O denotes rotation by θ and z is the unit vector in direction $\theta + \pi/2$, then

$$M_B^\delta f(z) \leq \sup_x \left| \int f(y)\psi(x - O^{-1}y)dy \right|$$

$$\leq \int |\hat{f}(\lambda)||\hat{\psi}(O^{-1}\lambda)|d\lambda$$

$$\leq \int_0^{C/\delta} \int_0^{2\pi} |\hat{f}(rw)||\hat{\varphi}(r\cos(\psi - \theta))|rdrd\psi.$$

Now apply Hölder's inequality,

$$\|M_B^\delta f\|_{L^2(S^1)}^2 \leq \int_0^{C/\delta} \int_0^{2\pi} \int_0^{2\pi} |\hat{f}(rw)|^2 |\hat{\varphi}(r\cos(\psi - \theta))|r^2 drd\psi$$

$$\times \int_0^{C/\delta} \int_0^{2\pi} \hat{\varphi}(r\cos(\psi - \theta))|r^2 drd\psi$$

$$\leq \int_0^{C/\delta} \frac{dr}{1+r} \int_0^{C/\delta} \int_0^{2\pi} |\hat{f}(rw)|^2 rdrd\psi$$

$$\leq C \left(\log \frac{1}{\delta} \right) \|\hat{f}\|_2^2.$$

Hint 9.44: By writing the operator as a sum of eight operators, we can reduce to considering rectangles $\mathscr{R}$ with angle in $[0, \pi/4]$. By dividing the plane into a grid of unit squares, and writing f as a sum of functions, each supported in one square, show it suffices to assume f is supported in $Q_0 = [0,1]^2$. Divide $3Q$ into δ-squares and for each such square Q, choose a $1 \times \delta$ rectangle $R_Q \in \mathscr{R}$ that hits Q. Define

$$Tg(x) = \sum_Q \text{area}(R_Q)^{-1} \int_{R_Q} g(y)dy \cdot \mathbf{1}_Q(x).$$

It suffices to prove the bound for T with a constant independent of our choice of rectangles (this is the standard way to linearize a maximal function). Show that the adjoint of T is given by

$$T^*h(y) = \sum_Q \text{area}(R_Q)^{-1} \int_Q h(y)dy\mathbf{1}_{R_Q}.$$

Given $h \in L^2(3Q_0)$, write $h = h_1 + \cdots + h_M$, $M \simeq 3/\delta$ by restricting h to disjoint vertical strips of width δ in $3Q_0$. Decompose each $h_k = h_{k1} + \cdots$ by restricting to each $\delta \times \delta$ square Q_{kj} in the vertical strip. Then

$$|T_f^*h(x)| \leq \sum_Q \frac{1}{\delta} \|h_Q\|_2 \delta \mathbf{1}_{R_Q} \leq \sum_Q \|h_Q\|_2 \cdot \mathbf{1}_{R_Q}(x).$$

Thus

$$\int |T^*h_k(x)|^2 dx \le \int \Big(\sum_j \|h_{jk}\|_2 \mathbf{1}_{R_{jk}}(x)\Big)^2 dx$$

$$\le \sum_{jk} \|h_{kp}\|_2 \|h_{kq}\|_2 \, \mathrm{area}(R_{kp} \cap R_{kq}).$$

Now use $\mathrm{area}(R_Q \cap R_{Q'}) \le C\delta^2/(\mathrm{dist}(Q,Q') + \delta)$ to deduce

$$\int |T_f^* h_k(x)|^2 = O(\delta) \sum_{p,q} \frac{\|h_{kp}\|_2 \|h_{kq}\|_2}{1 + |p - q|} = O\Big(\delta \log \frac{1}{\delta} \|h_k\|_2^2\Big),$$

and then

$$\|T^* h\|_2 = O\Big((\delta \log \frac{1}{\delta})^{1/2} \sum_k \|h_k\|_2\Big) = O\Big((\log \frac{1}{\delta})^{1/2} \|h\|_2\Big).$$

Hint 9.49: Take f, φ smooth and compact support on $\mathbb{R}^{d-k}$ and g, ψ smooth and compact support on $\mathbb{R}^k$ and apply Exercise 9.46 and Fubini's Theorem to deduce

$$I(x) = \int m(x,y)\hat{g}(y)\,\hat{\psi}(y)\,dy$$

is a Fourier multiplier for $\mathbb{R}^k$ with norm $\le |m|_p \cdot \|g\|_p \cdot \|\psi\|_q$. By Exercise 9.48, $|I(x)|$ is bounded pointwise by the same quantity, proving the result.

Hint 10.4: Let s be the intersection of the diagonals. By considering the triangle bcs, show the set of lines hitting both bs and cs has measure $|s - b| + |s - c| - |b - c|$. Add the analogous result for the segments as and ds.

Hint 10.5: Think of the needle as fixed and consider a random line hitting the disk $D(0, \frac{1}{2})$.

Hint 10.31: If the set is not uniformly wiggly then there is a sequence of squares along which the βs tend to zero. The assumptions imply there are group elements that map the parts of the boundary in these squares to "most of" $\partial \Omega$ and properties of Möbius transformations imply the images are close to circular arcs.

Hint 10.32: If it is not wiggly, then there is a sequence of squares along which the βs tend to zero. Iterates of such a square map it univalently until a critical point is encountered, which does not happen until it reaches size $\delta = \mathrm{dist}(\mathscr{J}, U^c)$. The Distortion Theorem for conformal maps then shows $\mathscr{J}$ can be covered by a finite number of analytic arcs.

Hint 10.33: First show the conformal map onto a half-plane is not maximal Bloch. If $\partial\Omega$ is not uniformly wiggly, then Ω can be rescaled by linear maps so as to converge to a half-plane and this gives a sequence of Möbius transformations that violate the definition of maximal Bloch. The maximal Bloch condition is characterized geometrically by Jones (1989). The condition also comes up elsewhere, as in O'Neill (2000).

Hint 10.34: If $1 - |z| \simeq 2^{-n}$, show the nth term is larger than the parts of the series that precede or follow it.

Hint 10.35: Construct Ω so it has at least one boundary point where the domain can be rescaled to converge to a half-plane.

Hint 10.39: Cut the integral into pieces with $|y - x| \simeq 2^{-k}|Q|$ over the indices $k = 1, \ldots, |\log_2 \beta|$, and similarly around z and use the regularity of γ to estimate how much μ-mass each piece has.

Hint 10.40: Use the previous exercise and the fact (proven in the text) that every pair x, y is associated to at least one and at most a bounded number of Qs. This implies $c^2(\mu)$ is bounded by the β-sum over $\mathscr{D}^*$, which is bounded by $\mathscr{H}^1(\gamma)$ by Theorem 10.2.3.

References

Adelman, Omer. 1985. Brownian motion never increases: a new proof to a result of Dvoretzky, Erdős and Kakutani. *Israel J. Math.*, **50**(3), 189–192.

Adelman, Omer, Burdzy, Krzysztof, and Pemantle, Robin. 1998. Sets avoided by Brownian motion. *Ann. Probab.*, **26**(2), 429–464.

Aizenman, M., and Burchard, A. 1999. Hölder regularity and dimension bounds for random curves. *Duke Math. J.*, **99**(3), 419–453.

Ala-Mattila, Vesa. 2011. Geometric characterizations for Patterson–Sullivan measures of geometrically finite Kleinian groups. *Ann. Acad. Sci. Fenn. Math. Diss.*, 120. Dissertation, University of Helsinki, Helsinki, 2011.

Arora, Sanjeev. 1998. Polynomial time approximation schemes for Euclidean traveling salesman and other geometric problems. *J. ACM*, **45**(5), 753–782.

Arveson, William. 1976. *An Invitation to C^*-algebras*. Springer-Verlag, New York–Heidelberg. Graduate Texts in Mathematics, No. 39.

Athreya, Krishna B., and Ney, Peter E. 1972. *Branching Processes*. New York: Springer-Verlag. Die Grundlehren der Mathematischen Wissenschaften, Band 196.

Avila, Artur, and Lyubich, Mikhail. 2008. Hausdorff dimension and conformal measures of Feigenbaum Julia sets. *J. Amer. Math. Soc.*, **21**(2), 305–363.

Azzam, Jonas. 2015. Hausdorff dimension of wiggly metric spaces. *Ark. Mat.*, **53**(1), 1–36.

Babichenko, Yakov, Peres, Yuval, Peretz, Ron, Sousi, Perla, and Winkler, Peter. 2014. Hunter, Cauchy rabbit, and optimal Kakeya sets. *Trans. Amer. Math. Soc.*, **366**(10), 5567–5586.

Bachelier, L. 1900. Théorie de la speculation. *Ann. Sci. Ecole Norm. Sup.*, **17**, 21–86.

Balka, Richárd, and Peres, Yuval. 2014. Restrictions of Brownian motion. *C. R. Math. Acad. Sci. Paris*, **352**(12), 1057–1061.

Balka, Richárd, and Peres, Yuval. 2016. Uniform dimension results for fractional Brownian motion. preprint, arXiv:1509.02979 [math.PR].

Bandt, Christoph, and Graf, Siegfried. 1992. Self-similar sets. VII. A characterization of self-similar fractals with positive Hausdorff measure. *Proc. Amer. Math. Soc.*, **114**(4), 995–1001.

Barański, Krzysztof. 2007. Hausdorff dimension of the limit sets of some planar geometric constructions. *Adv. Math.*, **210**(1), 215–245.

379

Barański, Krzysztof, Bárány, Balázs, and Romanowska, Julia. 2014. On the dimension of the graph of the classical Weierstrass function. *Adv. Math.*, **265**, 32–59.

Barlow, Martin T., and Perkins, Edwin. 1984. Levels at which every Brownian excursion is exceptional. Pages 1–28 of: *Seminar on probability, XVIII*. Lecture Notes in Math., vol. 1059. Berlin: Springer.

Barlow, Martin T., and Taylor, S. James. 1992. Defining fractal subsets of $\mathbf{Z}^d$. *Proc. London Math. Soc. (3)*, **64**(1), 125–152.

Bass, Richard F. 1995. *Probabilistic Techniques in Analysis*. Probability and its Applications (New York). New York: Springer-Verlag.

Bass, Richard F., and Burdzy, Krzysztof. 1999. Cutting Brownian paths. *Mem. Amer. Math. Soc.*, **137**(657), x+95.

Bateman, Michael, and Katz, Nets Hawk. 2008. Kakeya sets in Cantor directions. *Math. Res. Lett.*, **15**(1), 73–81.

Bateman, Michael, and Volberg, Alexander. 2010. An estimate from below for the Buffon needle probability of the four-corner Cantor set. *Math. Res. Lett.*, **17**(5), 959–967.

Benjamini, Itai, and Peres, Yuval. 1992. Random walks on a tree and capacity in the interval. *Ann. Inst. H. Poincaré Probab. Statist.*, **28**(4), 557–592.

Benjamini, Itai, and Peres, Yuval. 1994. Tree-indexed random walks on groups and first passage percolation. *Probab. Theory Related Fields*, **98**(1), 91–112.

Benjamini, Itai, Pemantle, Robin, and Peres, Yuval. 1995. Martin capacity for Markov chains. *Ann. Probab.*, **23**(3), 1332–1346.

Berman, Simeon M. 1983. Nonincrease almost everywhere of certain measurable functions with applications to stochastic processes. *Proc. Amer. Math. Soc.*, **88**(1), 141–144.

Bertoin, Jean. 1996. *Lévy Processes*. Cambridge Tracts in Mathematics, vol. 121. Cambridge: Cambridge University Press.

Bertrand-Mathis, Anne. 1986. Ensembles intersectifs et récurrence de Poincaré. *Israel J. Math.*, **55**(2), 184–198.

Besicovitch, A.S. 1919. Sur deux questions d'intégrabilité des fonctions. *J. Soc. Phys.-Math. (Perm')*, **2**(1), 105–123.

Besicovitch, A.S. 1928. On Kakeya's problem and a similar one. *Math. Z.*, **27**(1), 312–320.

Besicovitch, A.S. 1935. On the sum of digits of real numbers represented in the dyadic system. *Math. Ann.*, **110**(1), 321–330.

Besicovitch, A.S. 1938a. On the fundamental geometrical properties of linearly measurable plane sets of points (II). *Math. Ann.*, **115**(1), 296–329.

Besicovitch, A.S. 1938b. On the fundamental geometrical properties of linearly measurable plane sets of points II. *Math. Ann.*, **115**, 296–329.

Besicovitch, A.S. 1952. On existence of subsets of finite measure of sets of infinite measure. *Nederl. Akad. Wetensch. Proc. Ser. A.* **55** = *Indagationes Math.*, **14**, 339–344.

Besicovitch, A.S. 1956. On the definition of tangents to sets of infinite linear measure. *Proc. Cambridge Philos. Soc.*, **52**, 20–29.

Besicovitch, A.S. 1964. On fundamental geometric properties of plane line-sets. *J. London Math. Soc.*, **39**, 441–448.

Besicovitch, A.S., and Moran, P. A.P. 1945. The measure of product and cylinder sets. *J. London Math. Soc.*, **20**, 110–120.

Besicovitch, A.S., and Taylor, S.J. 1954. On the complementary intervals of a linear closed set of zero Lebesgue measure. *J. London Math. Soc.*, **29**, 449–459.

Besicovitch, A.S., and Ursell, H.D. 1937. Sets of fractional dimension v: On dimensional numbers of some continuous curves. *J. London Math. Soc.*, **12**, 18–25.

Bickel, Peter J. 1967. Some contributions to the theory of order statistics. Pages 575–591 of: *Proc. Fifth Berkeley Sympos. Math. Statist. and Probability (Berkeley, Calf., 1965/66), Vol. I: Statistics*. Berkeley, Calif.: Univ. California Press.

Billingsley, Patrick. 1961. Hausdorff dimension in probability theory. II. *Illinois J. Math.*, **5**, 291–298.

Binder, Ilia, and Braverman, Mark. 2009. The complexity of simulating Brownian motion. Pages 58–67 of: *Proceedings of the Twentieth Annual ACM–SIAM Symposium on Discrete Algorithms*. Philadelphia, PA: SIAM.

Binder, Ilia, and Braverman, Mark. 2012. The rate of convergence of the walk on spheres algorithm. *Geom. Funct. Anal.*, **22**(3), 558–587.

Bishop, Christopher J. 1996. Minkowski dimension and the Poincaré exponent. *Michigan Math. J.*, **43**(2), 231–246.

Bishop, Christopher J. 1997. Geometric exponents and Kleinian groups. *Invent. Math.*, **127**(1), 33–50.

Bishop, Christopher J. 2001. Divergence groups have the Bowen property. *Ann. of Math. (2)*, **154**(1), 205–217.

Bishop, Christopher J. 2002. Non-rectifiable limit sets of dimension one. *Rev. Mat. Iberoamericana*, **18**(3), 653–684.

Bishop, Christopher J. 2007. Conformal welding and Koebe's theorem. *Ann. of Math. (2)*, **166**(3), 613–656.

Bishop, Christopher J., and Jones, Peter W. 1990. Harmonic measure and arclength. *Ann. of Math. (2)*, **132**(3), 511–547.

Bishop, Christopher J., and Jones, Peter W. 1994a. Harmonic measure, L^2 estimates and the Schwarzian derivative. *J. Anal. Math.*, **62**, 77–113.

Bishop, Christopher J., and Jones, Peter W. 1994b. Harmonic measure, L^2 estimates and the Schwarzian derivative. *J. D'Analyse Math.*, **62**, 77–114.

Bishop, Christopher J., and Jones, Peter W. 1997. Hausdorff dimension and Kleinian groups. *Acta Math.*, **179**(1), 1–39.

Bishop, Christopher J., and Peres, Yuval. 1996. Packing dimension and Cartesian products. *Trans. Amer. Math. Soc.*, **348**(11), 4433–4445.

Bishop, Christopher J., and Steger, Tim. 1993. Representation-theoretic rigidity in PSL$(2, \mathbf{R})$. *Acta Math.*, **170**(1), 121–149.

Bishop, Christopher J., Jones, Peter W., Pemantle, Robin, and Peres, Yuval. 1997. The dimension of the Brownian frontier is greater than 1. *J. Funct. Anal.*, **143**(2), 309–336.

Bonk, Mario. 2011. Uniformization of Sierpiński carpets in the plane. *Invent. Math.*, **186**(3), 559–665.

Bonk, Mario, and Merenkov, Sergei. 2013. Quasisymmetric rigidity of square Sierpiński carpets. *Ann. of Math. (2)*, **177**(2), 591–643.

Bonk, Mario, Kleiner, Bruce, and Merenkov, Sergei. 2009. Rigidity of Schottky sets. *Amer. J. Math.*, **131**(2), 409–443.

382   *References*

Bourgain, J. 1987. Ruzsa's problem on sets of recurrence. *Israel J. Math.*, **59**(2), 150–166.

Bourgain, J. 1991. Besicovitch type maximal operators and applications to Fourier analysis. *Geom. Funct. Anal.*, **1**(2), 147–187.

Bourgain, J. 1999. On the dimension of Kakeya sets and related maximal inequalities. *Geom. Funct. Anal.*, **9**(2), 256–282.

Boyd, David W. 1973. The residual set dimension of the Apollonian packing. *Mathematika*, **20**, 170–174.

Brown, R. 1828. A brief description of microscopical observations made in the months of June, July and August 1827, on the particles contained in the pollen of plants; and on the general existence of active molecules in organic and inorganic bodies. *Ann. Phys.*, **14**, 294–313.

Burdzy, Krzysztof. 1989. Cut points on Brownian paths. *Ann. Probab.*, **17**(3), 1012–1036.

Burdzy, Krzysztof. 1990. On nonincrease of Brownian motion. *Ann. Probab.*, **18**(3), 978–980.

Burdzy, Krzysztof, and Lawler, Gregory F. 1990. Nonintersection exponents for Brownian paths. II. Estimates and applications to a random fractal. *Ann. Probab.*, **18**(3), 981–1009.

Cajar, Helmut. 1981. *Billingsley Dimension in Probability Spaces*. Lecture Notes in Mathematics, vol. 892. Berlin: Springer-Verlag.

Carleson, Lennart. 1967. *Selected Problems on Exceptional Sets*. Van Nostrand Mathematical Studies, No. 13. D. Van Nostrand Co., Inc., Princeton, N.J.–Toronto, Ont.–London.

Carleson, Lennart. 1980. The work of Charles Fefferman. Pages 53–56 of: *Proceedings of the International Congress of Mathematicians (Helsinki, 1978)*. Helsinki: Acad. Sci. Fennica.

Chang, S.-Y.A., Wilson, J.M., and Wolff, T.H. 1985. Some weighted norm inequalities concerning the Schrödinger operators. *Comment. Math. Helv.*, **60**(2), 217–246.

Charmoy, Philippe H.A., Peres, Yuval, and Sousi, Perla. 2014. Minkowski dimension of Brownian motion with drift. *J. Fractal Geom.*, **1**(2), 153–176.

Chayes, J.T., Chayes, L., and Durrett, R. 1988. Connectivity properties of Mandelbrot's percolation process. *Probab. Theory Related Fields*, **77**(3), 307–324.

Cheeger, Jeff, and Kleiner, Bruce. 2009. Differentiability of Lipschitz maps from metric measure spaces to Banach spaces with the Radon-Nikodým property. *Geom. Funct. Anal.*, **19**(4), 1017–1028.

Chow, Yuan Shih, and Teicher, Henry. 1997. *Probability Theory*. Third edn. Independence, Interchangeability, Martingales. Springer Texts in Statistics. New York: Springer-Verlag.

Christ, Michael. 1984. Estimates for the k-plane transform. *Indiana Univ. Math. J.*, **33**(6), 891–910.

Christensen, Jens Peter Reus. 1972. On sets of Haar measure zero in abelian Polish groups. Pages 255–260 (1973) of: *Proceedings of the International Symposium on Partial Differential Equations and the Geometry of Normed Linear Spaces (Jerusalem, 1972)*, vol. 13.

Ciesielski, Z., and Taylor, S.J. 1962. First passage times and sojourn times for Brownian motion in space and the exact Hausdorff measure of the sample path. *Trans. Amer. Math. Soc.*, **103**, 434–450.

Colebrook, C.M. 1970. The Hausdorff dimension of certain sets of nonnormal numbers. *Michigan Math. J.*, **17**, 103–116.

Córdoba, Antonio. 1993. The fat needle problem. *Bull. London Math. Soc.*, **25**(1), 81–82.

Cover, Thomas M., and Thomas, Joy A. 1991. *Elements of Information Theory*. Wiley Series in Telecommunications. New York: John Wiley & Sons Inc.

Cunningham, Jr., F. 1971. The Kakeya problem for simply connected and for star-shaped sets. *Amer. Math. Monthly*, **78**, 114–129.

Cunningham, Jr., F. 1974. Three Kakeya problems. *Amer. Math. Monthly*, **81**, 582–592.

David, Guy. 1984. Opérateurs intégraux singuliers sur certaines courbes du plan complexe. *Ann. Sci. École Norm Sup.*, **17**, 157–189.

David, Guy. 1998. Unrectifiable 1-sets have vanishing analytic capacity. *Rev. Mat. Iberoamericana*, **14**(2), 369–479.

Davies, Roy O. 1952. On accessibility of plane sets and differentiation of functions of two real variables. *Proc. Cambridge Philos. Soc.*, **48**, 215–232.

Davies, Roy O. 1971. Some remarks on the Kakeya problem. *Proc. Cambridge Philos. Soc.*, **69**, 417–421.

de Leeuw, Karel. 1965. On L_p multipliers. *Ann. of Math. (2)*, **81**, 364–379.

Dekking, F.M., and Grimmett, G.R. 1988. Superbranching processes and projections of random Cantor sets. *Probab. Theory Related Fields*, **78**(3), 335–355.

Drury, S.W. 1983. L^p estimates for the X-ray transform. *Illinois J. Math.*, **27**(1), 125–129.

Dubins, Lester E. 1968. On a theorem of Skorohod. *Ann. Math. Statist.*, **39**, 2094–2097.

Dudley, R.M. 2002. *Real Analysis and Probability*. Cambridge Studies in Advanced Mathematics, vol. 74. Cambridge University Press, Cambridge. Revised reprint of the 1989 original.

Dudziak, James J. 2010. *Vitushkin's Conjecture for Removable Sets*. Universitext. New York: Springer.

Duistermaat, J.J. 1991. Self-similarity of "Riemann's nondifferentiable function". *Nieuw Arch. Wisk. (4)*, **9**(3), 303–337.

Duplantier, Bertrand. 2006. Brownian motion, "diverse and undulating". Pages 201–293 of: *Einstein, 1905–2005*. Prog. Math. Phys., vol. 47. Basel: Birkhäuser. Translated from the French by Emily Parks.

Durrett, Richard. 1996. *Probability: Theory and Examples*. Belmont, CA: Duxbury Press.

Dvir, Zeev. 2009. On the size of Kakeya sets in finite fields. *J. Amer. Math. Soc.*, **22**(4), 1093–1097.

Dvir, Zeev, and Wigderson, Avi. 2011. Kakeya sets, new mergers, and old extractors. *SIAM J. Comput.*, **40**(3), 778–792.

Dvir, Zeev, Kopparty, Swastik, Saraf, Shubhangi, and Sudan, Madhu. 2009. Extensions to the method of multiplicities, with applications to Kakeya sets and mergers. Pages 181–190 of: *2009 50th Annual IEEE Symposium on Foundations of Computer Science (FOCS 2009)*. IEEE Computer Soc., Los Alamitos, CA.

Dvoretzky, A., Erdős, P., and Kakutani, S. 1950. Double points of paths of Brownian motion in *n*-space. *Acta Sci. Math. Szeged*, **12**(Leopoldo Fejer et Frederico Riesz LXX annos natis dedicatus, Pars B), 75–81.

Dvoretzky, A., Erdős, P., and Kakutani, S. 1961. Nonincrease everywhere of the Brownian motion process. Pages 103–116 of: *Proc. 4th Berkeley Sympos. Math. Statist. and Prob., Vol. II.* Berkeley, Calif.: Univ. California Press.

Dynkin, E.B., and Yushkevich, A.A. 1956. Strong Markov processes. *Theory Probab. Appl.*, **1**, 134–139.

Edgar, Gerald A. (ed). 2004. *Classics on Fractals.* Studies in Nonlinearity. Westview Press. Advanced Book Program, Boulder, CO.

Edmonds, Jack. 1965. Paths, trees, and flowers. *Canad. J. Math.*, **17**, 449–467.

Eggleston, H.G. 1949. The fractional dimension of a set defined by decimal properties. *Quart. J. Math., Oxford Ser.*, **20**, 31–36.

Einstein, A. 1905. Über die von der molekularkinetischen Theorie der Wärme geforderte Bewegung von in ruhenden Flüssigkeiten suspendierten Teilchen. *Ann. Physik*, **17**, 549–560.

Elekes, György, Kaplan, Haim, and Sharir, Micha. 2011. On lines, joints, and incidences in three dimensions. *J. Combin. Theory Ser. A*, **118**(3), 962–977.

Elekes, Márton, and Steprāns, Juris. 2014. Haar null sets and the consistent reflection of non-meagreness. *Canad. J. Math.*, **66**(2), 303–322.

Elekes, Márton, and Vidnyánszky, Zoltán. 2015. Haar null sets without G_δ hulls. *Israel J. Math.*, **209**(1), 199–214.

Ellenberg, Jordan S., Oberlin, Richard, and Tao, Terence. 2010. The Kakeya set and maximal conjectures for algebraic varieties over finite fields. *Mathematika*, **56**(1), 1–25.

Erdős, P. 1949. On a theorem of Hsu and Robbins. *Ann. Math. Statistics*, **20**, 286–291.

Erdős, P. 1961. A problem about prime numbers and the random walk. II. *Illinois J. Math.*, **5**, 352–353.

Erdős, Paul. 1940. On the smoothness properties of a family of Bernoulli convolutions. *Amer. J. Math.*, **62**, 180–186.

Evans, Steven N. 1992. Polar and nonpolar sets for a tree indexed process. *Ann. Probab.*, **20**(2), 579–590.

Falconer, K.J. 1980. Continuity properties of *k*-plane integrals and Besicovitch sets. *Math. Proc. Cambridge Philos. Soc.*, **87**(2), 221–226.

Falconer, K.J. 1982. Hausdorff dimension and the exceptional set of projections. *Mathematika*, **29**(1), 109–115.

Falconer, K.J. 1985. *The Geometry of Fractal Sets.* Cambridge Tracts in Mathematics, vol. 85. Cambridge: Cambridge University Press.

Falconer, K.J. 1986. Sets with prescribed projections and Nikodým sets. *Proc. London Math. Soc. (3)*, **53**(1), 48–64.

Falconer, K.J. 1988. The Hausdorff dimension of self-affine fractals. *Math. Proc. Cambridge Philos. Soc.*, **103**(2), 339–350.

Falconer, K.J. 1989a. Dimensions and measures of quasi self-similar sets. *Proc. Amer. Math. Soc.*, **106**(2), 543–554.

Falconer, K.J. 1989b. Projections of random Cantor sets. *J. Theoret. Probab.*, **2**(1), 65–70.

Falconer, K.J. 1990. *Fractal Geometry.* Chichester: John Wiley & Sons Ltd.

Falconer, K.J. 2013. Dimensions of self-affine sets: a survey. Pages 115–134 of: *Further Developments in Fractals and Related Fields*. Trends Math. Birkhäuser/Springer, New York.

Falconer, K.J., and Howroyd, J.D. 1996. Projection theorems for box and packing dimensions. *Math. Proc. Cambridge Philos. Soc.*, **119**(2), 287–295.

Fang, X. 1990. *The Cauchy integral of Calderón and analytic capacity*. Ph.D. thesis, Yale University.

Fefferman, Charles. 1971. The multiplier problem for the ball. *Ann. of Math. (2)*, **94**, 330–336.

Feller, William. 1966. *An Introduction to Probability Theory and its Applications. Vol. II*. New York: John Wiley & Sons Inc.

Ferguson, Andrew, Jordan, Thomas, and Shmerkin, Pablo. 2010. The Hausdorff dimension of the projections of self-affine carpets. *Fund. Math.*, **209**(3), 193–213.

Ferrari, Fausto, Franchi, Bruno, and Pajot, Hervé. 2007. The geometric traveling salesman problem in the Heisenberg group. *Rev. Mat. Iberoam.*, **23**(2), 437–480.

Folland, Gerald B. 1999. *Real Analysis*. Pure and Applied Mathematics (New York). New York: John Wiley & Sons Inc.

Ford, Jr., L.R., and Fulkerson, D.R. 1962. *Flows in Networks*. Princeton, N.J.: Princeton University Press.

Freedman, David. 1971. *Brownian Motion and Diffusion*. San Francisco, Calif.: Holden-Day.

Freud, Géza. 1962. Über trigonometrische approximation und Fouriersche reihen. *Math. Z.*, **78**, 252–262.

Frostman, O. 1935. Potential d'equilibre et capacité des ensembles avec quelques applications à la théorie des fonctions. *Meddel. Lunds Univ. Math. Sen.*, **3**(1-118).

Fujiwara, M., and Kakeya, S. 1917. On some problems for the maxima and minima for the curve of constant breadth and the in-revolvable curve of the equilateral triangle. *Tohoku Math. J.*, **11**, 92–110.

Furstenberg, H. 1967. Disjointness in ergodic theory, minimal sets, and a problem in Diophantine approximation. *Math. Systems Theory*, **1**, 1–49.

Furstenberg, H. 1970. Intersections of Cantor sets and transversality of semigroups. Pages 41–59 of: *Problems in Analysis (Sympos. Salomon Bochner, Princeton Univ., Princeton, N.J., 1969)*. Princeton Univ. Press, Princeton, N.J.

Furstenberg, H. 1981. *Recurrence in Ergodic Theory and Combinatorial Number Theory*. Princeton, N.J.: Princeton University Press.

Gabow, H.N. 1990. Data structures of weighted matching and nearest common ancestors with linking. In: *Proceedings of the 1st Annual ACM–SIAM Symposium on Discrete Algorithms*.

Garnett, John B. 1970. Positive length but zero analytic capacity. *Proc. Amer. Math. Soc.*, **24**, 696–699.

Garnett, John B. 1981. *Bounded Analytic Functions*. Academic Press.

Garnett, John B., and Marshall, Donald E. 2005. *Harmonic Measure*. New Mathematical Monographs, vol. 2. Cambridge: Cambridge University Press.

Gerver, Joseph. 1970. The differentiability of the Riemann function at certain rational multiples of π. *Amer. J. Math.*, **92**, 33–55.

Graczyk, Jacek, and Smirnov, Stanislav. 2009. Non-uniform hyperbolicity in complex dynamics. *Invent. Math.*, **175**(2), 335–415.

Graf, Siegfried, Mauldin, R. Daniel, and Williams, S.C. 1988. The exact Hausdorff dimension in random recursive constructions. *Mem. Amer. Math. Soc.*, **71**(381), x+121.

Guth, Larry. 2010. The endpoint case of the Bennett–Carbery–Tao multilinear Kakeya conjecture. *Acta Math.*, **205**(2), 263–286.

Hahlomaa, Immo. 2005. Menger curvature and Lipschitz parametrizations in metric spaces. *Fund. Math.*, **185**(2), 143–169.

Hahlomaa, Immo. 2007. Curvature integral and Lipschitz parametrization in 1-regular metric spaces. *Ann. Acad. Sci. Fenn. Math.*, **32**(1), 99–123.

Hahlomaa, Immo. 2008. Menger curvature and rectifiability in metric spaces. *Adv. Math.*, **219**(6), 1894–1915.

Hamilton, David H. 1995. Length of Julia curves. *Pacific J. Math.*, **169**(1), 75–93.

Har-Peled, Sariel. 2011. *Geometric Approximation Algorithms*. Mathematical Surveys and Monographs, vol. 173. American Mathematical Society, Providence, RI.

Hardy, G.H. 1916. Weierstrass's non-differentiable function. *Trans. Amer. Math. Soc.*, **17**(3), 301–325.

Harris, T.E. 1960. A lower bound for the critical probability in a certain percolation process. *Proc. Cambridge Philos. Soc.*, **56**, 13–20.

Hartman, Philip, and Wintner, Aurel. 1941. On the law of the iterated logarithm. *Amer. J. Math.*, **63**, 169–176.

Hausdorff, Felix. 1918. Dimension und äußeres Maß. *Math. Ann.*, **79**(1-2), 157–179.

Hawkes, John. 1975. Some algebraic properties of small sets. *Quart. J. Math. Oxford Ser. (2)*, **26**(102), 195–201.

Hawkes, John. 1981. Trees generated by a simple branching process. *J. London Math. Soc. (2)*, **24**(2), 373–384.

Hedenmalm, Haken. 2015. Bloch functions, asymptotic variance and zero packing. preprint, arXiv:1602.03358 [math.CV].

Hewitt, Edwin, and Savage, Leonard J. 1955. Symmetric measures on Cartesian products. *Trans. Amer. Math. Soc.*, **80**, 470–501.

Hewitt, Edwin, and Stromberg, Karl. 1975. *Real and Abstract Analysis*. New York: Springer-Verlag.

Hochman, Michael. 2013. *Dynamics on fractals and fractal distributions*. arXiv:1008.3731v2.

Hochman, Michael. 2014. On self-similar sets with overlaps and inverse theorems for entropy. *Ann. of Math. (2)*, **180**(2), 773–822.

Hochman, Michael. 2015. *On self-similar sets with overlaps and inverse theorems for entropy in* $\mathbb{R}^d$. In preparation.

Hochman, Michael, and Shmerkin, Pablo. 2012. Local entropy averages and projections of fractal measures. *Ann. of Math. (2)*, **175**(3), 1001–1059.

Housworth, Elizabeth Ann. 1994. Escape rate for 2-dimensional Brownian motion conditioned to be transient with application to Zygmund functions. *Trans. Amer. Math. Soc.*, **343**(2), 843–852.

Howroyd, J.D. 1995. On dimension and on the existence of sets of finite positive Hausdorff measure. *Proc. London Math. Soc. (3)*, **70**(3), 581–604.

Hsu, P.L., and Robbins, Herbert. 1947. Complete convergence and the law of large numbers. *Proc. Nat. Acad. Sci. U.S.A.*, **33**, 25–31.

Hunt, Brian R. 1994. The prevalence of continuous nowhere differentiable functions. *Proc. Amer. Math. Soc.*, **122**(3), 711–717.

Hunt, Brian R. 1998. The Hausdorff dimension of graphs of Weierstrass functions. *Proc. Amer. Math. Soc.*, **126**(3), 791–800.

Hunt, Brian R., Sauer, Tim, and Yorke, James A. 1992. Prevalence: a translation-invariant "almost every" on infinite-dimensional spaces. *Bull. Amer. Math. Soc. (N.S.)*, **27**(2), 217–238.

Hunt, Brian R., Sauer, Tim, and Yorke, James A. 1993. Prevalence. An addendum to: "Prevalence: a translation-invariant 'almost every' on infinite-dimensional spaces" [Bull. Amer. Math. Soc. (N.S.) 27 (1992), no. 2, 217–238; MR1161274 (93k:28018)]. *Bull. Amer. Math. Soc. (N.S.)*, **28**(2), 306–307.

Hunt, G.A. 1956. Some theorems concerning Brownian motion. *Trans. Amer. Math. Soc.*, **81**, 294–319.

Hutchinson, John E. 1981. Fractals and self-similarity. *Indiana Univ. Math. J.*, **30**(5), 713–747.

Ivrii, Oleg. 2016. Quasicircles of dimension $1 + k^2$ do not exist. Preprint, arXiv:1511.07240 [math.DS].

Izumi, Masako, Izumi, Shin-ichi, and Kahane, Jean-Pierre. 1965. Théorèmes élémentaires sur les séries de Fourier lacunaires. *J. Analyse Math.*, **14**, 235–246.

Jodeit, Jr., Max. 1971. A note on Fourier multipliers. *Proc. Amer. Math. Soc.*, **27**, 423–424.

Jones, Peter W. 1989. Square functions, Cauchy integrals, analytic capacity, and harmonic measure. Pages 24–68 of: *Harmonic Analysis and Partial Differential Equations (El Escorial, 1987)*. Lecture Notes in Math., vol. 1384. Berlin: Springer.

Jones, Peter W. 1990. Rectifiable sets and the traveling salesman problem. *Invent. Math.*, **102**(1), 1–15.

Jones, Peter W. 1991. The traveling salesman problem and harmonic analysis. *Publ. Mat.*, **35**(1), 259–267. Conference on Mathematical Analysis (El Escorial, 1989).

Jordan, Thomas, and Pollicott, Mark. 2006. Properties of measures supported on fat Sierpinski carpets. *Ergodic Theory Dynam. Systems*, **26**(3), 739–754.

Joyce, H., and Preiss, D. 1995. On the existence of subsets of finite positive packing measure. *Mathematika*, **42**(1), 15–24.

Juillet, Nicolas. 2010. A counterexample for the geometric traveling salesman problem in the Heisenberg group. *Rev. Mat. Iberoam.*, **26**(3), 1035–1056.

Kahane, J.-P. 1964. Lacunary Taylor and Fourier series. *Bull. Amer. Math. Soc.*, **70**, 199–213.

Kahane, J.-P. 1969. Trois notes sure les ensembles parfait linéaires. *Enseignement Math.*, **15**, 185–192.

Kahane, J.-P. 1985. *Some Random Series of Functions*. Second edn. Cambridge Studies in Advanced Mathematics, vol. 5. Cambridge: Cambridge University Press.

Kahane, J.-P., and Peyrière, J. 1976. Sur certaines martingales de Benoit Mandelbrot. *Advances in Math.*, **22**(2), 131–145.

Kahane, J.-P., Weiss, Mary, and Weiss, Guido. 1963. On lacunary power series. *Ark. Mat.*, **5**, 1–26 (1963).

Kakeya, S. 1917. Some problems on maxima and minima regarding ovals. *Tohoku Science Reports*, **6**, 71–88.

Kakutani, Shizuo. 1944. Two-dimensional Brownian motion and harmonic functions. *Proc. Imp. Acad. Tokyo*, **20**, 706–714.

Kamae, T., and Mendès France, M. 1978. van der Corput's difference theorem. *Israel J. Math.*, **31**(3-4), 335–342.

Karatzas, Ioannis, and Shreve, Steven E. 1991. *Brownian Motion and Stochastic Calculus*. Second edn. Graduate Texts in Mathematics, vol. 113. New York: Springer-Verlag.

Karpińska, Boguslawa. 1999. Hausdorff dimension of the hairs without endpoints for $\lambda \exp z$. *C. R. Acad. Sci. Paris Sér. I Math.*, **328**(11), 1039–1044.

Katz, Nets Hawk, and Tao, Terence. 2002. New bounds for Kakeya problems. *J. Anal. Math.*, **87**, 231–263. Dedicated to the memory of Thomas H. Wolff.

Katznelson, Y. 2001. Chromatic numbers of Cayley graphs on $\mathbb{Z}$ and recurrence. *Combinatorica*, **21**(2), 211–219.

Kaufman, R. 1968. On Hausdorff dimension of projections. *Mathematika*, **15**, 153–155.

Kaufman, R. 1969. Une propriété métrique du mouvement brownien. *C. R. Acad. Sci. Paris Sér. A-B*, **268**, A727–A728.

Kaufman, R. 1972. Measures of Hausdorff-type, and Brownian motion. *Mathematika*, **19**, 115–119.

Kechris, Alexander S. 1995. *Classical Descriptive Set Theory*. Graduate Texts in Mathematics, vol. 156. Springer-Verlag, New York.

Kenyon, Richard. 1997. Projecting the one-dimensional Sierpinski gasket. *Israel J. Math.*, **97**, 221–238.

Kenyon, Richard, and Peres, Yuval. 1991. Intersecting random translates of invariant Cantor sets. *Invent. Math.*, **104**(3), 601–629.

Kenyon, Richard, and Peres, Yuval. 1996. Hausdorff dimensions of sofic affine-invariant sets. *Israel J. Math.*, **94**, 157–178.

Khinchin, A.Y. 1924. Über einen Satz der Wahrscheinlichkeitrechnung. *Fund. Mat.*, **6**, 9–20.

Khinchin, A.Y. 1933. *Asymptotische Gesetze der Wahrscheinlichkeitsrechnung*. Springer-Verlag.

Khintchine, A. 1926. Über eine Klasse linearer diophantischer Approximationen. *Rendiconti Circ. Math. Palermo*, **50**(2), 211–219.

Khoshnevisan, Davar. 1994. A discrete fractal in $\mathbf{Z}_+^1$. *Proc. Amer. Math. Soc.*, **120**(2), 577–584.

Kinney, J.R. 1968. A thin set of circles. *Amer. Math. Monthly*, **75**(10), 1077–1081.

Knight, Frank B. 1981. *Essentials of Brownian Motion and Diffusion*. Mathematical Surveys, vol. 18. Providence, R.I.: American Mathematical Society.

Kochen, Simon, and Stone, Charles. 1964. A note on the Borel–Cantelli lemma. *Illinois J. Math.*, **8**, 248–251.

Kolmogorov, A. 1929. Über das Gesetz des iterierten Logarithmus. *Mathematische Annalen*, **101**(1), 126–135.

Körner, T.W. 2003. Besicovitch via Baire. *Studia Math.*, **158**(1), 65–78.

Kruskal, Jr., Joseph B. 1956. On the shortest spanning subtree of a graph and the traveling salesman problem. *Proc. Amer. Math. Soc.*, **7**, 48–50.

Kuipers, L., and Niederreiter, H. 1974. *Uniform Distribution of Sequences*. Wiley-Interscience [John Wiley & Sons], New York–London–Sydney. Pure and Applied Mathematics.

Łaba, Izabella. 2008. From harmonic analysis to arithmetic combinatorics. *Bull. Amer. Math. Soc. (N.S.)*, **45**(1), 77–115.

Lalley, Steven P., and Gatzouras, Dimitrios. 1992. Hausdorff and box dimensions of certain self-affine fractals. *Indiana Univ. Math. J.*, **41**(2), 533–568.

Lamperti, John. 1963. Wiener's test and Markov chains. *J. Math. Anal. Appl.*, **6**, 58–66.

Larman, D.H. 1967. On the Besicovitch dimension of the residual set of arbitrary packed disks in the plane. *J. London Math. Soc.*, **42**, 292–302.

Lawler, Gregory F. 1991. *Intersections of Random Walks*. Probability and its Applications. Boston, MA: Birkhäuser Boston Inc.

Lawler, Gregory F. 1996. The dimension of the frontier of planar Brownian motion. *Electron. Comm. Probab.*, **1**, no. 5, 29–47 (electronic).

Lawler, Gregory F., Schramm, Oded, and Werner, Wendelin. 2001a. The dimension of the planar Brownian frontier is 4/3. *Math. Res. Lett.*, **8**(4), 401–411.

Lawler, Gregory F., Schramm, Oded, and Werner, Wendelin. 2001b. Values of Brownian intersection exponents. I. Half-plane exponents. *Acta Math.*, **187**(2), 237–273.

Lawler, Gregory F., Schramm, Oded, and Werner, Wendelin. 2001c. Values of Brownian intersection exponents. II. Plane exponents. *Acta Math.*, **187**(2), 275–308.

Lawler, Gregory F., Schramm, Oded, and Werner, Wendelin. 2002. Values of Brownian intersection exponents. III. Two-sided exponents. *Ann. Inst. H. Poincaré Probab. Statist.*, **38**(1), 109–123.

Le Gall, J.-F. 1987. The exact Hausdorff measure of Brownian multiple points. Pages 107–137 of: *Seminar on Stochastic Processes, 1986 (Charlottesville, VA., 1986)*. Progr. Probab. Statist., vol. 13. Birkhäuser Boston, Boston, MA.

Lehmann, E.L. 1966. Some concepts of dependence. *Ann. Math. Statist.*, **37**, 1137–1153.

Lerman, Gilad. 2003. Quantifying curvelike structures of measures by using L_2 Jones quantities. *Comm. Pure Appl. Math.*, **56**(9), 1294–1365.

Lévy, Paul. 1940. Le mouvement brownien plan. *Amer. J. Math.*, **62**, 487–550.

Lévy, Paul. 1948. *Processus stochastiques et mouvement Brownien. Suivi d'une note de M. Loève*. Paris: Gauthier–Villars.

Lévy, Paul. 1953. La mesure de Hausdorff de la courbe du mouvement Brownien. *Giorn. Ist. Ital. Attuari*, **16**, 1–37 (1954).

Lindenstrauss, Elon, and Varju, Peter P. 2014. *Random walks in the group of Euclidean isometries and self-similar measures*. arXiv:1405.4426.

Lyons, Russell. 1989. The Ising model and percolation on trees and tree-like graphs. *Comm. Math. Phys.*, **125**(2), 337–353.

Lyons, Russell. 1990. Random walks and percolation on trees. *Ann. Probab.*, **18**(3), 931–958.

Lyons, Russell. 1992. Random walks, capacity and percolation on trees. *Ann. Probab.*, **20**(4), 2043–2088.

Lyons, Russell, and Pemantle, Robin. 1992. Random walk in a random environment and first-passage percolation on trees. *Ann. Probab.*, **20**(1), 125–136.

Lyons, Russell, and Peres, Yuval. 2016. *Probability on Trees and Networks*. Cambridge University Press.

Makarov, N.G. 1989. Probability methods in the theory of conformal mappings. *Algebra i Analiz*, **1**(1), 3–59.

Mandelbrot, Benoit. 1974. Intermittent turbulence in self-similar cascades: divergence of high moments and dimension of the carrier. *J. Fluid Mechanics*, **62**, 331–358.

Markowsky, Greg. 2011. On the expected exit time of planar Brownian motion from simply connected domains. *Electron. Commun. Probab.*, **16**, 652–663.

Marstrand, J.M. 1954. Some fundamental geometrical properties of plane sets of fractional dimensions. *Proc. London Math. Soc. (3)*, **4**, 257–302.

Marstrand, J.M. 1979. Packing planes in $\mathbf{R}^3$. *Mathematika*, **26**(2), 180–183 (1980).

Mattila, Pertti. 1990. Orthogonal projections, Riesz capacities, and Minkowski content. *Indiana Univ. Math. J.*, **39**(1), 185–198.

Mattila, Pertti. 1995. *Geometry of Sets and Measures in Euclidean Spaces*. Cambridge Studies in Advanced Mathematics, vol. 44. Cambridge: Cambridge University Press. Fractals and rectifiability.

Mattila, Pertti. 2015. *Fourier Analysis and Hausdorff Dimension*. Cambridge Studies in Advanced Mathematics, vol. 150. Cambridge University Press.

Mattila, Pertti, and Mauldin, R. Daniel. 1997. Measure and dimension functions: measurability and densities. *Math. Proc. Cambridge Philos. Soc.*, **121**(1), 81–100.

Mattila, Pertti, and Vuorinen, Matti. 1990. Linear approximation property, Minkowski dimension, and quasiconformal spheres. *J. London Math. Soc. (2)*, **42**(2), 249–266.

Mattila, Pertti, Melnikov, Mark S., and Verdera, Joan. 1996. The Cauchy integral, analytic capacity, and uniform rectifiability. *Ann. of Math. (2)*, **144**(1), 127–136.

Mauldin, R. Daniel, and Williams, S.C. 1986. On the Hausdorff dimension of some graphs. *Trans. Amer. Math. Soc.*, **298**(2), 793–803.

McKean, Jr., Henry P. 1961. A problem about prime numbers and the random walk. I. *Illinois J. Math.*, **5**, 351.

McKean, Jr., Henry P. 1955. Hausdorff–Besicovitch dimension of Brownian motion paths. *Duke Math. J.*, **22**, 229–234.

McMullen, Curtis T. 1984. The Hausdorff dimension of general Sierpiński carpets. *Nagoya Math. J.*, **96**, 1–9.

McMullen, Curtis T. 1998. Hausdorff dimension and conformal dynamics. III. Computation of dimension. *Amer. J. Math.*, **120**(4), 691–721.

Melnikov, Mark S., and Verdera, Joan. 1995. A geometric proof of the L^2 boundedness of the Cauchy integral on Lipschitz graphs. *Internat. Math. Res. Notices*, 325–331.

Milnor, John. 2006. *Dynamics in One Complex Variable*. Third edn. Annals of Mathematics Studies, vol. 160. Princeton, NJ: Princeton University Press.

Mitchell, Joseph S.B. 1999. Guillotine subdivisions approximate polygonal subdivisions: a simple polynomial-time approximation scheme for geometric TSP, *k*-MST, and related problems. *SIAM J. Comput.*, **28**(4), 1298–1309 (electronic).

Mitchell, Joseph S.B. 2004. Shortest paths and networks. Chap. 27, pages 607–641 of: Goodman, Jacob E., and O'Rourke, Joseph (eds), *Handbook of Discrete and Computational Geometry (2nd Edition)*. Boca Raton, FL: Chapman & Hall/CRC.

Montgomery, Hugh L. 2001. Harmonic analysis as found in analytic number theory. Pages 271–293 of: *Twentieth Century Harmonic Analysis – a Celebration (Il Ciocco, 2000)*. NATO Sci. Ser. II Math. Phys. Chem., vol. 33. Kluwer Acad. Publ., Dordrecht.

Moran, P.A.P. 1946. Additive functions of intervals and Hausdorff measure. *Proc. Cambridge Philos. Soc.*, **42**, 15–23.

Mörters, Peter, and Peres, Yuval. 2010. *Brownian Motion*. With an appendix by Oded Schramm and Wendelin Werner. Cambridge Series in Statistical and Probabilistic Mathematics. Cambridge: Cambridge University Press.

Muller, Mervin E. 1956. Some continuous Monte Carlo methods for the Dirichlet problem. *Ann. Math. Statist.*, **27**, 569–589.

Nazarov, F., Peres, Y., and Volberg, A. 2010. The power law for the Buffon needle probability of the four-corner Cantor set. *Algebra i Analiz*, **22**(1), 82–97.

Nevanlinna, Rolf. 1936. *Eindeutige Analytische Funktionen*. Grundlehren der Mathematischen Wissenschaften in Einzeldarstellungen, vol. 46. J. Springer, Berlin.

Newhouse, Sheldon E. 1970. Nondensity of axiom A(a) on S^2. Pages 191–202 of: *Global Analysis (Proc. Sympos. Pure Math., Vol. XIV, Berkeley, Calif., 1968)*. Amer. Math. Soc., Providence, R.I.

Nikodym, O. 1927. Sur la measure des ensembles plan dont tous les points sont rectalineairément accessibles. *Fund. Math.*, **10**, 116–168.

Oberlin, Richard. 2010. Two bounds for the X-ray transform. *Math. Z.*, **266**(3), 623–644.

Oh, Hee. 2014. Apollonian circle packings: dynamics and number theory. *Jpn. J. Math.*, **9**(1), 69–97.

Okikiolu, Kate. 1992. Characterization of subsets of rectifiable curves in $\mathbf{R}^n$. *J. London Math. Soc. (2)*, **46**(2), 336–348.

O'Neill, Michael D. 2000. Anderson's conjecture for domains with fractal boundary. *Rocky Mountain J. Math.*, **30**(1), 341–352.

Pajot, Hervé. 2002. *Analytic Capacity, Rectifiability, Menger Curvature and the Cauchy Integral*. Lecture Notes in Mathematics, vol. 1799. Berlin: Springer-Verlag.

Pál, Julius. 1921. Ein minimumproblem für ovale. *Math. Ann.*, **83**(3-4), 311–319.

Paley, R.E.A.C., Wiener, N., and Zygmund, A. 1933. Notes on random functions. *Math. Z.*, **37**(1), 647–668.

Parry, William. 1964. Intrinsic Markov chains. *Trans. Amer. Math. Soc.*, **112**, 55–66.

Pemantle, Robin. 1997. The probability that Brownian motion almost contains a line. *Ann. Inst. H. Poincaré Probab. Statist.*, **33**(2), 147–165.

Pemantle, Robin, and Peres, Yuval. 1994. Domination between trees and application to an explosion problem. *Ann. Probab.*, **22**(1), 180–194.

Pemantle, Robin, and Peres, Yuval. 1995. Galton–Watson trees with the same mean have the same polar sets. *Ann. Probab.*, **23**(3), 1102–1124.

Peres, Yuval. 1994a. The packing measure of self-affine carpets. *Math. Proc. Cambridge Philos. Soc.*, **115**(3), 437–450.

Peres, Yuval. 1994b. The self-affine carpets of McMullen and Bedford have infinite Hausdorff measure. *Math. Proc. Cambridge Philos. Soc.*, **116**(3), 513–526.

Peres, Yuval. 1996a. Points of increase for random walks. *Israel J. Math.*, **95**, 341–347.

Peres, Yuval. 1996b. Remarks on intersection-equivalence and capacity-equivalence. *Ann. Inst. H. Poincaré Phys. Théor.*, **64**(3), 339–347.

Peres, Yuval, and Schlag, Wilhelm. 2000. Smoothness of projections, Bernoulli convolutions, and the dimension of exceptions. *Duke Math. J.*, **102**(2), 193–251.

References

Peres, Yuval, and Schlag, Wilhelm. 2010. Two Erdős problems on lacunary sequences: chromatic number and Diophantine approximation. *Bull. Lond. Math. Soc.*, **42**(2), 295–300.

Peres, Yuval, and Shmerkin, Pablo. 2009. Resonance between Cantor sets. *Ergodic Theory Dynam. Systems*, **29**(1), 201–221.

Peres, Yuval, and Solomyak, Boris. 1996. Absolute continuity of Bernoulli convolutions, a simple proof. *Math. Res. Lett.*, **3**(2), 231–239.

Peres, Yuval, and Solomyak, Boris. 2000. Problems on self-similar sets and self-affine sets: an update. Pages 95–106 of: *Fractal Geometry and Stochastics, II (Greifswald/Koserow, 1998)*. Progr. Probab., vol. 46. Birkhäuser, Basel.

Peres, Yuval, and Solomyak, Boris. 2002. How likely is Buffon's needle to fall near a planar Cantor set? *Pacific J. Math.*, **204**(2), 473–496.

Perron, Oskar. 1929. Über stabilität und asymptotisches verhalten der integrale von differentialgleichungssystemen. *Math. Z.*, **29**(1), 129–160.

Prause, István, and Smirnov, Stanislav. 2011. Quasisymmetric distortion spectrum. *Bull. Lond. Math. Soc.*, **43**(2), 267–277.

Przytycki, F., and Urbański, M. 1989. On the Hausdorff dimension of some fractal sets. *Studia Math.*, **93**(2), 155–186.

Quilodrán, René. 2009/10. The joints problem in $\mathbb{R}^n$. *SIAM J. Discrete Math.*, **23**(4), 2211–2213.

Ray, Daniel. 1963. Sojourn times and the exact Hausdorff measure of the sample path for planar Brownian motion. *Trans. Amer. Math. Soc.*, **106**, 436–444.

Revuz, Daniel, and Yor, Marc. 1994. *Continuous Martingales and Brownian motion*. Second edn. Grundlehren der Mathematischen Wissenschaften [Fundamental Principles of Mathematical Sciences], vol. 293. Berlin: Springer-Verlag.

Rippon, P.J., and Stallard, G.M. 2005. Dimensions of Julia sets of meromorphic functions. *J. London Math. Soc. (2)*, **71**(3), 669–683.

Rogers, C.A., and Taylor, S.J. 1959. The analysis of additive set functions in Euclidean space. *Acta Math.*, **101**, 273–302.

Rohde, S. 1991. On conformal welding and quasicircles. *Michigan Math. J.*, **38**(1), 111–116.

Root, D.H. 1969. The existence of certain stopping times on Brownian motion. *Ann. Math. Statist.*, **40**, 715–718.

Rudin, Walter. 1987. *Real and Complex Analysis*. Third edn. New York: McGraw-Hill Book Co.

Ruzsa, I.Z., and Székely, G.J. 1982. Intersections of traces of random walks with fixed sets. *Ann. Probab.*, **10**(1), 132–136.

Saint Raymond, Xavier, and Tricot, Claude. 1988. Packing regularity of sets in n-space. *Math. Proc. Cambridge Philos. Soc.*, **103**(1), 133–145.

Salem, R., and Zygmund, A. 1945. Lacunary power series and Peano curves. *Duke Math. J.*, **12**, 569–578.

Saraf, Shubhangi, and Sudan, Madhu. 2008. An improved lower bound on the size of Kakeya sets over finite fields. *Anal. PDE*, **1**(3), 375–379.

Sawyer, Eric. 1987. Families of plane curves having translates in a set of measure zero. *Mathematika*, **34**(1), 69–76.

Schief, Andreas. 1994. Separation properties for self-similar sets. *Proc. Amer. Math. Soc.*, **122**(1), 111–115.

Schleicher, Dierk. 2007. The dynamical fine structure of iterated cosine maps and a dimension paradox. *Duke Math. J.*, **136**(2), 343–356.

Schoenberg, I.J. 1962. On the Besicovitch–Perron solution of the Kakeya problem. Pages 359–363 of: *Studies in Mathematical Analysis and Related Topics*. Stanford, Calif.: Stanford Univ. Press.

Schul, Raanan. 2005. *Subsets of rectifiable curves in Hilbert space – the analyst's TSP*. Ph.D. thesis, Yale University.

Schul, Raanan. 2007a. Ahlfors-regular curves in metric spaces. *Ann. Acad. Sci. Fenn. Math.*, **32**(2), 437–460.

Schul, Raanan. 2007b. Analyst's traveling salesman theorems. A survey. Pages 209–220 of: *In the Tradition of Ahlfors–Bers. IV*. Contemp. Math., vol. 432. Providence, RI: Amer. Math. Soc.

Schul, Raanan. 2007c. Subsets of rectifiable curves in Hilbert space – the analyst's TSP. *J. Anal. Math.*, **103**, 331–375.

Schwartz, J.T. 1980. Fast probabilistic algorithms for verification of polynomial identities. *J. Assoc. Comput. Mach.*, **27**(4), 701–717.

Selberg, Atle. 1952. The general sieve-method and its place in prime number theory. Pages 286–292 of: *Proceedings of the International Congress of Mathematicians, Cambridge, Mass., 1950, vol. 1*. Providence, R. I.: Amer. Math. Soc.

Shannon, C.E. 1948. A mathematical theory of communication. *Bell System Tech. J.*, **27**, 379–423, 623–656.

Shen, Weixiao. 2015. Hausdorff dimension of the graphs of the classical Weierstrass functions. Preprint, arXiv:1505.03986 [math.DS].

Shmerkin, Pablo. 2014. On the exceptional set for absolute continuity of Bernoulli convolutions. *Geom. Funct. Anal.*, **24**(3), 946–958.

Shmerkin, Pablo, and Solomyak, Boris. 2014. *Absolute continuity of self-similar measures, their projections and convolutions*. arXiv:1406.0204.

Skorokhod, A.V. 1965. *Studies in the Theory of Random Processes*. Translated from the Russian by Scripta Technica, Inc. Addison-Wesley Publishing Co., Inc., Reading, Mass.

Smirnov, Stanislav. 2010. Dimension of quasicircles. *Acta Math.*, **205**(1), 189–197.

Solecki, Sławomir. 1996. On Haar null sets. *Fund. Math.*, **149**(3), 205–210.

Solomyak, Boris. 1995. On the random series $\sum \pm \lambda^n$ (an Erdős problem). *Ann. of Math. (2)*, **142**(3), 611–625.

Solomyak, Boris. 1997. On the measure of arithmetic sums of Cantor sets. *Indag. Math. (N.S.)*, **8**(1), 133–141.

Solomyak, Boris, and Xu, Hui. 2003. On the 'Mandelbrot set' for a pair of linear maps and complex Bernoulli convolutions. *Nonlinearity*, **16**(5), 1733–1749.

Spitzer, Frank. 1964. *Principles of Random Walk*. The University Series in Higher Mathematics. D. Van Nostrand Co., Inc., Princeton, N.J.-Toronto-London.

Strassen, V. 1964. An invariance principle for the law of the iterated logarithm. *Z. Wahrscheinlichkeitstheorie und Verw. Gebiete*, **3**, 211–226 (1964).

Stratmann, Bernd O. 2004. The exponent of convergence of Kleinian groups; on a theorem of Bishop and Jones. Pages 93–107 of: *Fractal Geometry and Stochastics III*. Progr. Probab., vol. 57. Basel: Birkhäuser.

Sullivan, Dennis. 1982. Disjoint spheres, approximation by imaginary quadratic numbers, and the logarithm law for geodesics. *Acta Math.*, **149**(3-4), 215–237.

Sullivan, Dennis. 1984. Entropy, Hausdorff measures old and new, and limit sets of geometrically finite Kleinian groups. *Acta Math.*, **153**(3-4), 259–277.

Talagrand, Michel, and Xiao, Yimin. 1996. Fractional Brownian motion and packing dimension. *J. Theoret. Probab.*, **9**(3), 579–593.

Taylor, S. James. 1953. The Hausdorff α-dimensional measure of Brownian paths in n-space. *Proc. Cambridge Philos. Soc.*, **49**, 31–39.

Taylor, S. James. 1964. The exact Hausdorff measure of the sample path for planar Brownian motion. *Proc. Cambridge Philos. Soc.*, **60**, 253–258.

Taylor, S. James. 1966. Multiple points for the sample paths of the symmetric stable process. *Z. Wahrscheinlichkeitstheorie und Verw. Gebiete*, **5**, 247–264.

Taylor, S. James, and Tricot, Claude. 1985. Packing measure, and its evaluation for a Brownian path. *Trans. Amer. Math. Soc.*, **288**(2), 679–699.

Taylor, S. James, and Wendel, J.G. 1966. The exact Hausdorff measure of the zero set of a stable process. *Z. Wahrscheinlichkeitstheorie und Verw. Gebiete*, **6**, 170–180.

Tolsa, Xavier. 2003. Painlevé's problem and the semiadditivity of analytic capacity. *Acta Math.*, **190**(1), 105–149.

Tolsa, Xavier. 2014. *Analytic Capacity, the Cauchy Transform, and Non-homogeneous Calderón–Zygmund theory*. Progress in Mathematics, vol. 307. Birkhäuser/Springer, Cham.

Tricot, Claude. 1981. Douze définitions de la densité logarithmique. *Comptes Rendus Acad. Sci. Paris*, **293**, 549–552.

Tricot, Claude. 1984. A new proof for the residual set dimension of the Apollonian packing. *Math. Proc. Cambridge Philos. Soc.*, **96**(3), 413–423.

Tricot, Claude. 1982. Two definitions of fractional dimension. *Math. Proc. Cambridge Philos. Soc.*, **91**(1), 57–74.

Tukia, Pekka. 1989. Hausdorff dimension and quasisymmetric mappings. *Math. Scand.*, **65**(1), 152–160.

Urbański, Mariusz. 1990. The Hausdorff dimension of the graphs of continuous self-affine functions. *Proc. Amer. Math. Soc.*, **108**(4), 921–930.

Urbański, Mariusz. 1991. On the Hausdorff dimension of a Julia set with a rationally indifferent periodic point. *Studia Math.*, **97**(3), 167–188.

Urbański, Mariusz. 1997. Geometry and ergodic theory of conformal non-recurrent dynamics. *Ergodic Theory Dynam. Systems*, **17**(6), 1449–1476.

van der Corput, J.G. 1931. Diophantische ungleichungen. I. Zur gleichverteilung modulo eins. *Acta Math.*, **56**(1), 373–456.

Volkmann, Bodo. 1958. Über Hausdorffsche Dimensionen von Mengen, die durch Ziffereigenschaften charakterisiert sind. VI. *Math. Z.*, **68**, 439–449.

Weierstrass, K. 1872. Über contiuirliche functionen eines reellen arguments, die für keinen werth des letzeren einen bestimmten differentialquotienten besitzen. *Königl. Akad. Wiss.*, **3**, 71–74. Mathematische Werke II.

Wiener, N. 1923. Differential space. *J. Math. Phys.*, **2**.

Wisewell, L. 2005. Kakeya sets of curves. *Geom. Funct. Anal.*, **15**(6), 1319–1362.

Wolff, Thomas. 1995. An improved bound for Kakeya type maximal functions. *Rev. Mat. Iberoamericana*, **11**(3), 651–674.

Wolff, Thomas. 1999. Recent work connected with the Kakeya problem. Pages 129–162 of: *Prospects in Mathematics (Princeton, NJ, 1996)*. Providence, RI: Amer. Math. Soc.

Xiao, Yimin. 1996. Packing dimension, Hausdorff dimension and Cartesian product sets. *Math. Proc. Cambridge Philos. Soc.*, **120**(3), 535–546.

Zdunik, Anna. 1990. Parabolic orbifolds and the dimension of the maximal measure for rational maps. *Invent. Math.*, **99**(3), 627–649.

Zhan, Dapeng. 2011. Loop-erasure of planar Brownian motion. *Comm. Math. Phys.*, **303**(3), 709–720.

Zippel, Richard. 1979. Probabilistic algorithms for sparse polynomials. Pages 216–226 of: *Symbolic and Algebraic Computation (EUROSAM '79, Internat. Sympos., Marseille, 1979)*. Lecture Notes in Comput. Sci., vol. 72. Berlin: Springer.

Zygmund, A. 1959. *Trigonometric Series*. Cambridge University Press.

Index

Printed in the United States
by Baker & Taylor Publisher Services